G. Paillotin (Ed.)

European Agricultural Research in the 21st Century

Springer
Berlin
Heidelberg
New York
Barcelona
Budapest
Hong Kong
London
Milan
Paris
Singapore
Tokyo

G. Paillotin (Ed.)

European Agricultural Research in the 21st Century

Which Innovations Will Contribute Most to the Quality of Life, Food and Agriculture?

Palais de l'Europe, Strasbourg, France 28–29 November 1996

With 35 Figures and 37 Tables

INRA EDITIONS

Springer

G. Paillotin
INRA
147 rue de l'Université
75007 Paris
France

ISBN 3-540-64414-8 Springer-Verlag Berlin Heidelberg New York

Library of Congress Cataloging-in-Publication Data

Recherche agronomique européenne dans le monde du 21e siècle. English: *European agricultural research in the 21st century* Guy Paillotin (ed.). p. cm. Papers from a seminar held at the European Parliament in Strasbourg, Nov. 28–29, 1996. „INRA editions." Includes bibliographical references. ISBN 3-540-64414-8 (hardcover) 1. Agriculture–Research–Europe–Congresses. I. Paillotin, Guy. II. Institut national de la recherche agronomique (France) III. Title. S542.E8R4313 1998 630'.7204–DC21 98-24947 CIP

Printed in Germany

Cover design: design & production GmbH, Heidelberg
Typesetting: Camera ready by the editor

SPIN 10674110 31/3137 5 4 3 2 1 0 - Printed on acid-free paper

CONTENTS

The Strasbourg Declaration IX

PLENARY SESSION

Foreword
Guy Paillotin 2

The New International Economic Context for European Agricultural Research
Joachim von Braun 5

TOMORROW'S DIET

Workshop 1 – Decisive Elements in the Consumer's Choice

Trends in Food Consumption in the European Union: Towards a Euro Diet?
Bruce Traill 14

Consumer Choices: Tastes, Preferences, Constraints
Pierre Combris 20

Effects of Media Discourse and Marketing
Francesco Bisceglia 30

Influence of Media and Marketing
Chantal Jacquet 34

Workshop 2 – Quality in the Food Sector

The Components of Quality
Gianfranco Valfré 37

The Strategy of the Food Industry
Christiane Mercier 43

Quality from a Consumer's Perspective
Arie van Genderen 48

Application of Biotechnology in the Food Industry
Hans Nieuwenhuis 53

The Role of Public Authorities at the National and International Level
Joachim Heine 60

Workshop 3 – Diet, Well-Being and Health

Food Choice and Health
Pierre Louisot 67

How Functional Foods in the Diet Contribute to Health
Marcel Roberfroid 71

Food Safety
Jean-François Girard 76

AGRICULTURE, ENVIRONMENT AND RURAL AREAS

Workshop 1a – Agriculture, Natural Resources and Environmental Constraints

Water Management for Agriculture
Carlos Tio 80

Soil Management and Agriculture
Christian de Kimpe 84

Man and Biodiversity
Robert Barbault 92

Workshop 1b – Production Improvement

The Genetic Control of Plant Parasites: „Un plat de résistance"
Ian R. Crute 102

Objectives and Challenges in Improving Biological Material for Animal Breeding
Pascal L. Leroy 114

Integrated Agriculture
Jean Thiault 123

Improvements in Production Systems
Dieter Smidt 130

Workshop 2 – Product Diversification

Non-Food Production: Economic, Environmental and Land Use Aspects
Philippe Chartier 138

Non-Food Production: Technological and Scientific Bottlenecks,
The Role of Bioenergy in Europe's Energy Strategies
Wolfgang Palz 146

Forest and Agroforestry
Jan-Erik Hällgren 165

Trends in Fisheries and Aquaculture Production in Europe
Robin L. Welcomme 174

Workshop 3 – Agriculture as Part of Rural and Regional Development

The Diversity of Rural Space and its Typologies
Elena Saracenono 190

Agriculture at the Local and Regional Level
André Valadier 207

Amenities in Rural Areas: Responding to a New Social Demand
Markus F. Hofreither 209

AGRICULTURAL RESEARCH, INNOVATION AND SOCIETY

Workshop 1 – Research, Technological Change and Economic Development

Lessons from a Macro-Historical Analysis: The Future of European Agricultural Research
Ad M. van der Woude 217

Agricultural Research and Technical Change in Agriculture: Lessons from Microeconomic Studies
Robert E. Evenson 223

The Agro-Food Industry and Innovation
Christian Godde 250

Agricultural Research in France. A Useful Tool in Effective, Pervasive Agriculture
Maurice Rigaud 254

Workshop 2 – The Evolution of Partnerships

First Panel Discussion: Scientific and Industrial Partnerships
Michel Callon, on behalf of *Jean Gabolde* 260

Second Panel Discussion: Future Relations Between Agricultural Research and Agricultural Extension Services
Michel SEBILLOTE 265

Workshop 3 – Research and Society: Scientific Expertise and Public Decisions

Why are Public Organisations Needed for Agronomic Research?
Patrick Cohendet, Pierre-Benoit Joly 270

Third Panel Discussion: Scientific Expertise and Public Decision
Noëlle Lenoir 279

Scientific Appraisals: The Organisation and Responsibility of Experts
Marie-Angèle Hermitte 287

PLENARY SESSION

European Agricultural Research, World Food Situation and Development. Agricultural Research after the World Food Summit
Stein W. Bie 305

European Agricultural Research, World Food Situation and Development
Michel Petit 313

Europe and Agricultural Research for Development
Alain Darthenucq 321

Closing Speech
François d'Aubert 329

THE STRASBOURG DECLARATION

The participants to the Strasbourg Symposium,

members of European public and private research organisations, representatives of producers, companies in the food industry, consumer organisations and environmental associations, after examining European agricultural research and the problems it has to face on the eve of the 21st century, after stating and comparing their viewpoints, hereby adopt the following declaration:

1 - Feeding humanity without depleting natural resources

The individual's right to food, and the satisfaction of world food requirements are priorities to which research on agriculture and food must respond. However, the prospects of population growth and climatic change are making the availability of natural resources such as water, arable soils, forests, and genetic diversity, a major concern. For the 21st century it is essential to strengthen sustainable production systems.

European research on agriculture, foods, and the quality of life should enable Europe to contribute to the growth and quality improvement of world food production, all the while protecting natural resources and the environment. It must strive to become more competitive without destabilising markets. Further, it must facilitate access to technology and strengthen cooperation with countries where there are problems of malnutrition by contributing to the development of their research systems.

2 - Improving the management of natural, cultivated and forest lands, and preserving diversity

During the next century the peoples of Europe will be increasingly attentive to sustainable management of rural lands. The diversity of these lands, production systems, and food consumption habits are part of the cultural heritage of European citizens and contribute to the quality of their living conditions.

European research on agriculture, foods, and the quality of life preserve and optimise the diversity of geographical, ecological, and cultural conditions in Europe, especially considering the prospects of an enlarged European Union.

3 - Supporting job creation through the diversification of productions and activities

It is imperative to create jobs in the food sector and in rural areas. Innovation, in particular in the small- and medium- size enterprises, and diversification into (food) production with higher added value, non-food productions and non-agricultural activities will contribute to these goals.

European research on agriculture, foods, and the quality of life should be based on appropriate partnerships with producers, industry, and local and regional authorities to contribute to job creation through the diversification of productions and activities both within and outside of agriculture.

4 - Improving quality and guaranteeing food security

Consumers increasingly seek high quality and healthy foods. Food security is absolutely essential. Consumers must be able to trust the products which they buy, and be informed in keeping with their expectations (composition, origin, etc.).

European research on agriculture, foods, and the quality of life must contribute to improving the quality, safety, and nutritional value of food products. Reference to scientific knowledge should be increased. All the parties involved in this industry must show cooperation and solidarity.

5 - Developing the use of biotechnologies

The application of biotechnology to agriculture and food holds great promise and will advance progress in understanding the living environment, and innovations in processes and productions, but in Europe public opinion shows some concern about these new technologies.

European research on agriculture, foods, and the quality of life must help those in positions of responsibility to make scientifically informed decisions. Its duty is to shed light on discussions on technological choices, the benefits thereof, and potential risks by providing validated knowledge. It must contribute actively to educating the public, and to informing and engaging a dialogue with the media, the consumers, and the citizens.

6 - Enhancing the role of researchers as experts by providing them with the means to be independent

Scientific research on agriculture, foods and the quality of life concerns the citizens in their daily life. Faced with problems of society, researchers are increasingly solicited to provide their expert opinions in various situations: commercial conflicts, public health problems, differences of opinion, risk evaluation, etc.

European researchers must discharge their role as experts with adequate means and complete independence. They must be attentive to questions of ethics. It is their duty to show vigilance and responsibility towards the socio-economic players, the public authorities, and the citizens.

This colloquium was supported by the European Commission, Fruit d'Or Recherche and Rhône-Poulenc

FRUIT D'OR
RECHERCHE

RHÔNE-POULENC

- Rhône-Poulenc Secteur AGRO
- Rhône-Poulenc AGRO France
- Rhône-Poulenc Animal Nutrition
- Rhône-Poulenc Alimentaire
- Rhône Mérieux

The Strasbourg City Council, the Regional Alsace Council, the European Council, the European Parliament, the association of Céréaliers de France - Unigrains, Pernod-Ricard, Roquette and the news paper "Les Dernières Nouvelles d'Alsace also contribued to the success of this colloquium.

FOREWORD

Guy Paillotin
President of INRA
147, rue de l'Université
F-75007 PARIS, France

During the last fifty years, agricultural research has been keeping apace of the deep human and technological changes in the agriculture and food sectors.

Agricultural research, all the while, has also been going through profound transformation as it strived towards new objectives: after production came the transformation of agricultural and food products, food security, consumption, ecological issues and land use planning. Research has reached new heights in biology and is exploring other disciplines. It is forever changing, as are the needs of society.

The first change must be considered together with the success of agricultural research which was designed in the middle of the 20th century at the national level. Spurred, as it was, by dynamic incentive policies, research generated large yield increases for markets that were too narrow. Since that time, Europe has developed its agriculture so mightily that the continent is too small to absorb its enormous agricultural potential.

The recent Uruguay Round agreements open new prospects for a global economy hitherto governed by protectionist policies that allowed costly destruction of surplus output to coexist with malnutrition in certain parts of the world. Trade in the agriculture and food sector is becoming more liberal. In this new context, research should help farmers and agro-food industries find new competitive advantages. This is the overriding challenge to European agricultural research which has often contributed to increasing production but has paid little attention to outlets and markets.

The second challenge concerns the new demands from consumers and members of developed societies. The voice of these demands is especially loud because, serendipitously, the fear of food shortage has disappeared, as we can see from the emphasis placed on:

- the desire for better environmental protection;
- preference for high quality products, with information on their origin and conditions of production;
- health in relation to the consumer's diet;
- respect for certain ethics such as the well-being of farm animals.

Europeans are returning to «what's natural» and are eyeing scientific progress more critically. The recent bovine spongiform encephalopathy crisis, the debates on genetically modified organisms and on cloning stand out as evidence that goes well beyond any borders.

Agricultural research is being put to question, and must find the right answers.

What foods will we eat tomorrow? How can quantitative and qualitative food requirements be met? How can agriculture be productive without mismanaging natural resources and lands? How can we give European agricultural research the strength and efficiency needed to meet these challenges? What guidance can be given to partnerships and scientific cooperation? How can we increase people's confidence in research and innovation?

These are the main questions discussed at the seminar «European Agricultural Research in the World of the 21st Century: which innovations will contribute most to the quality of life, food and agriculture?», organised by INRA at the European Parliament in Strasbourg on 28 and 29 November 1996 to commemorate its fiftieth anniversary.

For two days, 400 participants with an interest in research discussed these issues. The participants included scientists from both the public and private sector, academics, public and private research administrators, politicians, government officials, and leaders from agricultural organisations, agro-industries and various associations.

The seminar brought together representatives from close to thirty countries, including most of Europe, North America, Japan, Africa, and the major international organisations such as FAO, World Bank and the European Commission.

It ended with the adoption of the «Strasbourg Declaration», a joint statement reasserting the importance of agricultural research, foods and the quality of life for the 21st century. Although many different opinions were expressed, six major points were recommended as priorities for European research. They can be referred to as a common framework for the future and equip INRA to start a preemptive review that will be followed through by its partners. Our second rendez-vous will be organised by DLO in 1998 in the Netherlands.

This event could not have taken place without the efforts of the scientific committee and the organising committee, which brought us together to discuss these very vital issues for the future of agricultural research and for society. I would like to express our gratitude for their remarkable work.

We are also obliged to the European Commission for the support it has given us via its directorate general for agriculture, DG VI, and its directorate general for research and technological development, DG XII.

I would also like to thank two companies that have been with us since we first started planning the seminar: Fruit d'Or Recherche-Astra Calvé and Rhône Poulenc which participated via its subsidiaries, Rhône-Poulenc secteur Agro, Rhône-Poulenc Agro France, Rhône-Poulenc Animal

Nutrition, Rhône-Poulenc Food Additives. We owe them a lot. Without them, the seminar could not have taken place.

Several other institutions and partners gave us support in a variety of ways. There was the city of Strasbourg, the Regional Council of Alsace, the Council of Europe, the European Parliament, the Céréaliers de France-Unigrains, Pernod-Ricard, Ets Roquette and the newspaper, *les Dernères Nouvelles d'Alsace*. I wish to acknowledge them all.

THE NEW INTERNATIONAL ECONOMIC CONTEXT FOR EUROPEAN AGRICULTURAL RESEARCH

JOACHIM VON BRAUN
Christian Albrecht University
Olshausenstrasse, 40
D-24098 Kiel, Germany

When a person celebrates his or her 50th birthday, reflections tend to be directed toward the past five decades. When an organisation celebrates its 50th anniversary, by reflecting on the future and its potential role in shaping the future, it is a sign of vitality. I commend INRA for taking on this difficult task. I congratulate INRA on its birthday and wish it success in the 21st century for the benefit of consumers and producers of food and agricultural goods in France, Europe and beyond.

THE NEW INTERNATIONAL ECONOMIC CONTEXT

This conference started with reflections on the international economic context which shapes the challenges, adjustment needs, and opportunities for European agricultural research in the coming century. However, there are not just one-way relationships between the economic context and agricultural research. There are also powerful linkages between agricultural research and the long-term international economic context.

What do we mean by "the new international economic context"?

We mean the new institutional, market and price environment that provides new chances and challenges for agriculture, the food industries, and consumers. Its key components are: the steps toward a more liberal international trade regime under GATT/WTO and the related agricultural policy changes in EU and other OECD countries; the redirection of former planned economies toward market orientation in Central and Eastern Europe, China and elsewhere; the formation of regional trade and integration zones or their expansion, such as the EU's eastward expansion, NAFTA, APEC etc.

GLOBALISATION AND THE FOOD SITUATION

Regarding globalization of the world economy, we must note that many countries are left out and much happens in regional trading blocks. Regional trading blocks are being rapidly formed or consolidated, and are stimulating regional integration in trade and knowledge transfers within them. It will be an issue of global governance whether further globalization of the economic system or whether confrontation and conflict will grow out of the regional trade blocks. Regional integration may take the lead for some decades to come before global policy co-ordination can make much further progress. Still, global policy co-ordination is needed in a number of areas related to natural resources and agriculture; the 1992 Earth Summit and the 1996 Food Summit have achieved substantial progress. Agricultural research systems within regional integration blocks will grow into larger clusters and can exploit positive economies of scale wherever they exist. There is a need for transnational co-operation within regional formations such as the European Union. In the long term, agricultural research systems will not only provide public goods but will increasingly be called upon to provide services to the private sector. It follows that in a more globalized world economy, agricultural research systems will be exposed to increased competition as well.

The economic context in which the future of the world food and agriculture system is going to be placed will be determined by the extent to which poverty will be reduced. It is the buying power and nutritional well-being of today's poor and their children which will determine the challenges and opportunities of agriculture and the food system of the future. If the rich get richer, agriculture will not be confronted with much added market demand. There may be a bit more demand for higher service content in foods and for more landscape in our regions.

In crude economic terms, the hunger and malnutrition problem is probably one of the largest, world-wide wastes of potential 'economic resources' - the lives of billions of potentially productive people now and over the coming decades - and is probably the biggest failure of market functioning yet to be overcome. Any consideration of the economic context of food and agricultural research in the 21st century must therefore also take this fact into account and address the related issues. The size of the problem has been well publicised by the recent World Food Summit (FAO 1996).

In a world increasingly integrated through trade and politico-economic ties among nations, the availability of sufficient food throughout the world is of increasing importance for household food security and nutrition. Thus far, world production has kept pace with population growth. However, because of continuing population growth, increasing land scarcity, and mounting difficulties in achieving sustainable increases in food-crop yields, global food availability cannot be taken for granted over the long run. In view of low food storage levels in the 1990s, the world food situation may remain quite unstable for some time. The private sector will not invest in storage to the extent industrial countries did before the recent GATT/WTO agreements. How will food and agricultural policy react to fluctuating and possibly higher world prices? How will supply respond to prices in the short and in the long run? Apparently, at least in the short run, we recognise that the policy response is very different by regions. The USA responded to increased world prices with new legislation called "freedom for farming", which entailed reduced production regulations and set aside programmes, furthermore export opportunities were seized. On the other hand in the EU, policies continued to hamper transmission of international price changes to the domestic market. If sustained in the long run, these policies will be inefficient, will not be fair to low-income countries and will not be coherent with development policy. In an integrated international food system, first world agriculture and food policies have to pay ever more attention to the de-stabilisation effects of their actions.

The prospects for meeting future food needs will depend on both the supply and the demand for food, and are likely to vary enormously by region. Most projections agree that food trade from today's main exporters, that is North America and Europe, will increase further in the coming decades (Alexandratos, 1995). Some developing countries will be reasonably well placed to manage their food needs, but the outlook is very worrisome for Africa and South Asia.

Population growth in Africa will outstrip growth in food production for a long time to come unless much more is done to accelerate agricultural growth. It is extremely unlikely that the region will have the necessary foreign exchange to import large amounts of food. The real prices of Africa's traditional export crops are low and declining, whereas the non-agricultural sector is tiny and will take a long time to expand sufficiently to generate enough foreign exchange to purchase the necessary food. It is equally unlikely that African governments will be able to count on enough food aid to make up the difference. World cereal food aid quantities declined 50% between 1993 and 1996. If current trends continue, it is estimated that by the year 2020, Africa will have an annual market demand for cereals of 158 million tons, and regional production of 132 million tons (Rosegrant *et.al.*, 1995). However, many of the needy will be priced out of the market. All indicators concur that poverty, malnutrition, and hunger will increase rapidly in Africa in the coming years, unless serious action is taken to avoid them. This in turn will accelerate the further degradation of vital natural resources, and threaten the sustainability of any production increases that are achieved.

Essentially, all estimates concur that South Asia, and particularly India and Bangladesh, holds the largest proportion of the developing world's underweight children. While in terms of food outlooks, although not as badly off as Africa today, there are disquieting trends in Asia. Yields are increasing at a slower rate than they did in the past three decades. Growth in the rice yields, for instance, has slowed from an annual rate of 3% in the late 1970s and early 1980s to less than 2% during the 1980s. China's food policy will largely determine the future of world food markets. An expansion of import demand is almost certain, but by how much? Discussions often focus on the supply side, which is indeed quite a constraint. Demand so far follows in the tracks of other cast Asian countries, and this mean rapid rise in the consumption of animal products. A cereals import demand of 60 to 100 million tons per annum by 2020 does not seem unlikely according to accumulating comprehensive analyses.

Developments in Central and Eastern Europe are particularly important to Europe. Due to malfunctioning markets, lack of safety nets, and underemployment, the early 1990s have seen a substantial increase in the food-insecure population groups of some economies in transition. The significance of access to land for household food security increased. In Russia, for instance, 25 million households derive much of their staple foods from garden plots. The income earned in cash and kind from the household plots is, for instance, about 26% in western Russia and the contribution to household calories is large (von Braun *et al.*, 1996). Rapid changes and adjustments are still taking place in transforming market and pricing systems, in the system of subsidies to output and input markets and the credit market, and in the process of privatisation and other structural reforms. As part of the still greater economic changes taking place in the former planned economics of Central and Eastern Europe, these transformations may have extensive ramifications for international food markets. Will Russia, for instance, continue to be a major net importer of food commodities, a marginal exporter, or even a significant exporter of food? When authority was shifted to the regional level, the increased segmentation of food and agricultural policy made it unlikely that trade opportunities would be exploited rapidly. Furthermore the state of infrastructure and friction in the marketing system are not conducive to trade. It seems likely that the large metropolitan areas, such as Moscow and St. Petersburg, will become increasingly dependent on food imports instead of stimulating faster development of domestic food industries. These import demands will be partly for high-quality foods for a society characterised by an increasingly skewed distribution of income and a higher demand for "convenient" foods that take little time to prepare. Russian agriculture will probably remain "in transition" for a long time.

SELECTED KEY AREAS OF POLICY ATTENTION

The discussion of policy for improving nutrition must not be limited to direct food- and agriculture-related policies. *Non-agricultural and economy-wide policies* such as industrial protection and fiscal policies are highly relevant to food and agricultural prices, income, and the employment of the poor, and thus also for food security over the short term and long term (Krueger, Schiff and Valdés, 1988). Long-term effects of alternative development strategies for poverty reduction and growth have shown the striking relevance of the choice of economic strategy.

Appropriate technological innovations in agriculture reduce unit costs of production and marketing, and induce economic gains by stimulating agricultural growth, improving employment opportunities, and expanding for supplies, all of which involve and benefit poor producers and consumers, and help to reduce food insecurity. *National and international agricultural and research systems*, in particular, are the forces driving the technological innovation required to achieve the sustainable agricultural growth that will make the needed food available to the world's growing population (Pinstrup-Andersen, 1994). Renewed actions are needed to accelerate technological innovation in many smallholder-dominated regions of the world in order to meet the nutritional goals directly and indirectly.

Commercialisation and market integration of agriculture in low income countries contributes to improving food security and the poor benefit from the increased income and employment generated by such activities. Gains in real income from commercialisation typically translate into gains in food consumption and nutritional welfare (von Braun and Kennedy, 1994). Those affected can acquire more food, reduce their workloads and thus improve child care, enhance their household sanitation and housing environments so reduce their exposure to infectious diseases, and strengthen the effective demand for both preventive and curative health care. Smallholders often strive to maintain subsistence food production along with new commercial production, despite higher returns to land and labour from cash crops. The poor are forced to adopt this strategy, more than anyone else. Given their risky economic environments and the lack of insurance markets, maintaining their own food supplies may be a sound economic strategy. Agricultural policy can effectively support it by promoting technological change in the production of subsistence foods. This also provides further latitude for specialisation at the farm level, and thereby permits smallholders to derive further gains from market integration.

In urban areas, household food security is primarily a function of the real wage rate (*i.e.* the rate relative to food prices) and of the level of employment. The prevalence of food deficiency tends to be lower in urban areas than in rural areas. But the miserable sanitation environment in poor urban locations and certain aspects of urban lifestyles sometimes make the urban nutritional situation qualitatively worse than in the rural situation. Urban food insecurity and malnutrition will become an increasingly important problem in the future, as both rates of urbanisation and problems with urban sanitation, diet quality and food safety grow. By the year 2025, 57% of Africa's population may be urban, as opposed to only 34% in 1990.

The problem of undernutrition is paralleled by extensive and growing public-health problems with *overweight and obesity*, especially in urban areas, not only in rich countries, but also low- and middle-income countries. Agricultural research systems will have to move closer to the consumption and nutritional issues in the future to address these complex matters of nutritional well being. Food industries serving low income consumers in the mega-cities of developing countries in the future must be seen as partners of agricultural and food research systems.

IMPORTANCE OF FOOD RELATED PARADIGMS

In the long run the context of food and agriculture will be determined largely by how we think about the relationships shaping nutritional well being and how society values nutritional welfare. Actually, the food and nutrition problem may be viewed from three different perspectives, representing three different paradigms: first, as a basic human right, second, as a symptom of broader poverty and development problems and, third, as a cause of these poverty and development problems.

Human Right

Considering nutritional well-being to be a basic human right of every individual means that, in principle, there can be no compromise. The World Food Conference re-confirmed the right to food as a *basic human right*, which had already been mentioned in the Universal Declaration of Human Rights of the United Nations in 1948. Although states continue to endorse the right to food, they have not translated it into law or developed national or international mechanisms to supervise its implementation. This does not imply that the stated human right is meaningless. The consensus and its codification provide a foundation for advocacy and political pressure in the countries which signed the related declarations. Establishing new rights may have an important effect in the long run. But in this case rhetoric and reality are still far appart.

Symptom of poverty

Considering malnutrition to be a *symptom of poverty and development problems* (*i.e.* an outcome) suggests that food availability and access (being mainly functions of structural conditions and changes in income, agriculture, and trade) interact with health, the sanitation environment, human behaviour and knowledge in giving rise to nutritional outcomes. Policy is then called upon to rectify constraints in any of these domains. This is the most accepted view. *While food availability* may be a problem for many people when availability declines and prices rise, the problem assumes crisis proportions mostly for the poor. This is why food availability needs to be evaluated within the context of poverty when availability problems turn into access problems (*e.g.* when prices rise), be it at the national or household level.

Precondition of Development

Taking the third view that nutritional well-being is a *pre-condition for development* argues that lack of productivity (in a broad sense) is a result of malnutrition. The nutritional well-being of the poor is thus not merely an outcome of development, but a pre-condition for it. The linkages between the two are of both a direct, short-term nature and an indirect, long-term one, whereby the latter also relates closely to population growth. Improved nutrition leads to higher physical productivity and higher economic productivity in the labour market. The policy implications are clearly radical and argue for a strong emphasis on 'nutrition first'. Clearly, there are strong positive relationships between improved nutrition and economic development (Fogel, 1994).

The position proposed here is to accept each of the three views (human right, symptom of poverty, precondition of development) and, in that context, give more emphasis to the generally under-valued third perspective, "food as the precondition to development view".

CONCLUSION

The *new international economic context* of agriculture is revealed by new trade opportunities, increased economic differences between world regions, and new emerging legal and policy conditions for agriculture and natural resources. Further globalization in trade and knowledge exchange puts agriculture production and scientific systems to test with respect to their national and regional comparative advantage.

Food and agricultural policy must give due consideration to a *long-term perspective* for a world without hunger, for dynamic development of the world's rural areas, and for high quality and safe food. The long-term challenges in world food and agriculture, and the acute food problems of the poor, must be addressed simultaneously. Europe has a large responsibility in that respect. Coherence between agricultural policy and development policy has not yet been achieved.

The *changing diets* in high income and in rapidly developing countries increase demand for a broad range of food choices. Increasing market segmentation in response to demand requires renewed attention to demand, and rapid urbanisation calls for consideration of efficient rural-urban linkages and co-operation between food industry and research systems.

Agriculture is increasingly constrained by *land scarcity and endangered natural resources*, and the improved recognition of the real value of these resources (soils, water, genetic resources). Appropriately dealing with the long-term challenges of agricultural demand and the resource constraints requires constant *innovation of technologies and institutions.* Rapid changes in technology, both biological and informational, present new opportunities for agricultural production, processing, and marketing in the 21st century.

However, *agricultural research* must not just take a position of passive response to changes in the international context. Research can and has to act as an active force of change: the innovations growing out of it will shape the future of peoples' well-being. European agricultural research systems need to respond by forming effective alliances in Europe and internationally, and by searching new links with public and private research. In any case, the accumulated scientific abilities and know-how in European agricultural research systems are an asset that must be made available to meet European, global, and low-income countries' food and agricultural challenges.

SELECTED REFERENCES

ALEXANDRATOS N.ed, 1995. *World Agriculture: Towards 2010. An FAO Study.* Baltimore, Md., and London: Johns Hopkins University Press, for the Food and Agriculture Organization of the United Nations.

VON BRAUN J. ed. 1995. *Employment for poverty reduction and food security.* Washington, D.C.

VON BRAUN J., and KENNEDY E., eds. 1994. *Agricultural commercialization, economic development, and nutrition.* Baltimore, Md., and London: Johns Hopkins University Press.

VON BRAUN J., SEROVA E., THO SEETH H., and MELYUKHINA O.1996. *Russia's food economy in transition: Current policy issues and long-term consumption and production perspectives.* Discussion Paper. Washington, D.C.: International Food Policy Research Institute.

FOGEL R. W. 1994. Economic growth, population theory, and physiology: The bearing of long-term processes on the making of economic policy. *The American Economic Review* 84 (3): 369-395.

FOOD AND AGRICULTURE ORGANIZATION OF THE UNITED NATIONS (FAO), 1996. *Technical background documents,* World Food Summit, Rome,. *p . 1-5.*

IFPRI (International Food Policy Research Institute). 1995. *A 2020 Vision for food, agriculture, and the environment.* Washington D.C., International Food Policy Research Institute.

INTERNATIONAL MONETARY FUND, World Economic Outlook - A Survey by the Staff, *World Economic and Financial Surveys*, Washington, D.C., May 1995.

INTERNATIONAL MONETARY FUND, World Economic Outlook - A Survey by the Staff, *World Economic and Financial Surveys*, Washington, D.C., October 1995.

KRUEGER A. O., SCHIFF M. W., and VALDÉS A. 1988. Agricultural incentives in developing countries: Measuring the effect of sectoral and economy-wide policies. *The World Bank Economic Review* 2 (September): 255-271.

MCRAE H., 1994, *The World in 2020 - Power, Culture and Prosperity: a vision of the future .* Harper Collins Publishers.

ORGANIZATION FOR ECONOMIC CO-OPERATION AND DEVELOPMENT (OECD),1996. Globalization and Linkages to 2020: Challenges and Opportunities for OECD Countries, *OECD Working Papers*, Vol.IV, No.76, Paris 1996.

ORGANIZATION FOR ECONOMIC CO-OPERATION AND DEVELOPMENT (OECD), *Science, Technology and Industry Outlook*, Paris, 1996.

PINSTRUP-ANDERSEN P. 1994. *World food trends and future food security.* Food Policy Report. Washington, D.C.: International Food Policy Research Institute.

QURESHI Z., 1996. *Global Capital Supply and Demand - Is There Enough to Go Around ?* The International Bank for Reconstruction and Development / The World Bank, Washington, D.C.,.

ROSEGRANT M. W., AGCAOILI-SOMBILLA M., and PEREZ N. D. 1995. *Global Food Projections to 2020: Implications for Investment.* Food, Agriculture, and the Environment Discussion Paper 5. Washington, D.C.: International Food Policy Research Institute.

THE WORLD BANK, 1996. *Global Economic Prospects and the Developing Countries,* The World Bank Book, Washington, D.C..

WORLD TRADE ORGANIZATION, 1995. *International Trade - Trends and Statistics* (prepared by the Economic Research and Analysis Division and the Statistics and Information Systems Division).

CHAIRMAN: MARION GUILLOU, *MINISTRY OF AGRICULTURE, FRANCE*

VICE CHAIRMAN: PIERRE COMBRIS, *INRA, IVRY, FRANCE*

THEME 1

TOMORROW'S DIET

Workshop 1
Decisive elements in the consumer's choice

TRENDS IN FOOD CONSUMPTION IN THE EUROPEAN UNION: TOWARDS A EURO DIET?[1]

BRUCE TRAILL
Centre for Food Economics Research (CeFER)
Department of Agricultural and Food Economics
The University of Reading
Reading RG6 6AR, UK

ABSTRACT

Economics, demographics, life styles and consumer attitudes to food are all tending to converge within Europe and beyond. The extent to which this process progresses has important implications for the strategies of firms and for the organisation and management of food systems throughout Europe. It is also important for the orientation of agricultural and food research programmes in individual countries and the EU. This paper explores some of the issues and reviews some of the recent literature and evidence on convergence. Clearly food consumption patterns have become more similar across countries, but some big questions remain unanswered: how far will the process go and how quickly? It is suggested that rather than look at countries in aggregate, it would be better to consider segments of consumers with similar consumption behaviour and ask to what extent such segments will cross national borders and to what extent the size of the segments will converge across countries.

INTRODUCTION

The intention of this short paper is to consider whether food consumption patterns in Europe are converging and what might be the final state of any convergence process. More detail of some of the forces for change will be covered in the other papers in this session.

The issue of convergence is of evident importance to providers of goods and services: the more alike markets are, the more alike can be the marketing mix which firms use to target those markets; no need for costly product modifications to satisfy local tastes; no need to develop different advertising/promotion strategies; no need to develop products suitable both for the corner shop and the hypermarket; and no need to price differently in each market to reflect consumers' perception of the positioning of the product in the luxury/basic spectrum. Stated like this, the benefits to firms of convergence appear unequivocal, but it should be born in mind that differing demand characteristics (as

---

1. This paper is being printed with the permission of the European Review of Agricultural Economics

well as the ability to separate markets) is a precondition for price discrimination, so firms with market power in a number of markets may prefer that demand does not converge and that markets remain separated.

EVIDENCE FOR CONVERGENCE

Connor (1991, 1994) argues that Europe is moving towards the US in food consumption. In the earlier article (an NC-194 working paper: Connor 1991 p2) he states his belief that 'all consumers are basically alike' in the sense that they 'with the same incomes and socio-demographic characteristics, facing the same relative prices, and holding the same information, will tend to choose the same basket or array of goods'.[2] Thus, as incomes, prices and demographic factors in Europe catch up with developments in North America, so food consumption patterns will converge. Connor shows that for a range of processed foods, European consumption correlates strongly with American consumption 5 and 10 years earlier, but not with consumption in the same time-period.

The same logic applied within Europe predicts that as the European nations converge, so will their food consumption patterns. There is some evidence to support the view that convergence has been taking place. Using the broad product categories of FAO, Table 1 indicates the coefficient of variation in consumption across 29 European countries in 1961 and 1990. In all cases it is lower in 1990, implying that for all of these products, consumption has become more similar across European countries during the last 30 years. However, this is not to say that they have all become the same. Hermann and Röder (1995) and Gil *et al.* (1995) apply different statistical methodologies to, respectively, OECD and EU food consumption data, in both cases concluding that convergence is occurring, though in the latter case concluding that the speed of convergence is diminishing. Gil *et al.* (1995) demonstrate that significant dietary differences remain: using cluster analysis, they identify 7 Western European country groupings: Portugal and Spain; Greece and Italy; Benelux, France, Ireland and the UK; Austria, Germany and the Netherlands; Finland; Denmark; and Norway and Sweden.

Table 1- Coefficients of variation of food consumption across 29 European countries (kg per cap per year). *Source: Computed from FAO food balance sheet data*

	1961	1990
Alcohol	70.0	52.5
Cereals	31.6	30.1
Eggs	47.3	31.3
Fruit	58.5	42.4
Meat	39.8	28.8
Milk	43.4	31.2
Pulses	99.1	80.5
Starchy roots	45.0	43.1
Sugar	41.4	21.8
Vegetables	43.5	42.6

[2] This somewhat controversial statement had been expunged by the time the article emerged in journal form (Connor, 1994).

CULTURE AND INDIVIDUAL PERSONALITY DIFFERENCES MATTER

One reason that food consumption patterns should not be expected completely to converge among countries even if socio-economic and demographic factors do, is that culture is an important influence on behaviour, and cultural diversity has proved resistant to pressures from foreign travel, global media and telecommunications. Another is that individuals differ, both within and between countries, and that individual personal characteristics are another determinant of food consumption (see *e.g.* Steenkamp, 1996, for a general discussion of these issues). Different individuals have different 'values' and, though psychologists may disagree about precisely how many fundamental values there are, they accept that different people place different emphasis on different values. For example, Hofstede (1984), uses four dimensions to categorise personality, (i) individualism *vs* collectivism (importance attached to individual freedoms *vs* society) (ii) power distance (tolerance of inequality in wealth and power/centralisation of authority) (iii) risk (the extent to which risk is avoided through laws, rules, religion) (iv) masculinity *vs* femininity (emphasis placed on masculine values of performance, aggression, visible achievement). In a study based on questionnaires with 116,000 IBM employees world-wide, the responses were used to group similar countries. Some of the groups are shown in Table 2: The *Anglo-American* group consists of the U.S., U.K. Australia, New Zealand, Ireland and South Africa; the *Nordic* group has all the Scandinavian countries; the *Germanic* includes Germany, Austria and Switzerland; the *Latin-European* contains France, Spain, Italy, Portugal, Belgium and Latin America. These country groupings are intuitively more logical than those obtained by clustering according to economic and demographic variables (*e.g.* Krause *et al.,* 1995), which puts the U.S., Japan and Sweden in the same cluster.

Table 2- Characteristics of Country Groupings
Source: Adapted from Hofstede, 1984.

Group	Power Distance	Uncertainty Avoidance	Individualism	Masculinity
Anglo-American	LL	LL	HH	H
Nordic	LL	H	L	L
Germanic	LL	HH	HH	H
Latin-European	H	HH	H	L
Japan	H	HH	LL	HH
Far East	HH	L	L	M

LL=Very Low, L=Low, M=Medium, H=High, HH=Very high.

Country groupings themselves, however, represent only an average of their populations, whereas any businessman or marketing academic will testify to the fact that food markets within individual developed Western countries are becoming more finely segmented (*e.g.* Hughes, 1994; Barkema *et al.,* 1993), and the trend is very firmly away from mass markets. Recognising that consumers within countries are not all the same, the most sensible approach is to understand consumer markets as 'groups of buyers that share the need and desire for a product and the ability to pay for it rather than those who share a common border. Buyers in a segment seek similar benefits from and exhibit similar behaviour in buying a product' (Blackwell *et al.*, 1994, p221.). According to this approach, demographic and economic considerations remain important, but so do psychosometric, attitudinal, cultural and lifestyle characteristics, and the process of convergence is best viewed as the growing importance of homogeneous segments of consumers which cross national boundaries. Sometimes such groupings are known in the international marketing literature as strategically equivalent segments because they respond to the same marketing mix.

The nearest that exists so far to a tool for cross-country segmentation along these lines for food consumption is the 'food-related life styles' work of Grunert *et al.* (1993) which clusters consumers according to characteristics associated with their attitudes to food and their behaviour towards food, in terms of shopping and eating (which of course are related to economic and demographic as well as cultural and psychosometric characteristics). Thus far the work has concentrated mainly on the validation of the instrument for cross-cultural analysis (*i.e.* do consumers in different cultures interpret the questions in the same way?) (Grunert *et al.* 1993) and its application to separately develop food related life style segments in three countries, Germany, France and UK (*e.g.* Grunert *et al.* 1995), but not in detail to the search for cross-country segments. However, as Table 3 indicates, efforts to look at trans-national segments for these countries has begun.

Table 3- Food-related life style segments in three countries
Source: Grunert et al. (1996).

Segment	Germany	France	Great Britain
The uninvolved food consumer	21%	18%	9%
The careless food consumer	11%		27%
The moderate food consumer		16%	
The conservative food consumer	18%	13%	19%
The rational food consumer	26%	35%	33%
The hedonistic food consumer		18%	
The adventurous food consumer	24%		12%

Another study funded by the European Union is underway, using somewhat different methodology (Laddering, via Means-End Chains. See Steenkamp *et al.*, 1995) based on consumer interviews in 11 EU states. At the end of that study we will know more about the extent of cross-European market segments, though there remains one problem even if such segments are 'uncovered': do such methods of segmentation provide good predictions of food consumption? The suspicion remains that similar food related life styles in different countries could lead to different food consumption.

CONCLUSION

In conclusion, the evidence for either globalization or Europeanisation of food consumption is rather limited. There clearly are forces at work, as described by Connor (1994), which tend to encourage convergence in consumption. There is evidence that this is happening within Europe. However, it is simplistic to assume that all consumers are basically the same. Even within countries marketers have long used psychosometric as well as economic and demographic variables to segment markets. Between countries, cultural differences magnify these effects, though there are signs that some global segments of consumers are emerging. Ongoing research should shed light on whether trans-European segments of consumers already exist, and the economic, demographic and other characteristics of those segments should be useful in deducing the extent to which they are likely to become more important over time and/or can be targeted by firms as strategically equivalent segments.

SELECTED REFERENCES

BARKEMA A., DRABENSTOTT M., COOK M. L., 1993. The Industrialization of the U.S. Food System, pp 3-20. *In* Padberg, D I (ed) *Food and Agricultural Marketing Issues for the 21st Century*, FAMC 93-1, Texas.

BLACKWELL R.D., AJAMI R., STEPHAN K. 1994. Winning the Global Advertising Race: planning Globally and Acting Locally, pp 209-232. *In* Hassan, S.S. and Kaynak, E. (eds). *Globalization of Consumer Markets: Structures and Strategies*, International Business Press, New York, 333 p.

CONNOR J., 1991. North America as a Precursor of Changes in Western European Food Purchasing Patterns, NC-194 *Occasional Paper* OP-22, 56 p.

CONNOR J., 1994. North America as a Precursor of Changes in Western European Food Purchasing Patterns, *European Review of Agricultural Economics*, pp 155-173.

GIL J. M., GRACIA A., PÉREZ Y PÉREZ L., 1995. Food Consumption and Economic Development in the European Union, *European Review of Agricultural Economics*, 22, pp 385-399.

GRUNERT K.G., BRUNSO K., BISP S., 1993. Food-Related Life Style: Development of a Cross-Culturally Valid Instrument for Market Surveillance, *MAPP Working Paper* 12, Aarhus.

GRUNERT K.G., BRUNSO K, BISP S., 1995. Food-Related Life Style in Great Britain, *MAPP Project Report* 14, Aarhus.

GRUNERT K.G., BAADSGAARD A., HARTVIG LARSEN H., MADSEN T.K., 1996. *Market orientation in food and agriculture*. Boston, MA: Kluwer.

HERMANN R. and RÖDER C., 1995. Does Food Consumption Converge Internationally? Measurement, Empirical Tests and Determinants, *European Review of Agricultural Economics*, 22, pp 400-414.

HOFSTEDE G., 1984. *Culture's Consequences*, Sage Press, London.

HUGHES D. (ed), 1994. *Breaking with Tradition: Building Successful Partnerships and Alliances in the European Food Industry*, Wye College Press.

KRAUSE J.H., WILSON W.W., DOOLEY F.J., 1995. Global Market Segmentation for Value-Added Agricultural Products, *Agribusiness*, 11(3), pp 195-206.

STEENKAMP J-B, AVLONITIS G., GRUNERT K.G., TRAILL W.B., VANDEN ABEELE P., WEDEL M., AUDENAERT A., 1995. *Consumer-led approach to foods in the EU: development of market-*

oriented strategies based on pan-European segments, Proceedings of the 24th Conference of the European Marketing Academy (EMAC), pp 2009-2019, Paris.
STEENKAMP J-B., 1996. Dynamics in Consumer Behaviour with respect to Agricultural and Food Marketing, *in* Wieringa, B, Grunert, G, Steenkamp, J-B, Wedel, M and Van Tilburg, A (eds) *Agricultural Marketing and Consumer Behaviour in a Changing World*, Proceedings of the 47th Seminar of the Association of Agricultural Economists, Wageningen, pp 15-38.

CONSUMER CHOICES : TASTES, PREFERENCES, CONSTRAINTS

Pierre Combris
INRA, Laboratoire de Recherche sur la Consommation
63, rue Brandebourg
94205 Ivry-sur-Seine

Knowledge of what determines consumer choices is essential for anyone who wants to have an idea of the foods of tomorrow. Basic determining factors include individual tastes and preferences, but because tastes, and even moreso product preferences are usually heterogeneous, and sometimes unstable, they are difficult to determine and predict. Behaviour, at least to a certain degree, is made predictable by the existence of nutritional, economic, temporal, social and other constraints.

Fortunately for the consumers, but unfortunately for the forecasters, some of these constraints gradually subside and lead to increased diversification in food consumption. We are not necessarily entering an era of total unpredictability, even though there are many signs hinting that we are indeed at the beginning of a new phase of food history. The coming era will of course have features that are already appearing rather clearly, and do not risk being called into question since they are connected to both fundamental consumer expectations and to some well-anchored socio demographic trends.

After these elements create the structure, there are all sorts of possibilities. Research in the field of food consumption, as it should, will focus on the formation and evolution of this vast domain, marked by freedom of choice, but it will also contribute to providing it with tangible contents. The challenge is mighty.

THE THREE PHASES IN THE EVOLUTION OF FOOD CONSUMPTION: QUANTITATIVE GROWTH, STRUCTURAL EVOLUTION, OVERALL DIFFERENTIATION

A long-term analysis of the combined effects of nutritional and economic constraints gives an idea of changes in the level and structure of food consumption in most developed countries. Very schematically, these changes can be divided into three distinct phases: an overall quantitative food consumption growth phase, an evolutionary phase for the nutritional structure of the diet, and finally a generalised products differentiation phase.

Figure 1: Energy supplied by major food groups in France during the last two centuries

Source: from J.C. Toutain, op. cit.

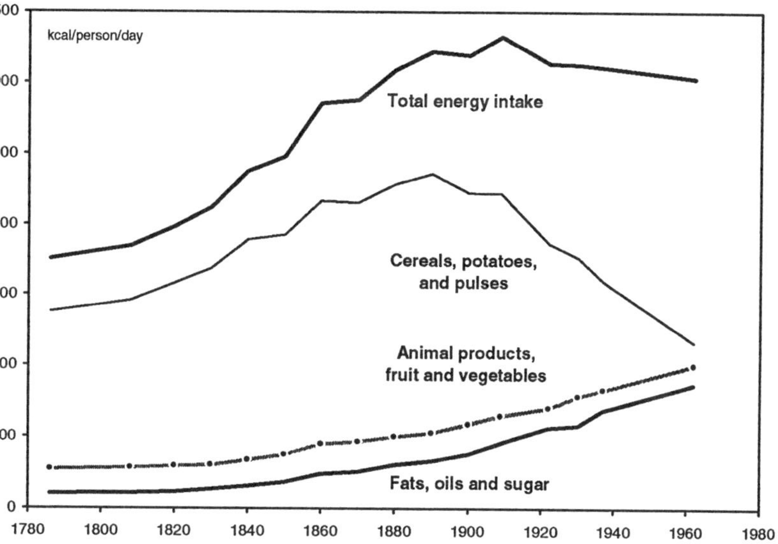

Figure 2: Percentage of energy from carbohydrates, fat and protein in France since 1900

Sources: adapted from J.C. Toutain, op. cit., and OECD, Food Consumption Statistics.

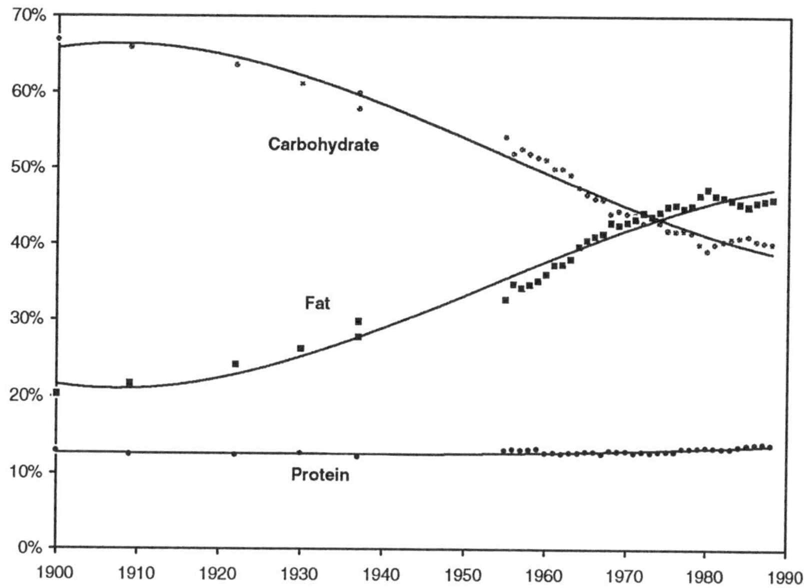

Figure 1, for instance, shows the trend of this evolution in France over the last 200 years. During the 19th century, the total calorie intake rose steadily as a result of increased consumption of all foods. Between 1880 and 1890, when the overall saturation level had been reached, the consumption of staples (cereals, starches, pulses) started dropping off as the consumption of other products (products of animal origin, fruits, vegetables, fats and sugar) steadily rose. This type of change occurred in almost all the countries as money and supplies became available, and reflects a strong similarity in basic tastes which, depending on the country, are expressed by special preferences for specific foods connected to what is locally available and, thus, to national and regional cultures. In nutritional terms, this trend translated into continuous, radical changes in the structure of the average diet (Fig.2), a structural evolution that strongly characterised food consumption patterns in most developed countries during the greater part of the 20th century.

Hence, for slightly more than a century, per capita food consumption measured in terms of total caloric intake has been levelling off, a trend that, however, has not triggered a major crisis in the production end since it has been largely counterbalanced by the increased demand created by population growth and by greater intermediary consumption (cereal grains that are no longer consumed direct are processed into animal protein). In other words, the structural evolution in consumption throughout this period has been the main source for increasing market opportunities.

Apparently, at least in the developed countries, the process is coming to an end. Global demand has no reason to grow except to keep up with the - increasingly slower - population growth rate. Furthermore, there is no group of foods that is especially slated to develop at the cost of other groups. This new situation will gradually and profoundly change the food offer.

The fact that the nutritional structure of the diet stopped changing in the beginning of the 1980s in France (Fig.3), and in many other developed countries, apparently was the first sign of a new stationary phase, although it is too early to consider it as a new long-term trend. This structural stabilisation is now commonly viewed as the result of dietary recommendations to reverse trends observed during the years of economic growth, and most of all, to reduce the intake of lipids, especially the ones that are high in saturated fatty acids, *e.g.* products of animal origin. (Buss 1991, Dupin 1981).

If the new situation is really the outcome of a more widespread awareness of the link between diet and health, and as such represents a permanent change in consumer preferences and not an effect of slower economic growth, we must be coming to the end of a long transition phase and entering a new era of our food history.

The combination of foreseeably slight increases in global demand and doubts about the priority growth for certain food groups will lead to across the board competition among food products. In the era before us, the consumption of one food will increasingly mean a corresponding decrease in the consumption of another. This differential growth, which already affects declining food groups, is going to impact the whole household basket. Because all food products, to a certain extent, can supplant each other, suppliers will have to make their products more distinctive, a task that will require new and more systematic study on food feature mixes: quality, taste, processing sophistication, nutritional characteristics (nutritional enrichment or reduction). This explains why the coming phase will probably be marked by pronounced food differentiation and increasingly segmented markets.

Figure 3: Percentage of energy from carbohydrates, fat and protein in France since 1955

Source: OECD, Food Consumption Statistics.

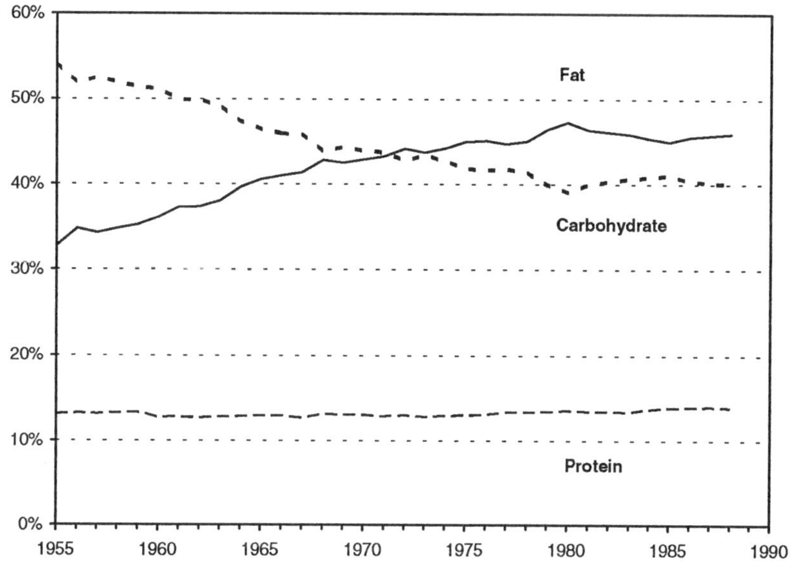

In a universe of massive competition and substitution, the objects of economics become more complex. Since consumers have to choose from among numerous multi-feature products, proper information and labelling become very important, and hence, the creation, validation and dissemination of scientific information on the virtues and dangers of a food or any of its components become strategic challenges as a counterpart to this new step in the evolution of food consumption.

THE CONSUMER: BASIC EXPECTATIONS AND NEW DEMANDS

Recognising the existence of forces that encourage supply differentiation is not enough to give specific traits to foods that manage to break into the market. It is no easier to predict the result of a "societal" selection process than a "natural" selection process. Although lacking in tangible content, careful scrutiny of consumer expectations can give us an idea of the direction in which products may probably develop.

What consumers basically expect from food can be described in six sub-sets related to food characteristics. Expectations focus on the sensorial appeal, food safety, health, availability, convenience and social acceptability.

• Sensory appeal includes all observable physical characteristics indicative of the individual's appreciation of a food: first of all its looks, colour, and smell (and, increasingly, packaging also), then and of special importance, organoleptic qualities, in other words, texture, aroma and taste.

• Safety concerns short or longer term harmlessness. It is not present at first sight, but the least doubt, be it related to appearance, storage conditions, contamination risks, absence of information or guarantees will suffice to doom a product.

• Health includes the expected medium- and long-term effects foods will have on one's physical conditions. Unlike food safety, where uncertainties are not tolerated, the health factor can stand a certain degree of compromise because the diet is more important than the individual food item as far as the health factor is concerned. This said, each food can claim it has the nutritional virtues that one consumer may seek...which another consumer seeks to avoid.

• Availability concerns the facility with which a commodity can be obtained. Before it can be consumed, a food item has to be physically accessible and financially affordable! It should be easy to find and buy (in a pinch, we could add "regardless of place or time") at bottomline prices, of course.

• Convenience in transport, storage, unpackaging, preparation, and waste elimination cover all the factors that ease constraints to household food procurement. The development of these characteristics has been (and for long will still be) a major avenue to explore for new food products.

• The social acceptability of a food item is the result of many factors, connected to both culture and ethics. Depending on the nature and strength of constraints, a product will be actively sought, easily accepted, or, on the contrary, totally excluded from a population's or particular social group's food repertoire. De luxe foods and ethnic foods are examples of the first two categories. Under "rejects" we find foods that are banned for reasons of religion, ethics, politics and principles (rejection of foods judged to be produced under conditions hostile to nature, animals, or people).

Since consumers' food expectations are very general, there is no reason for them to change. Consequently, the differentiated strategies mentioned above boil down to a search for a competitive advantage in one of the six domains of customer expectations or another.

These domains occupy a more or less central position, depending on the products and the times. At present, focus to some degree or another is on health, and ranges from respect for a balanced diet to a preventive or even a quasi-medical diet which is considered to be blessed with curative effects or be able to improve one's intellectual and physical performance. The concerns which the public authorities have about foods, as reflected in the media's interest in the subject, slowly but surely have brought about changes in eating habits, with popularity fading for some foods like red meat, alcohol and animal fats in favour of the "healthier" foods such as fresh fruits and vegetables, dairy products and plant oils. This tendency may gain ground in the coming years through the demand for products guaranteed to be nutritious and rich in micro-nutrients. Importance to health will probably continue to be a major factor in choosing foods, but it should not be forgotten that the products that benefit or suffer from this trend are always at the mercy of new well-circulated piece of information. Here again we see that providing scientific information is an extremely important strategic challenge.

Today's consumer, partly for reasons related to the present day context (*e.g.* the "mad cow" or BSE crisis, the use of growth hormones, the development of genetically modified products) is extremely sensitive to the question of food safety and ethics. We can expect a very sharp increase in the demand for information and regulatory guarantees in the future. The absolute refusal to accept any risks of foods being unsafe or unclean could spread to the field of nutrition and taste, especially as concerns fresh foods. Along similar lines, the increased opposition to certain production and trade conditions could lead a significant part of the consumers to demand fuller guarantees concerning the origin of foods, and systematically reject products suspected to be "unethical". Depending on the case, in other words, depending on the level of competition, on the producers' strategic choices and on the consumer's "tolerance threshold", new requirements may lead to the production of specific ranges of products that appeal to certain groups of consumers who are willing to pay the price, or that bring about a higher level of minimum guarantees for all products to meet.

The differences in domains of consumer expectations may provide ample opportunities for differentiating products to attract corresponding segments of the population. These segments will also be formed on the basis of the characteristics of consumer households. Are there any significant trends in this field?

SOCIO-DEMOGRAPHIC TRENDS AND CHANGES IN FOOD CONSUMPTION

As restrictions to food purchases such as product price and availability, household income, etc. gradually slacken, the range of choices expands, bringing about more and more behavioural diversity. The result is that the relation between food consumption and socio-demographic household characteristics becomes less intrinsic, thereby making behaviour less predictable.

Economic determinants provide a good example. As income and price constraints eased, meat consumption gradually rose. The increase was much greater for the meats with the sharpest price cuts *i.e.* pork and poultry. Most households can buy as much of these meats as they please, which means that the connection between consumption level and income is no longer apparent; this is not the case for beef (Fig.4).

As indicated above, the position of a household on the revenue scale can no longer be used to forecast pork and poultry consumption levels. It can still be used, to a certain extent, to predict beef consumption, but since the range of beef consumers is bound to grow broader with time, it will become increasingly difficult to tie predictions to income.

This means that the search is on for other variables that have a systematic effect on consumption. There are some socio-demographic characteristics, *e.g.* age, educational level, profession, but it is becoming increasingly necessary to work on a case by case and product by product basis.

Figure 4: Influence of income on quantity purchased and price paid in France in 1991

Source: *National Food Survey*, INSEE, 1991.

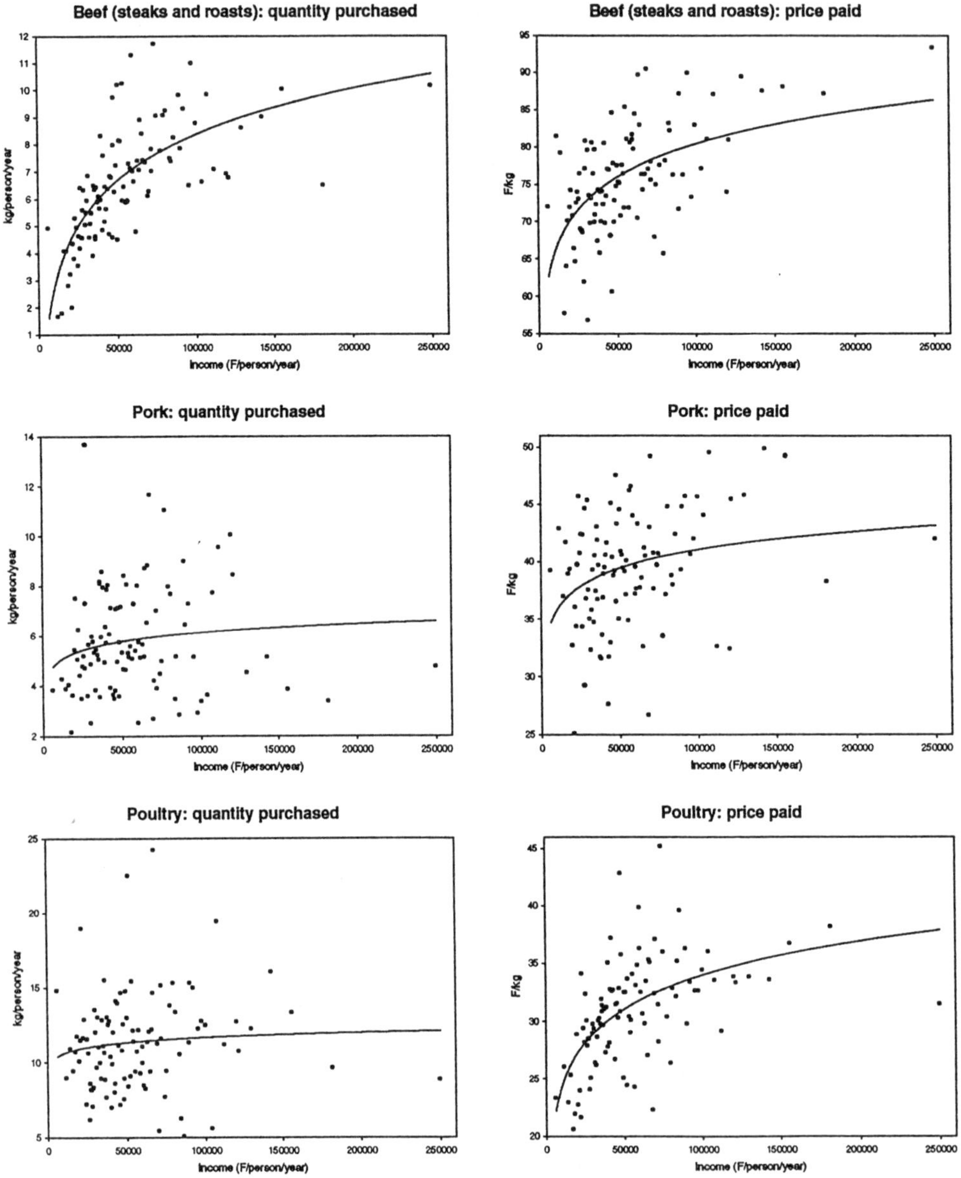

Yet there are some general tendencies that are still predictable since they are linked to deep-rooted socio-demographic trends which are not subject to sudden change. To anticipate the future changes in demand, it seems reasonable to count on longer life spans, older average population age, smaller households, increasingly more wage-earning women, specialisation per phase of the life, higher education levels, and differing value systems.

• People will live longer and the average age of the population will go up, thus moulding and developing the market for "seniors", with its attendant concern for health (sodium level, fat content, calcium, vitamins, etc.).

• The size of the household will continue to grow smaller: between 1962 and 1995 the average number fell from 3.1 to 2.5. This will be accompanied by an increased demand for products sized for one-person alone and adapted for consumption under various social circumstances. (In 1995 close to 28% of the households were composed of one person.)

• One of the major trends is the substantial increase in the number of women on the job. (Over 75% of the women in the 25 to 40 age group work now, as against slightly over 40% at the end of the 1960s.) This implies a very strong demand for preprocessed foods and the development of carry-away ready-to-eat dishes.

• There will be a sharper contrast between the various phases of the life cycle (longer schooling, later marriages, fewer children, longer active period without children, longer period of retirement) which will also encourage the development of products adapted to the constraints of each category of household.

• Last, the higher educational levels and the greater differences in value systems will be accompanied by a stronger desire for diversity and novelty. A taste for variety, exoticism and crosscultural eating has already started and will continue to grow. This trend, however, will not curtail the consumption of traditional and special regional foods.

Alongside these major trends there are evolutions that are more distinctly connected to the economic context, and therefore have scope and direction that are more difficult to anticipate. This particularly applies to increased social inequality (more rich consumers, and more financially weak or even poor consumers) which could cause the market to split, with one part being accompanied by the specific nutritional problems created by the overconsumption of very cheap foods. Remember that food is one of the main items in the family budget, and that its contract-free character and its flexibility make it a natural financial shock absorber.

The development of out-of-home consumption is also closely connected to economic growth. Between 1971 and 1991 the number of people who ate five lunches a week outside their home doubled (from 6 to 12%). France is considered to have a real potential for further growth since 45% of the people still eat their noon meal at home. The ready-to-eat foods constitute an inexpensive substitute for restaurant services, which again leads to the question of ensuring a balanced diet when the meal time and structure are less regulated.

IMPLICATIONS FOR RESEARCH ON CONSUMPTION

The paths that research must explore in order to improve our diet can be plotted by analysing the most significant economic, socio-demographic and cultural trends, together with consumers' requirements and expectations. This type of an approach uses expectations as an analytical grid to identify the strengths and weaknesses of a product in a given socio-economic context. It is useful in designing new products or, more often, in finding new attributes for existing products.

In the far-reaching competition to be expected in the coming years, public research on consumption, obviously, can contribute in particular to sectors that do not invest heavily in product innovation or research in marketing. But it will also have to consolidate its own orientations which concern both basic behavioural studies, and a view of the long term and of the global value of applied research.

First and foremost, the increasing behavioural differentiation implies the need to develop tools for observing customers (continuous recording of purchases and consumption of a large enough sample, ethnography of micro-behaviour, etc.). This is the basis of all research on consumption. Any retrospective study of trends, so badly needed in order to understand prevailing phenomena and define new behaviour, requires this type of observation. Greater differentiation is also important when making detailed studies and quantifying the effects of economic and socio-demographic variables on consumption. Furthermore, it constitutes the basis of the sociological analyses needed to explain consumer attitudes and the underlying social mechanisms.

The second important aspect of development-targeted research comes from a depth analysis of behavioural heterogeneity and consumer typologies. The stress that marketing research has laid on consumer typologies (psycho-cultural or lifestyles, that, in general combine the characteristics of the life style and the value systems) points to one of the possible paths to explore. The robustness and relevance of a tool such as this can be tested by trying to validate it in a transcultural analysis. Typology research is still an open problem. Including eating habits (regularity/irregularity of eating, relative importance of snacks, casual and formal meals, associations of foods, variety of diets, etc.) in this type of research could provide an interesting path in trying to understand behaviour and could be relevant in identifying populations at risk.

Last, by studying the mechanisms of "choice-making" in real situations, and in particular by analysing information "processed" by the consumers (general information on nutrition and foods, nutritional allegations concerning food products, etc.) it should be possible to simultaneously consider various components of the choosing process which are usually analysed separately (price, *brand name*, tastes, etc.). Thanks to the aforementioned nutritional typologies, such research may lead to a better understanding of the consumer's selection and, perhaps, ensure that the increased competition (which seems unavoidable) will lead to a true expansion of choice, rather than to a volatile loss of preferences and to behaviour that is harmful to the consumer's long-term well-being.

SELECTED REFERENCES

BERTRAND M., INSEE, 1993, Consommation et lieux d'achat des produits alimentaires en 1991, *INSEE Résultats*, no 262-263, *Consommation-Modes de vie*, no 54-55.

BUSS D., 1991, "The Changing Household Diet", in *Fifty Years of the National Food Survey 1940-1990*, Ministry of Agriculture, Fisheries and Food, HMSO, London.

COMBRIS P., 1986, "L'évolution de la consommation alimentaire des Français", *Problèmes Politiques et Sociaux*, no 544, La Documentation Française, pp. 9-12.

COMBRIS P., 1990, "L'évolution du modèle alimentaire en France de 1949 à 1988 : continuité et ruptures", *Annales de Gembloux*, 96, pp. 279-304, and *in* Eizner N. (ed.), *Voyage en alimentation*, ARF Editions, Paris, 1995.

COMBRIS P., 1996, "Mangeurs et aliments : que nous apprend l'analyse économique ?" dans *Identités des mangeurs, images des aliments, in* Giachetti I. (ed.), CNERNA-CNRS, Polytechnica, Paris.

DUPIN H., 1981, *Apports nutritionnels conseillés pour la population française*, CNRS-CNERNA, Technique et Documentation, Lavoisier, Paris.

RITSON C., GOFTON L., MCKENZIE J. (eds.), 1986, *The Food Consumer*, John Wiley & Sons.

GRIGNON C., GRIGNON CH., 1996, Long-term Trends in Food Consumption: a French Portrait, *Working Paper INRA-CORELA*, no 9610.

SENAUER B., ASP E., KINSEY J. (eds.), 1991, *Food Trends and the Changing Consumer*, Eagan Press.

TOUTAIN J.C., 1971, "La consommation alimentaire en France de 1789 à 1964", *Economies et sociétés, Cahiers de l'ISEA*, Tome V, no 11, pp. 1909-2049.

EFFECTS OF MEDIA DISCOURSE AND MARKETING

FRANCESCO BISCEGLIA
Soremartec, Group Ferrero
Drève de l'Arc en Ciel, 102
6700 Arlon Schoppach, Belgium

A long time ago a certain Mr Karl Marx said that demand generated supply. His vision of the market was simple and schematic: there "needs" felt by a mass of people who, today, we call "consumers". And the production world was expected to respond by providing the required goods and services. Yet Marx did not - and could not - know about the impact of products such as Coca Cola. And more than that, he had no idea of the incredible efficiency of our modern mass media. A successful product and strong media coverage form an irresistible couple that upset all the theories and changed the rules of the market game. Look at events (it may be more accurate to speak of phenomena) that have occurred during the last few decades.

Who needs a "need"? The game only requires an idea, a company line and a communications strategy. And if there are a few professionals around to lead the game, the consumer will be a sure catch. Or will he be trapped, as the more malicious (and there are many of them) would say, because, frankly speaking, this new scenario is puzzling, and even frightening. I seriously wonder whether an efficient marketing service and a well orchestrated advertisement campaign could succeed in selling useless, or even dangerous products. And to a certain extent, corporate communications, that very powerful, devouring god, Moloch, was held responsible for society's overconsumption.

In this type of a situation there is always a risk of being too partial. In the end, you always exaggerate, regardless of whether you are defending or attacking the way the company does things. I want to avoid being too categorical because I am convinced that this would not be the best approach to use in trying to understand the various steps in the changes that have taken place during the last few years. And yet, I feel that this exaggeration is a decisive element in trying to understand the world we live in and work out hypotheses for a future that is just around the corner. In sum, what I am trying to say is that when we talk about the market as a superpower and, even more so, the tremendous influence-wielding power of the media, it is as if we were justifying the great mutations that have turned the market upside down and radically changed the consumer's behaviour.

This is a mistake. There have been, and still are other, relevant mutations.

Don't forget that the last few decades have not all been the same.

The first undeniable fact is so elementary that we forget to think of its importance, and forget it altogether, *viz.* in modern society primary needs are usually satisfied and the consumer is attracted to new products he never thought he would "need". To paraphrase the famous words of Queen Marie Antoinette, we have been able to offer cake to people who already have had enough bread, and because they might already have had enough (or too much!) they are psychologically ready for something else. This was what happened during the post-war years when all of Europe was enjoying good economic health. Greater wealth and security led to a higher purchasing power and, thus, a higher consumption rate, especially for food. Furthermore, the modern household basket had more room for both quantity and quality. The question was what new products, or even what products would the consumer buy.

As mentioned above, I think it would be better to show the connection between the various stages of the evolution that started thirty years ago.

During the first decade, products were sold at an ex-factory stage of development. They were often the fruit of technological innovation, especially in the processing/transformation stage, and were better than what was offered during the preceding ten years. Industrial interests created a process that corporate strategy implemented. The consumer, with all due respect, was nothing more than the end of the line, not a very glorious role and, more importantly, he was exposed to many risks. He could only hope that the producer had a sense of responsibility and professionalism. But this was a hope not supported by any possibility to intervene or exercise control. The only control was the effect of market popularity, which would only become visible in the long term.

The next decade was marked by relatively clear confirmation of the fragility of certain equilibria. In absolute terms, we can't say that the 1970s were years lived on the brink of poverty. Yet for several countries and for several social classes these were times of minimum well-being. Above that, they were years that insidiously ushered in a certain insecurity that grew alarmingly until the middle of the 1980s. Just think back to the social conflicts, the crisis that hit much of the industrial sector and, with it, the ugly head of unemployment, and the oil shock. Most (but not all) corporate strategies could do no more than try to contain the risk. Contrary to most expectations, product renovation and investments inspired little interest. We can even go as far as to say that the product as such, received little attention. This was even felt at the communications end. This second decade can be considered as a long period betwixt and between, a sort of intermission affected by myriad negative external factors in the field of marketing and company communications.

The 1990s are proving to be very different. At present it is recognised that attention has to be showered on the consumer and his needs, a message that the media constantly repeats. Results are positive, and that is what really counts. At this stage, there is no certainty that producer awareness came from spontaneous choice rather than from the need to avoid losing contact with the consumer who, in the meantime, had become better informed and more mature. The difficult years were very instructive. Fears and uncertainties were pondered and led to more receptive behaviour. Even now, a traditional product can have consumer appeal if it is priced right. There is no happenstance approach in advertisement campaigns that bank on the promotional sales pitch to win over families with short budgets. But the economic angle is not the only new, important question. In marketing and communications strategies during the last few years the "why you spend" has become just as important as the "how much you spend". This is especially true in certain sectors such as the market for foods and environment-affecting products. Think of the attention given to low calorie foods, dietetic products, etc. All this is being done (or should be) to meet ever more modern and precise consumer demands.

In sum, and in tribute to Marx, forceful demand is moving in to stimulate supply. Businesses are again aiming at consumer needs which constitute a baseline reference not only for production and research, but also for marketing and communications.

We could ask whether old doctrines are beginning to take revenge. Has some mutation changed the rules of the game? Mistake you not - there are no rules. Or, in any case, if there are any, they are continually changing and following an evolution that could almost be called organic. Rules change as societies change, as consumer behaviour changes, as technological innovation progresses, and it invariably does. And this will always be the case. Trying to imagine how the next script will read requires a genuinely "pedestrian" approach, although certain tendencies that are becoming clear and convincing will give us an idea. Maybe the "protagonists" in our story will be called to play in a bigger "cast", thanks to something new and important.

Up to now, we have been thinking in terms of a dual relation, and, even, an almost antagonistic relationship. On the one side we have the consumers who are more or less capable of defending their interests. On the other side, we have industry with its marketing function and its impact on communications via the media. This gives us an antithesis with the "good" on the one side and the "wicked Moloch" on the other. Going a step further, we have to remember that tomorrow will be different because the context of right now, today, is already changing. The seller stands arrogantly between the producer and the buyer. The role of the retailer, especially the large retailer is becoming more important. Retailing is more than mere transmission, it is an increasingly independent and powerful part of marketing policies and communications strategies. In some cases, there is a direct, although not obligatory, proprietary link between the large retailer and the mass media. Almost all major department stores and associations, even trade associations, have the financial data and health to invest heavily in communications.

Advertisements in newspapers and on television lead to higher profits, and this is not expected to change. These "inroads" must not be underestimated because they also mean that the consumer can obtain information from an additional source. This is where the public can learn about a more advantageous product which would probably not have been included in the big communications circuits if its producers' arguments had been weak. Thank goodness for the existence of so many different media! But we might ask whether truly impartial information can actually exist. I would say that, to a certain extent, it already exists. Information that may not be completely impartial, but at least is spontaneous, has led the way to some newly opened major markets in the food sector. Just think of the many articles and services presented on T.V. a few years ago in support of the so-called Mediterranean diet. Readers and viewers showed some interest, and several companies took advantage of this interest to introduce new industrial processes.

Everything we have said is sort of a cliché for present day reality, or as the Romans in Antiquity said, "*sic stantibus rebus*" (that is how things go). But things change all the time, even right now while I am preparing this line of discourse. The most obvious proof is a look at the mutations of these last few decades, the world of mass media, communications, the sadly famous sacrifice-seeking Moloch that companies use. All of that is part of a more and more tight meshed system. Television has recently enjoyed exponential growth, unfortunately to the detriment of the written press. This is an inexorable process because a new, highly effective medium cannot be brought to a halt. Food products, from yoghurt to biscuits, or from fruit to nuts have been competing for television time. But during the last few years, we have learned that the snapshot of the moment cannot represent the reality of tomorrow. In sum, it would be a mistake to think that television and the written press are the only media leaders.

They won't disappear of course. Company information will be transmitted through advertisement for a long time to come. But advertisements will also use the new media produced by modern technology. And the revolution that for years has been shaking the telematic world will certainly provide possibilities for circulating much more information using supports that are not yet available for use or hardly even exist in the mind's eye. Imagine the impact of a phenomenon such as Internet with the limitless information it can provide. The individual person can use his PC to access sources of information that formerly were way beyond his grasp. In parallel to "official communications", which have always been with us, the world stage is beginning to welcome what we could call an alternative, or in any case a new, "spontaneous" form of communication. Consumers, henceforth, will be able to share their opinions and interact with associations that are defending their interests. In sum, they will have a redoubtable weapon to voice opinions and stand up for their rights. Internet will not only be universally accessible, it will also be proportionately less expensive.

Up to now, it was only the industrial Goliaths who, because of their size and financial breadth, could invest massively in communications, which is far more than just advertisement. Their investment capacity empowered them to call the cards, and even, in some cases, make the media bow to their influence. These privileges are on their way out and, as they exit, little David, the consumer, is gaining might. This is a most reassuring thought.

INFLUENCE OF MEDIA AND MARKETING

CHANTAL JACQUET
Carrefour France
BP 75
91002 Evry Cedex

The supermarkets are accused of every evil under the sun, but they have the tremendous advantage of being near their customers, and at Carrefour we learn more about our customers every day. Our aim is to satisfy their new emerging needs. The *Larousse* dictionary defines quality as "the capacity of products to respond regularly to the consumer's explicit and implicit needs". The explicit quality as perceived by the consumer means taste, performance, practicality, and price. The implicit needs include health, safety, and, since the mad cow crisis, we have seen a new sense of awareness cropping up, the consumer's recognition of the fact that as far as consumption is concerned, nothing is totally riskfree. Since our society shied away from disease and death, and lived with this illusion, this awareness is new.

The future, explicit needs include a very great demand for information, as we see every day in the mail we receive. The customer demands guarantees and transparency as concerns health and safety, he also wants to pay a just price in return for promises that are kept and that ensure a certain quality. Because the customer wants to be informed we try to give useful, clear answers. This means clear data on our labels, especially as concerns the presence of GMOs, and the identification of allergenic ingredients such as groundnut oil.

Being above board about safety and healthy requires knowledge of the whole animal and plant sector, and, therefore we carefully study what breeds and varieties to choose. A finished product will only be good when finished if the initial raw material is good. We also keep informed about the feed the animals eat, and the fertilisers being used. Our vigilance shows up through *selective* lists of microtoxin detection operations and soil analysis to limit the use of mineral nitrogen or prohibit the spreading of urban waste. We want our products to be traceable from the field to the plate, and require our manufacturers to abide by very precise technical specifications so that we can go upstream to examine all stages of the sector. We also carry out factory audits to identify the origin and nature of the raw materials, the additives and the technological *auxilliaries* being used. Without being "ayatollahs" we feel that certain additives, such as taste *exhausters* are useless. And after ascertaining that the raw material is good quality, we go on to demand that certain risk-bearing products used during processing be eliminated and that foreign matter be accurately detected.

We are also attentive to new demands from our customers. In another vein, we show our concern for environmental problems by studying how waste is treated, and we respond to people who complain, for instance, about the pollution caused by wastewater or the use of freon in industry.

For a product to be healthy it has to be protected by healthy packaging, which explains why we have worked closely with the packaging trade associations on the elimination of dangerous substances. We are trying to decrease the weight of the packaging and ensure maximum recycling.

In conclusion, although food in France is truly safety:
- each product has a price, and that price corresponds to a certain level of quality, and,
- the customer is entitled to the information he needs to be fully free in making a choice.

It is the duty of Carrefour to incorporate customer demand and anticipate the customers' implicit and explicit expectations. The quality/transparency approach is an important asset for French companies, both at the production and the distribution end. As Colbert said in 1664, "If our factories, through great care, make superior quality products, foreigners will prefer buying in France and their money will flow into the kingdom."

CHAIRMAN: JUAN **GALVEZ MORROS**, *FACULTY OF AGRICULTURE, MADRID, SPAIN*

VICE CHAIRMAN: LOUIS-AIMÉ **AUMAITRE**, *INRA, RENNES, FRANCE*

THEME 1

TOMORROW'S DIET

Workshop 2

Quality in the food sector

THE COMPONENTS OF QUALITY

GIANFRANCO VALFRÉ
Faculty of Veterinary Medicine
Institute of Animal Science
University of Milan
Via Celoria, 10
20133 Milan, Italy

INTRODUCTION

The "Workshop on Quality Control and Requirements for Quality of Food of Animal Origin" organised by the FAO, FEZ, and the RAISA-CNR Project for the Task Force on Livestock Production in Eastern Europe (1994), made the following observation: "Livestock production is undoubtedly characterised by quality uniformity for products of animal origin at the different levels. Guaranteed minimum hygiene will enable the various countries to meet the terms of the Marrakech Agreements and, thus, easily overcome the customs barriers. The criteria which *Codex Alimentarius* considers valid represent the minimum starting point for trade."

Moreover as far as food is concerned, the consumer feels most strongly about its being:
- healthy and reliable;
- rich in "noble" nutrients and, at the same time, low in calories;
- practical to prepare and easy to digest;
- adapted to processing and conservation techniques.

After decades of trying to win the "quantity race", now the concept of "Total Quality" has become the main objective for all productive processes in animal production. Foods must not only correspond to specific conditions or have specified characteristics, they also have to be the end result of a chain of operations that include several fields, such as:
- genetics;
- management (which includes animal rearing, the use of pharmaceutical drugs,...);
- food factors (the use of nutrients, additives, the presence of residues or contaminants,...);
- environment.

Specific, critical points must be controlled at every step of the way after having checked that the prerequisite health-and-hygiene conditions have been respected.

Good Practices criteria are applied to the processing and conservation systems used on products of animal origin. But, except for some minor measures to improve animal nutrition in the

Animal Production Sector, nothing has been done to improve on the specific standards and limits imposed for livestock feed.

The situation has become more difficult because plans for animal production have been imposed nearly everywhere in the world, not only in the EU countries, (except as concerns certain aspects of milk quality), and have been based almost exclusively on productivity factors, as can be seen from:

- he methods of genetic improvement;
- the use of hard technologies (use of pharmaceutical products and additives);
- the management of animals, which have been bred for one purpose only (morphological changes or metabolic/respiratory shortcomings);
- the organisation of intensive herding (minimum physiological conditions, sub-standard hygiene,...).

QUALITY PARAMETERS FOR PRODUCTS OF ANIMAL ORIGIN

We feel that total quality should be defined in terms of a series of variables such as:

- health and hygiene, with emphasis on the elimination of undesirable residues (pharmaceutical products or their metabolites, additives, pesticides, mycotoxins, agents of sporadic diseases, etc);
- chemico-bromatological factors;
- nutrition;
- organoleptic traits;
- technology.

These parameters, given by order of importance, are affected by food intake, genetics, management, animal health, and environment, which all affect the animal in a different manner, and are sometimes synergetic.

Consider the following elements in any example:

- health and hygiene quality, state of health, diets are above all connected to the animal rearing environment and to herd management;
- the chemico-bromatological and the nutritional quality are aboveall affected by genetics and diet;
- the organoleptic quality depends aboveall on diet, management and environment;
- the technological quality is connected to genetics and diet, but also to collection (milk, eggs) and slaughter (meat) conditions.

The main points in each of these factors are described below.

Diet

Components (of different levels, when present) have antinutritional, toxic, hormonal, antihormonal, antivitaminic effects that come from plants, mycotoxins, and industrial contaminants (polycyclic hydrocarbons, aromatic amino-acids, enantiomorphic forms) as the leading culprits.

Contaminants stemming from cross-reactions, through dust, poor food technology or substandard technology (BSE, botulism), and fraudulent use of pharmaceutical products, whether the product be authorised or banned, (hormones, beta-antagonists) and contaminated water constitute a second diet-related risk.

Genetics

Except for certain parameters in milk, (amount of butterfat, fats, fat globules, casein, serum proteins), little is known about the genetic basis of parameters that make up the quality of products of animal origin. The paucity of knowledge on the relationship between genotype and dietary needs, leads to considerable wastage of food resources. Last, the relation between genotype and disease- or stress resistance has only recently been thoroughly researched.

Management

The human factor plays the main role, in particular as concerns maintaining health and hygiene standards, *e.g.* incompatibility of certain types of animal productions, such as poultry and swine on the same litter bed, or the use of adjuvants and drugs. Further, production infrastructure are often not adapted to needs, and both dust and environmental controls are left essentially up the producer.

Animal health

It is essential to eradicate diseases such as tuberculosis, brucellosis, bovine leucinosis and salmonellosis, but the vaccination system has to be made more rational in order to avoid transferring the live virus together with the vaccine and to respect the international agreements on human and plant health contained in World Trade Organisation enactments.

Environment

There are very clear differences in environmental policies from one country to the next, even among the EU countries. These differences do not only concern soil and water protection, but also the control of dust, both in the macroenvironment and in the on-farm microenvironment. This recommendation targets the persistent effect of dust on the animal bronchial system, and the possibility that drugs and other active substances used in animal production might have residual effects.

The search for high quality products, must include attention to consumer demand and needs, *e.g.* need for cow, ewe or goat milk that is closer to human milk or free of allergy stimulators or closer to the food needs of the elderly. In certain cases, Guidelines for Good Practices in Animal Production should be prepared and enforced to curtail dangers created by lack of respect for critical recommendations. But that would not be enough, because the actions mentioned above may also relate to:

- the diet, with focus on certain legal aspects connected to the quality of cattle feed, conditions of use for additives and pharmaceutical drugs, and repression of illegal use;
- management of animals specially bred to increase productivity, and considered as veritable technological robots;

- animal health, that is managed with maximalistic technology as concerns vaccinations and stamping-out, although the latter is not always appreciated because of the dictates of demand;

- environmental management, methods which, in certain cases, make animals suffer the brunt of erroneous strategies designed for other sectors. This creates a situation much similar to that of catalytic pipes for automobiles.

A diversified strategy is being imposed which, on the one hand, focuses on the role of intensive animal production (with all the hard technologies that characterise it), and on the other, encourages extensive animal farming, a system which derives maximum benefit from **biodiversity**. It is fortunate that the useful micro-organisms in the plant and animal world are still present through biodiversity, because **biological diversity** not only represents different levels of quality, it also provides the only possible response from an animal world that is becoming increasingly uniform. Genetic resources that have been insufficiently exploited or have been eliminated purely because they are not productive enough must be reassessed. **Research** and research alone, at this juncture, and on the eve of the third millenium, can produce the desired results.

For 50 years, we have focused research on developing adapted adjuvants, adequate structures, effective disease-fighting drugs, readily digestible concentrated foods, and, last but not least, quality. Unfortunately, we neglected research on molecular genetics and physiology, and on the phenotypic control of the genetic basis of animals destined for human consumption.

RESEARCH

Research scheduled for the third millenium will have to start by more thoroughly studying the genetic basis of animal biodiversity:

- The **nutritional requirements** (energy, proteins, aminoacids, vitamins, macro- and oligo-elements) of breeds reared for economic gain, together with the chemico-bromatological and nutritional quality of the products obtained, and positive repercussions on environmental pollution must be studied. Overeating unbalanced diets and the improper use of additives and drugs harms the animal.

- **Disease resistance**, with due attention to prophylactic measures, should be given priority attention. Neither direct nor indirect approaches used for the major diseases have solved the problem. The most alarming diseases are lentivirus (scrapies, visna-maedi, BSE, etc.) for which information is urgently needed on the molecular activity that transforms the prion protein from its normal form into a pathological form. It is also important to know the distribution, within the animal populations, of genes that code for normal proteins. Healthy animals not only produce more, but the fact that they do not receive drugs or additives solves the problem of health quality connected to the presence of xenobiotic residue. In the case of vaccines with active viruses, the GATT agreement on "health and phytosanitary measures" must be guaranteed, even when derogatory measures are justified.

- **Hereditary diseases**: knowledge and eradication of diseases, both metabolical (PSE and DFD in pigs, for instance) and morphofunctional, which interfere badly with Total Quality Management, plead in favour of the most modern production systems.

- **Stress**. This syndrome represents an abnormal neuro-endocrine response to physical, chemical, climatological, and behavioural aggression. It can be further aggravated by various forms of management. Morphological variations introduced in certain animal species (abnormally fast weight gains, oxygen shortage, changes in endocrinal regulation, for instance) have interfered with animal health, and increased individual and group reactions.

- **Variability in neuro-endocrine responses**: it is important to know the site and level of the hormone receivers before scheduling or eliminating hormonal performance stimulators. Here our thoughts go to the oestrogens, the bST, the pST, and especially to the beta-agonistic substances.

- **Quality indicators**: we are looking for biochemical markers that will facilitate marker-assisted breeding and provide indices of guaranteed quality for the consumer.

The second important factor (after genetics) concerns the acquisition of the most appropriate scientific data on **food**, in order to achieve Total Quality, which is the primary aim in each production line, and includes the product's nutritional content. The science applied to all the technologies used in preparing cattle feed needs to be reviewed. This does not only apply to the raw materials, but also to the concentrates, the adjuvants, and the by-products whose numbers and uses are constantly increasing. Just think of the overwhelming effects of a minor change in the meat-and-bone meal production technique! The structure of the scrapie agent in ewes changed and was transmitted to cattle and, downstream thereof, created justified concern for human health. The thermal food purification treatment caused isomerisation of certain aminoacids which were no longer biologically available and could become highly toxic. Industrial processing of various products could also lead to the formation of dangerous chemical compounds such as aromatic polycyclic hydrocarbons in natural phosphates treated at high temperatures.

CONCLUSION

Animal production, which includes feed intake of animals bred especially to produce good meat, milk, and eggs, involves an almost infinite number of clones. This represent a dream come true for the 18th century French philosopher Jean d'Alembert, cited in Diderot's Encyclopaedia. It ensures diversity and quality for products of animal origin, and immediate pleasure for people in the most highly developed countries. There are more specific breeding choices that also contribute to certain quality-related conditions.

But management, food controls and legal enactments are not enough for the concept underlying the total quality system. The concept of "total quality" management has generated all the results it can, because it does not sufficiently embrace the genetic side of the question. The system can only be turned into a real "Total Quality System" if "Genetic Quality" becomes an applicable concept, of greater importance than all other parameters. Better care, greater attention to animals, and improved technologies are also very important. All aspects of biodiversity in animals need to be thoroughly understood, for this biodiversity constitutes the basis of many of our most familiar everyday products. This must be done before biotechnologies are used to make d'Alembert's prophetic dream come true.

SELECTED REFERENCES

Anania, G., 1995. *L'agricoltura nell'accordo conclusivo dell'Uruguay Round del GATT.* RAISA-CNR, Rome.

BOYAZOGLU, J., and RENEAUD J., 1991. *The livestock production in Eastern Europe as affected by current changes*. Pudoc, Wageningen.

CIHEAM, FAO, FEZ, 1996, International Symposium, *Les Fondements de la qualité des produits typiques méditerranéens*, Badajoz.

POLIDORI P., and RENEAUD J., 1995. *Quality control and requirement of food of animal origin*, FAO, Rome, Publication REUR no 40.

VALFRÉ, F., and MORETTI, V. M., 1991. *Characteristic, quality and control of animal products for human consumption*. Pudoc, Wageningen, 144-148.

THE STRATEGY OF THE FOOD INDUSTRY

CHRISTIANE MERCIER
GROUPE DANONE, Direction Scientifique
7 rue de Téhéran
75381, Paris Cedex 08, France

ABSTRACT

The challenge facing the food industry is going to be how to meet the requirements of consumers in the developed countries. The strategy will be to commercialise foods that first and foremost fulfil the following five criteria:

. **organoleptic satisfaction**: taste, colour, smell, texture.

. **safety**: microbiological hygiene, tracability of ingredient and food from "hayfork to table fork", control of the sector from "seed to plate", brand name commitment.

. **health**: serve health by developing balanced foods (*e.g.* Mediterranean diet) and more specific foods called "health foods". Dietary education backed by industry has become vital in order to demonstrate that **Good and Healthy** are inseparable.

. **service and practicality**: time-saving; easy to open, use and transport; new packaging.

These challenges can only be met if industry works steadily on the prices of consumer foods. This means improving productivity to finance innovation and make your presence felt on the world competitive market where, if you're not ahead of the game, you're late!

INTRODUCTION

At the World Food Summit proposed by the FAO Director General, Jacques Diouf from Senegal, and organised in Rome from 13 to 17 November 1996, a declaration of principle and, more importantly, a "universal food security plan of action" was adopted. All the participating parties - governments, international institutions, the civil society - must now go to work. The question is how to feed the world's population which, despite the "demographic transition", will continue to grow until at least the middle of the next century. How can it be done sustainably so that each person will be able to fulfil his food needs?

In the developing countries, priority goes to the yield increase, the improvement of seeds, and the storage of agricultural production. In the developed countries, the food industry already processes between 75% and 80% of the agricultural output, and sells the foods which, in the main, meet consumer demand. But the consumer of tomorrow will be more and more demanding and will no longer be seen as an archetype, a "standard consumer". On the contrary, according to Promar International 1996, by the year 2005, there will be different types of consumers that must be taken into consideration. They can be defined as: rich, poor, worrywart, aged, enamoured with well-being and good shape, single parent, family, multi-job worker (fig.1).

Figure 1: New consumer lifestyles - The fragmentation of the "meal "
(PROMAR International, February 1997)

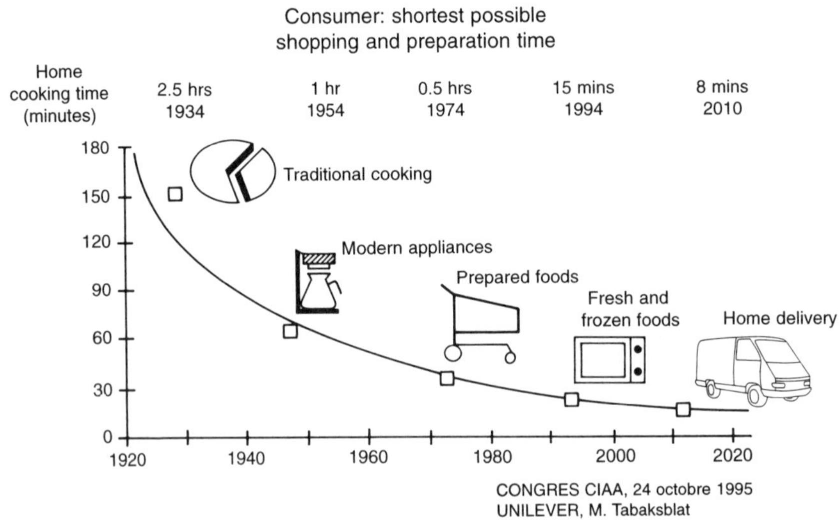

The consumer of the year 2005 will demand food that, first and foremost, has the following characteristics:

- safety and satisfaction, with ready-to-eat mini-meals;
- balanced meals made of healthy foods,
- as practical as can be,
- out-of-home consumption (restaurant, fast food)
- low cost.

The fact that the consumer has already reduced shopping and food preparation time and will reduce them even further in the near future will incite the food industry to market foods that have the aforementioned characteristics (Fig.2).

This will make industry adopt a research-innovation strategy that meets the following priority criteria:

Figure 2: Consumer shortest possible shopping and preparation time
(Congress CIAA, october 1995, UNILEVER, M.TABAKSBLAT)

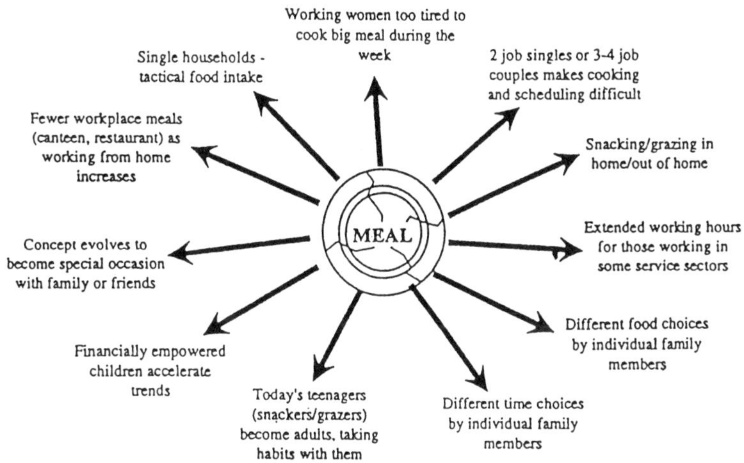

ORGANOLEPTIC SATISFACTION

All five senses are involved in the brain perception of what we call "taste", the taste organs themselves of course, but also smell, sight, touch and hearing. There are phenomena other than this combination of senses which contribute to pleasure and aversion such as the internal state of the body which determines hunger or satiety, the nutritional state, previous experiences, memory, and socio-cultural factors. That is why the food industry took the initiative of creating the European Centre for Taste Sciences in Dijon. Its research institute has a multidisciplinary structure that covers from chemistry to psychology, and from neurobiology to ethnology and history. In the developed countries, one eats with the brain rather than the body!

FOOD SAFETY

The growing demand for foods that are "natural", "fresh", "smooth", "practical" and "quick" has prompted industry to look for "soft technologies" that can produce food that is "good" and "safe". The so-called "ultra-clean" technologies make it possible, *inter alia*, to reduce culinary parameters which means minimal destruction of raw materials, in other words, keeping the food close to its natural condition and, at the same time, ensuring that the product respects human health.

Software now available on microbiological methods can simulate the development of micro-organisms caused by non-respect of storage conditions, and thus can help prepare for new formula and storage conditions.

Industry needs well adapted evaluation methods so that it can meet European and international health regulations. Molecular biology techniques such as the polymerisation chain reaction, applied to foods, constitute infallible tools for detecting, identifying and quantifying micro-organisms (be they pathogenic or not) and possible spoilage.

Swift, non-invasive physical methods such as image analysis, X-ray, NMR, can and should be adapted for use as "on-line sensors" in food production factories. For certain foods, these methods are indispensable in judging whether foods conform with rules, *e.g.* NMR can be used to detect the geographical origin of certain raw materials and ingredients, and their natural place of origin.

Food safety cannot be guaranteed unless the raw materials and ingredients are traceable, foods are monitored during processing for health and toxicological problems, and, - be it not forgotten, - new techniques are harmless.

HEALTH

Strive for health. This means,

- on the one hand, developing foods with the right balance in macro- and micro-nutrients by applying nutritional recommendations from the medical sector (such as certain oven-ready dishes, children's foods, drinks for athletes), the Mediterranean-type diet;

- on the other hand, bringing the more specific, "health foods" to the market for certain categories of consumers. A well-documented scientific file, with data on results in the healthy person, should be submitted to the regulatory authorities for allegations that they are "nutritious" or "healthy" before being used in "claim marketing". In both cases, before reaching the market these foods require thorough research in human nutrition with finalisation, *inter alia,* of specific nutritional tracers *e.g.* marking durum wheat with ^{13}C to study the digestibility of pasta, or marking milk with ^{15}N to evaluate the digestibility of bovine casein. The development of these health foods will only be possible if industry, research (in human nutrition) and the medical body form a strong partnership. Dietary education, thus, has become a necessity in order to show that Good and Healthy are inseparable; public authorities must work closely together to provide such education.

SERVICE AND PRACTICALITY

The food industry processes and conditions foods in various types of packaging (paper, cellophane, plastic, aluminium, glass, etc.), for different forms of storage, (bottling, pasteurisation, vacuum packing, freezing, etc.) and in various formats (brick, cartons, soft bags, boxes, pots, jars, bottles, etc.). It has to meet both the consumer's and retailer's requirements.

The consumer wants time-saving, easy-to-open-and-use, single-product units with legible labels. The retailer is interested in form, bulkiness, ease of transport, and shock resistance. The ecologist demand a reduction of packaging which incites international research to concentrate on ecological packaging that is compactable and recyclable (plastic), biodegradable and incinerable (PVC), and on new packaging materials (hybrid glan and plastic solutions).

The so-called "active" packaging (liquid and gas barriers, O_2 and odour absorbent, active substance releasers) helps improves shelf-life storage and, thus, food quality, and service.

PRODUCTIVITY/PRICE

To win over market shares and protect its image, industry's main strategy is to meet the consumer's and the retailer's needs. But these challenges can only be successfully met if constant attention is given to the price: lower cost/higher quality. This means that productivity must be improved, both as concerns the price of the raw material (in relation to its technical specifications) and the processes used. Automation of production lines, on-line quality monitoring by (physical, chemical or biochemical) sensors, modelling and digital simulation of certain steps are indispensable tools in improving production, without forgetting the importance of transparency in ways of product preparation.

CONCLUSION

To maintain the place of a respected brand name on the stage of international competition requires constant, offensive innovation coupled with a good consumer information policy. This is the life insurance of industry. If you're not ahead of the game, you're late!

QUALITY FROM A CONSUMER'S PERSPECTIVE

ARIE VAN GENDEREN
CONSUMENT & BIOTECHNOLOGIE (on behalf of Consumers International, CI)
P.O. Box 1000
2500 BA The Hague, The Netherlands

ABSTRACT

The introduction of new technologies in the food chain, such as biotechnology, raises ethical, religious, environmental and socio-economic questions with consumers, since such technologies may pose new and unknown risks.

In general, consumers expect foods to be safe, healthy, nutritious, tasteful and affordable. A growing number of consumers also show interest in the way a foodstuff is produced. They ask questions about the environmental impact production methods, about animal welfare and about ethical aspects of production techniques, such as modern biotechnology.

The question is whether modern biotechnology adds anything to the general quality aspects mentioned above.

INTRODUCTION

Consumers International, or CI for short, is a federation of consumer organisations dedicated to the protection and promotion of consumer rights world-wide. It co-ordinates consumer groups in some 65 countries and organises information networks, international seminars, workshops, etc.

CI also initiates research and action on global issues and represents consumer interests before international policy-making bodies. The central aims of CI's work is promoting the growth of a strong consumer movement and increasing protection for people in their role as consumers.

I work for the Consumer and Biotechnology Foundation (C&B) which was founded in 1991 to represent Dutch consumer organisations in the Netherlands. We are affiliated with the Consumentenbond, a large consumer organisation in the Netherlands with over 650,000 members.

The aim of our foundation is to map out the "pros" and "cons" of new technologies, biotechnology in particular, and inform and support the consumer organisations, the world over organised within CI.

WHO NEEDS MODERN BIOTECHNOLOGY

The primary function of food, or maybe I should say the primary function of eating, is to keep us alive. For a lot of people in this world staying alive is quite a burden, because they lack healthy foods in adequate quantities.

For other people, in different parts of the world, staying alive is also a burden, but not because they lack food, but because they have too much to eat. For those people, eating is more than satisfying basic needs. It is pleasure, fun and in some cases an addiction. (This doesn't mean of course that everybody has a problem!).

What those groups have in common is that they have a food and health problem, although the causes and solutions differ.

What the biotech companies want us to believe is that modern biotechnology is a necessary tool to solve both problems. It will help to increase agricultural yields and thus help us to overcome food shortages. And on the other hand it may help to create new products or products with new properties that may help reduce health problems. Food companies such as Unilever and Nestlé claim that further development of modern biotechnology will bring major benefits to society. It will assist us in delivering innovative food products which will meet consumer demands.

We, and I speak now for many environmental and consumer organisations all over the world, question those claims. I will try to explain why.

To begin with, I will comment on agriculture, then on quality issues, the general consumer concerns, and, finally the right to choose.

AGRICULTURE AND MODERN BIOTECHNOLOGY

Genetic modification of plants is sold to the public using various arguments. One argument is that we need genetic modification to create new plants or plants with special traits to feed the growing population of the world. Some advocates of biotechnology go as far as to predict even more famine in the near future if we do not use biotechnology to improve the major crops such as rice, corn, wheat, soy beans, etc. Others argue that we should build in resistance genes to make the plants less vulnerable to pests and diseases. With an expected positive spin off through reduced pesticide use. Another argument is that we can create crops with improved consumer traits, such as delayed ripening and better taste, green vegetables with a lower nitrate content and products with better fatty acid composition.

Although the researchers work hard, many of the promised/expected improvements are still science fiction. And the improvements that have been made, such as BT-cotton, BT-corn seed and BT-rapeseed and the herbicide tolerant plants, can expect a lot of opposition from various groups and individual consumers. Opposition which we think is justified.

SAFE, HEALTHY, NUTRITIOUS, TASTEFUL AND AFFORDABLE

In our rich western society consumers want their food to be safe, healthy, nutritious, tasteful and affordable (good price/quality relation). In general, the present supply of foodstuffs meets those demands. I know of no consumer surveys that give an indication that the majority of the buyers are dissatisfied with the present offer of products. Nor is there any indication that we need all kinds of novel foods, regardless of the production procedures. When the large food companies speak of upgrading products in order to meet consumer demands they mean in fact something else. To the industry modern biotechnology is a tool they can use to create new products or products with new traits. Not to meet consumer demands, but to create new markets or become more competitive by reducing production costs.

Of course, no consumer opposes the idea of making foodstuffs safer, better, healthier or cheaper. But the question is whether we need modern biotechnology to do the job. In our opinion there is no simple yes or no answer to this question. When you put this question before industry representatives the chances are that they will say 'yes'. But many environmentalists and consumer representatives will say 'no'.

Our position is somewhere in the middle. We suggest that it is better to work with a case-by-case approach: look at the products or product groups and identify any problems or bottlenecks concerning safety, nutrition, quality, etc. Subsequently try to identify the techniques or measures which are needed to solve the problem, if any. In some cases modern biotechnology may be part of the solution, in other cases it won't.

When we take a helicopter view of the food supply we encounter five major problem areas:

- food hygiene (salmonellas in meat products for example);
- loss of nutrients, colour and flavour as a result of processing;
- residues of pesticides and environmental pollution;
- unhealthy diets;
- adequate food supply (developing countries and growing population).

There is an English expression: 'An ounce of prevention is worth a pound of cure'. With this in mind, possible problems should, in our view, be tackled as follows:

- hygiene problems: better management techniques throughout the chain, starting at the base of the production;
- loss of nutrients: better, 'softer' processing techniques or in some cases, less processing and more emphasis on whole foods.
- residue problems: clean and sustainable production techniques that avoid environmental pollution and thus reduce or eliminate the residue problem.
- unhealthy diets: more consumer education, elimination or reducing the unhealthy components from the diet.
- adequate food supply: food shortages are caused by a range of factors, such as bad land management, type of crops (often the result of the economic situation in a region), civil wars, poor storage systems, etc. Solutions should begin with improving those conditions and systems.

The simple conclusion is that in none of the five areas is biotechnology automatically needed. And when used, in our view, it should always be part of a broader approach. In some cases modern biotechnology could be used for the production of processing aids, such as enzymes. But there seems to be no direct need for the production of major ingredients.

In relation to the residue problems modern biotechnology may indirectly help to prevent environmental pollution by reducing the need for pesticides and thus indirectly reduce the residue problem.

SAFETY ASPECTS AND CONSUMER CONCERNS

There is still no legal EU framework, no Novel Foods Regulation, for the assessment of foodstuffs produced with genetically modified organisms or ingredients thereof. Awaiting a European Novel Foods Regulation The Netherlands has set up its own regulatory systems. Some other countries like Denmark and Great Brittain have an assessment system for Novel Foods.

In The Netherlands we have a comprehensive Novel Foods Regulation, an Advisory Council on Novel Foods and a Scientific Committee on the Safety of Novel Foods. However, when we look at the situation in Europe as a whole, not to speak of the situation world wide, things become blurred and unclear, especially for outsiders. This lack of transparency is one of the causes for consumer concerns regarding novel foods. Even in The Netherlands, in spite of regulation, consumer concerns on safety are widespread and not to be ignored. And when there is concern about the safety of the product, there will be concern about the other aspects as well. Public acceptance of novel foods may be improved when the products show a "plus" on one or more of the quality aspects. So far, hardly any product has been marketed that bears such a "plus".

THE RIGHT TO CHOOSE

Consumer surveys in various EU countries indicate that a fair majority of the consumers wish to be informed about the genetic modification of foodstuffs or ingredients and/or the safety of the products. The reasons for this vary: ethical concerns, fear of the unknown, safety concerns, concern for the environmental and ecological impact. The consumer organisations have adopted these aspects as a key issue in their policy regarding novel foods. This right to information applies to all consumer products, including novel foods. In order to make an informed choice the consumer needs to be informed and educated. The product label plays a crucial part in this. So, to give consumers the opportunity to make a choice between genetically modified products and non-gmo-products, the true state of the product needs to be labelled.

CONCLUSION

Consumers International is not opposing the application of modern biotechnology as such. On the other hand, we do not see it as our task to promote this technique either. Consumer concerns are widespread. In part, the concerns have an ethical background, which is hardly disputable. For another part, consumers are concerned about the safety of the products.

In The Netherlands we assess the application on a case by case basis. Generally speaking, products have a better chance of acceptance by consumers and consumer organisations, if:

- they are safe (for man, animals and the environment)
- there is a direct benefit for the consumer (better quality, lower price)
- there is an indirect benefit such as less environmental burden, less waste, *etc*
- they are labelled.

APPLICATION OF BIOTECHNOLOGY IN THE FOOD INDUSTRY

HANS NIEUWENHUIS
Unilever Research Laboratorium Vlaardingen
Olivier van Noortlaan 120
3133 AT Vlaardingen, The Netherlands

ABSTRACT

Unilever is one of the world's largest producers of consumer goods. Needs for consumer goods change over time because of changes in consumer demands. In order to remain competitive we have to meet these demands, and therefore we need innovations. Traditional and modern. Biotechnology plays an important role in these innovations. Modern biotechnology provides routes to obtain highly functional ingredients, reduce the gap between plant and desired product properties, and deliver raw materials closer to the end product. In addition it allows for the development of new food concepts. Before introducing a food-product which contains genetically modified products on the market, it is important to inform our customers and consumers as well as our own people about biotechnology, in what type of products modern biotechnology already is or may be applied, and particularly, what the benefits are to our consumers and the company. It is Unilever's belief that the further development of modern biotechnology will continue to bring major benefits to our consumers and the company.

INTRODUCTION

Unilever is one of the world's largest producers of consumer goods. They are sold in three major groups: foods, detergents and personal care products. About 50% of the sales is foods. We sell our products under brand names such as Magnum and Cornetto ice cream, Surf washing powder, Dove soap, Lipton tea, Fruit d'Or, Astra, Benedicta, and Boursin, just to mention a few. These products are made and marketed by more than 500 operating companies in more than 80 countries.

Brand names are the core of our business. They stand for consistently high quality, safety and reliability. When choosing one of our brands, the consumer expects added value - either in quality, functionality or performance. Our branded products are market leaders, and we aim to keep this position.

However, consumers are not static in their purchasing behaviour; they are continuously looking for products that fulfil their needs. These needs change over time because of changes in the available time for household duties, personal care, indulgence behaviour, health attitude, available budget, family living patterns, social status, social acceptance or environmental concerns. Especially in the area of foods, consumers increasingly demand products that have certain organoleptic qualities, that are not just wholesome but have specific, built-in nutritional characteristics (*e.g.* low fat, more polyunsaturated fatty acids), contribute better to a healthy lifestyle, are more natural or contribute to reducing the impact on the environment. In order to remain competitive we have to meet these demands, and therefore we need innovations. And that's where bioscience and biotechnology come in.

TRADITIONAL AND MODERN BIOTECHNOLOGY

A large and ever-growing proportion of our products is made from biological raw materials, and intended for human consumer or use. It will therefore not come as a surprise to learn that both the biosciences and biochemistry form a cornerstone of our research effort.

Classical biotechnology has already been used with increasing success for thousands of years. Examples are beer, bread, wine, fermented products such as cheese, yoghurt, sauerkraut and meat; antibiotics; and flavours and fragrances produced by micro-organisms.

Modern biotechnology, with its genetic component, is no longer at the laboratory stage. It is already being used by many branches of industry to develop new and improved products derived from biological materials and processes. Unilever believes that the further development of modern biotechnology will continue to bring major benefits to society.

In many of our food markets, modern biotechnology will contribute to obtaining products that are optimally targeted to consumer wishes, have very specific functionalities, and are more natural or contribute to a health-conscious lifestyle. And it will enable milder processing and production methods that reduce the impact on the environment. It has become an indispensable tool in present-day research.

ROLES FOR MODERN BIOTECHNOLOGY

We all know that food is essential to keep us alive. However, in affluent societies consumers expect more of foods: they want foods that improve the quality of life. Our main task, therefore, is to upgrade raw materials, mainly of vegetable origin, to high quality products that are in line with consumer wishes (Fig 1).

Two of these wishes have clearly emerged over the last decades: more functional foods and foods that are as natural as possible. The products marketed under our Heart Health brand names, such as Becel, Flora or Fruit d'Or, are proof of how we are increasingly trying to meet the demand for more functional foods. With modern biotechnology, opportunities to create these can be enlarged (Figure 2). We will look at these consumer trends and how modern biotechnology may assist us in delivering innovative products which meet such demands.

Figure 1: Production of food products (20th century)

Proceedings of raw material results in

Existing raw materials and ingredients ——————————> Existing products

reduced quality | action to restore quality

Figure 2: Production of food products (21st century)

Bioprocessing

Genetically improved raw materials ——————————> Products of the future:
- better quality
- longer keepability
- environmental friendly
- natural

improvement of required properties | Elimination of unwanted properties

Functional ingredients

For many of our products we use fermentation processes to make stabilisers, thickeners, flavours, starter cultures or emulsifying agents, while enzymes are used for *e.g.* juice clarification. Being involved in food processing we will continue to add ingredients, preferably natural ones. And, increasingly, we will be turning to functional ones. What do we mean by that?

A food contains numerous ingredients, which are mixed to make a healthy, tasteful and attractive product. All these ingredients have a very specific function. If they were to be left out, the food would taste different, have an odd texture, be less nutritious, or be much more difficult to digest. Such characteristics are called functions, and we therefore talk of functional ingredients. Using modern biotechnology we may then opt to have such ingredients produced in the most suitable host organism. In that way we will obtain a wider range of natural, functional ingredients that may be added to our food products. Many of these new ingredients will add healthiness, convenience and improved quality to our products.

Reducing the gap between plant and desired product properties

Adding more natural ingredients is one way to obtain more natural foods. But our real aim is adding less and using more of a raw material's intrinsic properties. Or better still, we would like to have raw materials that contain more of the functional ingredients we are really interested in. Therefore,

Unilever's aim is to do as little as possible in terms of processing and additions. In other words, we would like to narrow the current gap between the plant (or other raw material) and the end product as much as possible by modifying the properties of the plants in the direction of the requirements of the end product. To do so requires expertise in plant modification (plant biotechnology, plant breeding) and knowledge of the metabolic routes involved (biochemistry).

Raw materials closer to the end product

Plant breeding in the food chain used to be directed towards improving the yield, resistance against viruses, pests or herbicides, fertilisation options, facilitated processing and cost savings. Nowadays, these objectives are shifting: plant breeding is increasingly geared to enhancing a raw material's qualities such as improved nutritional value, better taste, improved texture or a more pleasing appearance, while environmental concerns are also taken into account.

Identification and modification of genes may ultimately lead us to plants with new functional ingredients and unique flavours or flavour delivery systems. In addition, if we could adapt plants to facilitate processing, we would not only make more efficient use of the raw material but also lower the impact on the environment. Genetic modification also allows enhancement of desirable properties and elimination of undesirable ones. An example is the work being done to reduce the nitrate content in spinach by increasing the levels of the enzymes of nitrate metabolism. Modification of these enzymes would permit a more efficient use of nitrogen fertiliser with its associated benefits to the environment and the farmer, while the consumer would be able to buy products with a lower nitrate level. With conventional breeding techniques it has not yet been possible to achieve this target.

Novel concepts for foods

Finally, modern biotechnology is assisting in the development of new food concepts. Take, for example, the use of polyfunctional yeast (a baker's yeast that not only makes carbon dioxide to let the dough rise but simultaneously delivers bread improving enzymes). We are investigating the possibility of selecting bacteria that modulate the intestinal microbial flora for improved bowel function and uptake of nutrients (such a concept is called probiotics).

These latter examples are still far in the future, but the one mentioned earlier demonstrates how classical and modern biotechnology already contribute to a wider choice of more natural, improved quality functional foods and to more efficient production processes that can be better controlled in the technical and scientific sense. In addition, these technologies contribute to reducing the impact on the environment.

THE ACCEPTABILITY OF MODERN BIOTECHNOLOGY

Some consumers and their organisations are hesitant about accepting consumer products, notably foods, made with the aid of modern biotechnology. They often raise questions that have not yet been openly answered by industry and/or the authorities. Unilever's position on modern biotechnology is to encourage open discussion with those concerned about the impact of modern biotechnology on ethical, socio-economic, environmental and safety issues.

Serious questions deserve serious answers

New technologies appearing on the market place are received with resistance from certain segments of the population because of anxiety. History provides many fine examples of worries about the introduction of new technologies: the crossbow, the steam engine, personal and hospital hygiene, the automobile, the aeroplane, contraconceptives and food irradiation, just to name a few. These worries are very understandable and should be taken seriously, because they often come from people who have thought long and hard about particular issues. Industry has paid dearly for not paying enough attention to such concerns, and therefore we adhere to the principle that serious questions deserve serious answers.

People tend to be apprehensive about events or technologies they cannot comprehend or control. In the case of biotechnology, the specialised nature of the science makes it a remote, incomprehensible technology. Another factor that raises anxiety is the lack of adequate information. We try to address both factors by being open for discussion, and issuing publications. The recent discussion around soja has given a lot of supportive evidence to this approach.

The necessity of the technology

Existing technologies are good but not good enough. To reduce the number of processing steps when making or refining a consumer product, or refine them we need to selectively alter the sources of raw material or we carefully select, biological processing steps using specific enzymes. If we want more natural foods we will need more natural ingredients. And so on and on. For all these targets we need a technology that can do more than the existing ones. Modern biotechnology, through its genetic component, may provide the 'more'. Modern Biotechnology is one of the main breakthroughs of this century. It is my personal opinion that if all those concerned do not seriously look at the benefits of this new development, its loss will cause a great regret for the future.

On ethics

But we do have to recognise our own limits. A technique with such a far reaching potential as genetic modification may be misused like any other technology; this is a fear often expressed by those hesitating to accept products of modern biotechnology. We recognise these anxieties. Even though they fall completely outside Unilever's scope, we strongly advocate, also in this area, effective, clear and properly enforced regulations to avoid misuse or undesired use, and an open dialogue so that we may be informed and in turn inform others as well. In addition, we need to be aware of, and understand, ethical concerns which may reflect lifestyles and religious or regional traditions.

Socio-economic consequences

With the introduction of every new technology there are fears for job losses, socio-economic disturbances or undesirable use. Every new wave of technology will, over the years, wipe out obsolete technologies while creating new opportunities. We believe in the creative and not in the destructive power of technology, both for the industrialised and the developing world. And it is not so much the technology itself that requires safeguarding but the manner in which people intend to use it.

Environmental benefits

So, what about the environment? Reducing the impact on the environment is a major asset of modern biotechnology. Virtually all food examples are geared to reduce environmental pollution, either by reducing the amount of chemical additives, the number of processing steps or the harmfulness of the compounds used. Modern biotechnology has the potential to become an important ally of the environment and not an enemy.

The safety of modern biotechnology products

Now we address the question of safety. First let's look at safety in general. Most consumers are conscious of the quality of products and appreciate the diversity in products offered. Whether consciously or unconsciously, the consumer trusts the information provided by the manufacturer. We, as manufacturers of consumer products, see no reason at all why the consumer should stop trusting us now that a new technology is emerging. As with all other innovations, we will assess the consequences for health, workers safety and the environment, and adhere to the safety regulations of the various governments. In fact, our research/development projects are already being subjected to such assessments. These assessments are not only carried out by our own scientists but also by independent scientific experts, since we believe that an independent review is important for a technology as new as modern biotechnology.

Safety assessment procedures

In biotechnological safety assessment procedures we first need to demonstrate that the gene(s) involved or any additional genes that have been introduced do not give rise to unwanted side effects. We also have to demonstrate the safety of the host organisms. Then comes the next question: is the product produced in the host organism identical to an existing product (*e.g.* sugar, or amylopectin) or is it a completely new product (*e.g.* a newly designed enzyme)? And is it the only metabolite formed or are others present as well? In both cases we follow assessment procedures agreed to by the international community and translated into national regulation.

How will the consumer know?

When the product has obtained approval for market introduction, the consumer has to be informed. It has not yet been definitively decided to what extent modern biotechnology offspring should carry a compulsory label and what should be on it. Whatever the outcome of this international debate, we feel that, wherever possible, the consumer should be able to make an informed choice either by information on the pack or by information on request via consumer information services and care-lines. When the product is identical to an existing one we don't expect many questions from our consumers, when it concerns a new one we should provide comprehensible information and be ready to answer fully any questions that may arise.

Judging the product

We do not intend to persuade consumers or anybody else to become indiscriminate supporters of modern biotechnology. But we are sorry to see modern biotechnology's offspring being discriminated against solely on the basis of the technology used. We hope that you will share our conviction that modern biotechnology, side by side with other technologies, has substantial benefits to offer to society at large as well as to the individual consumer. Therefore, we advocate that society, and in particular the organisations in which consumers put their trust, judge each new product or application on a case by case basis and accept products, irrespective of the technology used, when convinced of their safety and the benefits they have to offer.

THE ROLE OF PUBLIC AUTHORITIES AT THE NATIONAL AND INTERNATIONAL LEVEL

JOACHIM HEINE
Commission Européenne - DG VI
200, rue de la Loi
B-1049 Brussels, Belgium

INTRODUCTION

Public opinion has always been concerned about food quality, a subject that regularly elicits state intervention. But objectives have changed to keep abreast of economic and social development. When the overriding concern was to make enough food available, quality requirements were restricted to guaranteeing that the food was healthy and fulfilled nutritional requirements. Since the simple need to eradicate hunger has been replaced by the need to sell surplus agricultural produce, quality food production and foods themselves are vital when facing sharp competition and in acquiring or defending market share. With these changes on the board, it is logical that consumers have become increasingly demanding and consider quality as the result of a whole series of heterogeneous issues, ranging from composition to methods (more or less traditional, biological, and "animal-friendly"), from quality food production to geographical origin.

As quality takes on more facets and importance (competition), the role of the public authorities needs to be defined. There are two fundamentally opposite orientations that seem to be emerging:

- one invites the lawmakers to help organise the increasingly divided, competitive market and detail requirements imposed on products;
- the other, on the contrary, considers that competition on a free market will be able to settle the problem since only high quality products (regardless of consumer opinion) will be able to survive economically. If this doctrine is applied, the lawmaker's role will be curtailed to creating a legally flexible framework dominated by market forces.

The European Union includes defenders of both, although in the present political climate, which favours deregulation and preaches subsidiarity, the pendulum tends to swing more towards the latter. Reality requires a more nuanced attitude, since a simplistic approach cannot solve the multitude of problems related to this issue.

WHAT QUESTIONS NEED ANSWERING?

The question at hand is "food quality". Some people might say that it is obviously in everyone's interest to have high quality products at low prices. There is no doubt about the economic logic to this statement, but in this situation, it would be difficult for the public authorities to intervene efficiently. The state could set prices and then impose objective criteria on those products. But would this mean banning products with other characteristics (which, therefore would be priced differently)? Considering the growing number of consumer requirements, the lawmaker will have to cope with an almost unlimited number of prices and criteria since, in principle, there are at least as many specific demands as there are consumers. The very definition of "quality" has to be revisited. Quality can easily be defined in terms of a precise objective, but a general definition can only be based on the reference to the desires of the consumers. So ISO standard 8402 defines quality as "all the properties and characteristics of a product (or a service) which give it the capacity to satisfy expressed or implicit needs".

This definition of quality is not useable in administrative practices. It covers all imaginable cases, while public authority has always favoured, and should continue to favour certain possibilities and prohibit others. The true imperatives of public authority emerge when efforts are made to determine the underlying reasons for their preferences. The basic objective should be protection for everyone involved in the food chain, from the producer to the consumer. This protection is not unilateral, *i.e.* meant only to protect the consumers against the producers/processors and traders. All the economic players have specific needs, and these needs merit equal attention from the public authority, whose duty is to make choices and, in case of conflict, establish priorities. Choices will inevitably depend on basic political orientations, and actions will be influenced by a concept of the state and its duties to its citizens. But this basic approach cannot be found in laws governing the food sector.

Action by the public authorities should always predicate on consideration of pending interests, and concern for keeping them balanced. In the context of this paper, these interests can be summarised as follows:

- protection of health (requirements concerning the healthiness and cleanliness of products, and product safety);
- adequate information (appropriate labelling);
- fair competition (prohibiting misleading information on the labels; reservation of certain denominations.

There are many different ways to accommodate these interests. Since interests may overlap or be contradictory, depending on the circumstances, the specific rules that can be established include a large range of legal instruments, as can be seen through examples in the national food laws. Circumstances influence choice according to the case. The lawmaker's basic orientation for or against a detailed regulation will also have a considerable effect.

By making an initial distinction between these instruments, we see that there are measures that target the properties of the product, in the broad sense of the term, and others that focus on its presentation (the two may be connected).

a) *The approach to product properties may cover many aspects.*

- As far as policy is concerned, health requirements are of overriding importance. They do not only apply to the final properties of the product, but also, whenever necessary, to the quality of the unprocessed product and the conditions in which it is processed. Just think of the ban on certain growth promoters and additives. Quality maintenance partly depends on how the product is handled during the marketing stage. This means that rules have to be drawn up on packaging and marketing conditions.

- In the same domain, there are rules on product safety, although agricultural products are not affected in the same way as industrial products.

- In the concept of consumer protection, prime importance is given to respecting physical health, but this does not exclude the possibility of adopting other measures that give priority to the economic benefits that can be obtained by providing high quality (and not only healthy) goods. In this case the legislation will provide quality requirements that have to be met before a product can be put on the market; they may be limited to making an outline of categories (Handelsklassen) on the basis of criteria (more or less strict) that are (or are not) fulfilled by the product in question.

The quality requirements that a product must meet before being sold to the intervention board does not fall within this group of measures. They will not prevent a product of a different quality from being sold on the market, and are seen more as elements of a sales contract, with the buyer seeking certain assurance of quality. They have no constraining effect on the market in general. At best they may serve as an incentive.

b) Rules on product labelling seem to serve the interests of all the economic partners.

It can be in the producer's interest if the labelling rules prevent a competitor from using terms reserved for a product or a special group of producers (*e.g.* intellectual property of country of a designation of origin); these rules are also useful in ensuring fair competition at all levels and in providing the consumer with precise, accurate information.

There are several types of regulations:

- general regulations that apply to all the products and essentially aim at ensuring that the presentation is correct and understandable ;
- rules that impose certain appellations (*e.g.* butter vs. margarine);
- rules that reserve certain names for products that have been produced under certain conditions (*e.g.* biological production);
- rules that reserve certain denominations for certain persons (*e.g.* designation of origin, geographical indication);

Reserved names may be connected, (and normally are) to certain characteristics of the product.

THE ROLE OF THE PUBLIC AUTHORITY

In compliance with the separation of functions which characterises modern democracies, the public authority intervenes through the legislative, executive (administration) and judicial branches. Most of its actions are carried out via the legislative branch, but actions via the other two branches are also important.

The judicial authority is the arbiter and, as such, essentially serves to settle conflicts (between individuals and between individuals and public authorities) Its role, therefore, seems generally limited. But since permanent, coherent jurisprudence stems from a wide range of non-formal rules, respected despite their informality, this role may be very important. In many EU member states, food law can serve as an example.

The role of the administration is mainly to verify legal standards. At present, there is an evolution to limit its role to monitoring checks carried out by the parties directly involved, but the intensity of its activity will have considerable effect on how well these standards are applied and, thus, on the fulfilment of objectives sought through legislation.

The action of the lawmaker will depend on three possible positions:
- absence of legislation,
- encouraging action by individual parties,
- adoption of constraining rules.

In no case can we say that legislation on the food sector does not exist. The question is how intensive legislative activity is and how well problems are covered. The present tendency is to have the parties directly involved work out the rules, as long as the interests of direct concern to the public are not too closely involved. This especially applies to product composition and preparation. Industry is very familiar with industrial standards (AFNOR, DIN, CEN, ISO, etc.), but only a few industrial standards serve in the food sector. This is because food production has traditionally been less attentive to the needs of mass consumption than to its own distinction and diversification. Whatever the case, according to certain groups, such as the enthusiastic defenders of subsidiarity, more efforts should be made to standardise the food sector, especially where quality is concerned.

The states and the European Community, in principle, can abstain from any action, pending changes in standardisation work. They can also encourage action by inviting qualified groups to study the question or by providing financial assistance through subsidies.

The member states, in any case, can encourage good quality production by dispensing national aid. But these aid payment can only receive Commission agreement (through the procedure set out in Article 93 of the Treaty of Rome) if they respect certain criteria: aid cannot be granted on a product quantity basis, but it can be granted to pay for training, information, the creation of control organisations by producer groups, and verifications.

Subsidies can only be granted by legal authorisation, but this authorisation can be relatively broad and vague, leaving the authorities in charge a large discretionary margin. The authorities will no doubt encourage certain products and qualities, but, like in the case of industry's adoption of standards, the recipient party is not bound by any obligations in accepting the aid that is awarded, *e.g.* he is free to adopt the standards or not, he can ask for aid (and accept the criteria applied) or not.

Laws will be needed if rules on quality are to become obligatory and binding. To a certain degree they are inevitable. As long as there is an economic advantage (this usually means a good price) to respecting the rules, (whether they be standards, recommendations, criteria for granting subsidies, or regulations), the stakeholders are inclined to do so. If, on the other hand, respecting these requirements does not generate any additional benefit, they will be considered burdensome and costly, and will be infringed.

This is largely the case concerning sanitation. The requirements have to be respected, although there is no special financial advantage to doing so. Private initiatives will probably not be very effective and will be kept down to the minimum. That is why the lawmakers have always felt that health required detailed legislation.

Within the European Union there are three levels of legislation: regional, national and community. In Community law, an intervention by a regional legislator can be equated to that of a national legislator. Community legislation has to respect international rules derived from the GATT agreements, which ends up meaning that there are three superimposed judicial levels: national community, and international.

Judicial priority is given to Community legislation.

- Community legislation can override national authority. The member states (under their own constitutions) and their regions can freely legislate on food quality, but only as long as the Community lawmakers have not acted.
- National and regional legislation, in any case, must respect the rules of the Treaty. Above all, it must not obstruct trade. By virtue of Articles 30 and 34 of the EC Treaty, these laws cannot include quantitative or other restrictions to intra-Community importation and exportation.

In sum:

Member states, in the absence of Community decisions, may regulate the production, composition, packaging, presentation and marketing of their own products, but not products imported from other member states.

Any product imported from a member state must, in principle, be admitted to the importing member states, if it has been legally produced (in other words, conforms to the fair, traditional production procedures and the regulations of the exporting country) and is marketed in the exporting country.

By accepting this principle, the members states, when preparing commercial or technical regulations that may have an effect on the smooth and free circulation of goods, agree not to adopt an exclusively national position, or a position that only accommodates the requirements of their own national products. For the common market to operate well, each member state must accommodate the legitimate demands of the other member states. The importation and commercialisation of a food commodity, legally produced and marketed in another member state, may not be impeded - pending harmonised rules at the Community level - unless the decision to do so:

- can be justified as being necessary to satisfy a very serious requirement (protection of public health, defence of the consumer, fair business practices, environmental protection);
- is proportionate to the targeted goal;
- and, constitutes the least aggressive way of reaching this target.

These principles come from jurisprudence rendered by the Court of Justice, in particular the judgement called the "Cassis de Dijon" decision (dated 20 February 1979, case 1220-78, Rec. 1979, p. 469).

But the residual authority of the member states can by superseded through the harmonisation of rules at the Community level. In the 1985 White Book (point 65) on the completion of the single market, the Commission defined the principles of harmonisation at the Union level :

- a clear distinction is made between aspects that absolutely need harmonisation and aspects that can be governed by mutual recognition of national rules and standards;
- harmonisation through Community legislation that is binding on the member states is limited to coverage of requirements vital to health and security, the purpose being to ensure the free movement of products that conform to these requirements;
- harmonisation of other qualitative aspects are to be promoted as much as possible by the preparation, (without the intervention of the lawmakers), of European standards, but the absence of European standards should not impede free movement. Pending finalisation of European standards, the guiding principle is mutual recognition of national standards by the member states.

As far as food products are concerned, in principle the Community only adopts harmonised rules, if they relate to the protection of public health, defence of consumers, fair trade or environmental protection. This usually entails "horizontal measures", *i.e.* measures globally applied to food commodities. But in certain cases, the Community also adopts sectoral measures, *e.g.* standards on the composition of food products, definition of methods used in biological production. Furthermore, to promote a product quality policy at the Community level, a Community framework for the recognition of quality products that bear a designation of origin or have been traditionally produced has been set up.

The Community, on the other hand, has to live up to the obligations stemming from the Uruguay Round. This entails the agreement on technical barriers to trade (TBT) and the agreement on sanitary and phytosanitary measures (SPS).

The two agreements basically cover the same ban on obstacles to trade as the Treaty of Rome but give more detail and apply to GATT members. The SPS agreement requires, in principle, (article 3, para. 3) respect for international standards, directives and recommendations, unless there is scientific justification for requiring a higher level of protection, or if a given GATT member feels that a certain level of protection is appropriate.

As far as standards, directives and recommendations are concerned, in Annex 3.3, special mention is made of the Codex alimentarius whose rules, henceforth, will be much more important than in the past, since previously a member state could decide not to adopt them.

Relevant international agreements also include the TRIPS agreement on the legal aspects of intellectual property rights related to trade. This agreement has some impact in the case of geographical designation of origin.

Opinions in this paper reflect the viewpoint of the author, and not those of the European Commission

CHAIRMAN: JOSEPH HAUTVAST, *AGRICULTURAL UNIVERSITY, WAGENINGEN, THE NETHERLANDS*

VICE CHAIRMAN: SERGE RENAUD, *INSERM, BORDEAUX, FRANCE*

THEME 1

TOMORROW'S DIET

Workshop 3
Diet, well-being and health

FOOD CHOICE AND HEALTH

PIERRE LOUISOT
Professor at the Faculty of Medicine
Laboratoire de Biochimie, INSERM U 189/CNRS
B.P 12
69921 Oullins Cedex, France

Keeping up good health, obviously, is connected to a healthy lifestyle. And choosing the right foods is a fundamental element in healthy living. **Health maintenance**, thanks to a very varied daily diet, and **possible prevention** of certain illnesses by optimal food choices are considered basic sales pitches in marketing high quality products.

It is very commonplace to say that man needs **a regular, balanced daily diet**, of biologically absorbable proteins, glucids, lipids, vitamins, mineral salts, water, etc. **"Biologically absorbable"** does not only have the habitually understood meaning of foods that contain a molecular conformation which adapted to the receiver cells in the gastro-intestinal tract. That would be too easy. The term implies an appealing presentation of the metabolites resulting from careful cooking, a pleasant environment, good company, due length of time. This is just to show the strong psychological content of the physiological context... and how great a risk there is of annihilating the benefits of foods that are "metabolically correct" if consumed in an environmental desert! In today's world, when we think of food and health we automatically think of **"the calories"**. That this notion is vulgar and perfectly **unadapted to our metabolical complexity**, and that it disregards the precious context in which we live, cannot be overemphasised. The explanation, however, is easy enough to understand: the caloric value of foods can be roughly quantified, while quantifying discreet metabolic interactions is nearly impossible. The basic question of **quantifying the metabolic regulation of our bodies** using non-invasive techniques such as stable isotopes and NMR has become a scientific priority.

The relationship between nutrition and health has always been controversial: **atheroma and cancer** give us excellent examples. Is a nutritious, well-balanced diet a source of health? Although not systematically proven, it is very likely. Is an insufficient or unbalanced diet a direct cause of illness? This is also probable. How can these suggestions be translated into daily reality?

Food is only one of the many factors that make up the human person. Our metabolical equilibrium is a very fragile composition, the result of numerous interactions:

- genetic factors that are still very poorly understood;
- family habits and childhood education (if it exists!);
- working conditions, and how demanding they are;
- financial ease, that is not becoming easier;

- leisuretime: a copious meal for an older person religiously ends with coffee sweetened with a sugar substitute, followed by.... after-dinner liqueurs!

As far as foods are concerned, some of our contemporaries have strong illusions about their links with health. There are many variations to the theme:

- Some believe in the **metabolic marker.** Cholesterol is a popular one. It is essential to keep the blood level down. To deduce that all is well (or the opposite, if it rises) is classically naïve. This simplistic viewpoint is not uncommon because it comforts the mind. No complete analysis, no difficult thought process, reject all thought of complexity - that's much more reassuring!

- Others are under the **influence of the good Doctor X** (who is perfectly unknown beyond the borders of his community or the reaches of the journal that stands behind him), who usually has a cure for everything thanks to a diet which he alone (by divine inspiration) knows and which should help women everywhere (comforting notion of salvation for all)!

- Some people bend to **exotic influences** and want to live like an indian in the Amazon (except for the high infant mortality rate and the low life expectancy in the region of course, but that is carefully ignored), feeding on a **rare** leaf or root (even more highly prized if it is Chinese). And then there is the run-of-the-mill Frenchman who mentally associates food-health-flexibility-longevity and sees himself deftly springing from tree to tree, or sprightly dancing all night in a bar under the effect of a dangerously exciting drink which, of course, is called an "energy-builder".

It is easy, and dangerous, to disorganise the metabolic equilibrium that is acquired normally through a diversified daily diet. Fast nutritional correctors are not blissfully innocent, and are more expedient in causing nutritional problems than in relieving them. Atheroma, which is multifactorial *par excellence,* is unquestionably connected to foods. But it is also connected to smoking, drinking, cell ageing, arterial hemodynamics, free radicles and, without doubt, to many other factors, yet unknown. Giving a patient the impression that he can be cured by adding, or eliminating, a specific metabolite is totally unacceptable. Thumbnail epidemiology, or seemingly appealing statistics cause simple correlated, but unconnected, events, to be mistaken for cause-effect relationships. This simplification is appealing to the media, but it is radical and dangerous.

There are abundant examples of this idiosyncratic taste. We used to speak of the "light" foods; now we have the vitamin- or mineral-enriched foods, and tomorrow will no doubt bring with it other types of illusions.

The case of vitamin overdoses is legend. Since a person is in good health when his needs for the usual vitamins have been satisfied (and this is true), wouldn't he be in even better shape if he got more of these vitamins?

Vitamins and minerals, just like water, proteins, glucids and lipids, are fundamental to living matter. They should be:

- introduced through food;
- or synthesised from simple sources by the human organism.

The description of vitamin-deficiency diseases dates long back in time, although the basic causes were not known: beriberi was described by the Chinese in the 26th century B.C, and scurvy (in the Eber Papyrus) in the twelve centuries B.C.

Precise knowledge of the importance of vitamins in the metabolism and their molecular role is more recent, and stems from biochemical engineering and quantic biochemistry. As concerns structural and metabolic biochemistry, vitamins have truly exceptional chemical characteristics:

- they are heterocycles (carbon, oxygen, nitrogen, hydrogen, phosphorus, etc.) with extraordinary **chemical reactivity** (high molecular resonance, astounding electronic mobility, etc.);
- this remarkable chemical reactivity is the very foundation of their biological properties.

Such fundamental molecules might have been programmed into the synthesis of human organisms, but, in general, this has not been the case. The so-called lesser organisms such as bacteria, yeasts, fungi, etc. are capable of synthesising vitamins, but the human being, in this particular case, suffers a metabolic insufficiency, since he cannot synthesise the ones he needs. Mutations in certain biosynthetic chains have caused him to lose many elementary capabilities.

As a result, vitamins have to be provided from the environment he lives in, and in particular, from his daily diet, failing which he suffers from a phenomenon called vitamin deficiency or, the earlier stage, called "sub-deficiency".

The opposite is also true. If vitamin intake exceeds daily need, the surplus vitamin molecules, like surplus molecules of any substance, are metabolised in organs and cells, and then eliminated through catabolism, a complex, energy-consuming process that, in quantitative terms, is still partly unknown.

The situation, thus, is clear:

- In the event of vitamin or mineral deficiency, should the deficiency be offset by appropriate food supplements, regardless of form? The answer, basically, is "yes" of course. But what about the form? Attention needs to be given to the problem of real absorption of these supplements, in other words, the bioavailability of the vitamin or mineral, a subject that still requires enormous work.
- The second question is much more important: in a person who is normally supplied with vitamins and minerals as part of his balanced metabolism, is it advisable to create a veritable overdose in order (according to a belief not scientifically substantiated) to improve his biology and his capacity to exert effort (for those who still do!), prevent cardio-vascular illnesses and cancers, increase immunity, strengthen bones, avoid foetal malformations, and soothe the pains of older and old age, etc.?

In normal dietetics, an overdose of water, glucids, lipids and proteins is only encouraged in well-defined physiological and pathological situations. Most curiously, our contemporaries - who tend to be suspicious of foods - are totally uncritical of vitamins and minerals. They light-heartedly confuse the beneficial effects of making up for a deficiency with illusions about surplus intake. Many of them have become faithful consumers of every vitamin known under the sun. And compared to the North American and North European, the French consumer is a paragon of wisdom. It is not because a certain nutritional laxism has gained ground elsewhere that we should allow it to take root here, and lead us, *inter alia,* into a situation of disastrous obesity like that found among certain populations of the Western world, with all the repercussions that that can have on pathologies entailing high cardio-vascular risks.

CONCLUSION

The present-day controversy on the use of vitamins and provitamins in hypothetical prevention of cardio-vascular risks and cancers is reaching new heights, especially in the United States, and in some cases the results have run totally counter the sought goals; this means that the problem of nutritional change must be treated without passion, without preconceived ideas, without blissful innocence, but with great scientific rigour and concern for the long-term safety of our contemporaries. The case of vitamins is a good example, but it is not the only one, far from it. Periodically, without any serious scientific verification or nutritional study, the health benefits of some wayward metabolite reaches the zenith. But its stardom does not last long. Although this practice is extremely unhealthy, our current mores in communications, unfortunately, make it impossible to avoid.

Human nutrition is dependent on a fragile metabolic equilibrium which, when disturbed by surpluses or deficiencies, can lead to very long-lasting health impairments. Teach prudence and sagacity, preach "metabolic humility", popularise only scientifically demonstrated benefits, avoid deleterious effects, and beware of excessive or restrictive nutritional impulses which may disrupt the harmony of daily living just a little too much.

HOW FUNCTIONAL FOODS IN THE DIET CONTRIBUTE TO HEALTH

Marcel Roberfroid
Université catholique de Louvain
Département des sciences pharmaceutiques
Unité de biochimie toxicologique et cancérologique
73, avenue Mounier
B-1200 Brussels, Belgium

ABSTRACT

The most recent data on human biology justify the hypothesis that certain nutrients consumed in quantities that exceed recommendations, and certain non-nutritive components, if properly consumed, have specific, beneficial interactions with various functions of the organism and thereby contribute to maintaining good health and preventing certain diseases. This is the context in which the concept of functional diets has been developed. A functional diet, through properly dosed functional foods, provides components that can have a beneficial effect on the functions of the organism, thereby justifying functional or health claims.

The following concept has deliberately been approached from a nutritional angle, and its application requires the development of a functional food science. Recognition of the functional character of a food requires: clear identification of its interaction with a specific function, understanding of the mechanism, evidence of the beneficial effect, a well-formulated hypothesis, and confirmation (through appropriate studies) of this effect in man.

Essential functions, like the ones that govern the gastro-intestinal tract, the cardiovascular system, the redox equilibrium, the basic metabolism, development and growth, cognitive performance and appetite, and the metabolism of xenobiotics are all targets for functional foods.

In most cases, the road from the identification of these interactions to the dissemination of claims on their benefits to human health is long. This long road is full of promise for 21st c. nutrition, and must be lined with rigourous scientific procedures, for they are the only guarantors of well-merited success, and, even more so, guarantors of the beneficial effects that such science can justly claim to have on human health.

INTRODUCTION

The primary role of food is to nourish us (with a balanced diet) and, at the same time, give us a feeling of pleasure and well-being. Recent research in biochemistry and physiology, and also in pathogenesis, support hypotheses that the diet controls and modulates many, varied functions in the body and, thereby, participates in maintaining a level of health that inhibits the outbreak of certain diseases. The concept of functional foods and a new discipline in nutrition called the science of functional foods are based on these hypotheses.

A functional food is a food which not only has macro- and micro- nutrients, but also contains components capable of targetted interaction with functions of the organisms and thus can contribute not only to good nutrition, but also to maintaining good health and preventing disease.

It is important to stress from the outset that a functional food is only valid if it is included in a balanced diet, and that its functional components can only be evaluated in this global context. This means that it is probably more accurate to speak about a functional diet, rather than a functional food.

THE PRODUCTION OF FUNCTIONAL FOODS

There are four ways to create a functional food, *viz:*

- free a food of a component, such as an allergenic protein, that may cause an undesirable effect on the consumer;
- increase the amount of a component found naturally in a food in order to reach a dose level that could have the desired effect, *e.g.* fortify a food in micronutrients to reach a consumption level that is higher than the recommended daily intake and is compatible with the "dietary guidelines for disease prevention" (Block 1993) and, at the same time, increase the amount of a non-nutritive component whose beneficial effects have been established;
- add a component which, usually, is neither a macro- nor a micronutrient but whose beneficial effects have been established, *e.g.* a bifidogenic oligosaccharide;
- replace a component, usually a macronutrient, known to have adverse effects, *e.g.* certain lipids, by a component whose beneficial effects have been established, *e.g.* inulin (Franck-Frippiata, 1992).

To demonstrate the aforementioned beneficial effects requires a rigorous scientific procedure stemming from a strategy rooted in the science of functional foods. This is the only way to guarantee relevance to public health.

THE SCIENCE OF FUNCTIONAL FOODS

The first step in any research designed to develop a functional food entails the identification of a specific, potentially beneficial interaction between components in this food and a genomic, biochemical and/or physiological function of the organism. This process is part of basic research that should lead to the formulation of scientifically justified hypotheses concerning the mechanisms of such interaction, and provides the basis for defining a functional effect which then has to be demonstrated by modelling. This step constitutes the experimental part of the science of functional foods and leads to the formulation of hypotheses on the relevance of these effects on human health. Such hypotheses have to be tested in rigorous nutritional studies with adequate numbers of volunteer participants.

The most promising fields of investigation for the science of functional foods are:

- gastrointestinal functions, including functions connected to a balanced colonic flora, functions mediated by the organ's endocrinal activity, functions that depend on its immune activity, functions that govern the absorption of macronutrients and the bioavailability of ions, functions that control the transit of the food bolus and the mucosal motility, and functions that modulate the proliferation of its epithelium. The main food components that may have beneficial effects on the gastrointestinal system are the probiotics, the non-digestible oligosaccharides, resistant starches, non starch polysaccharides, and certain peptides, (Roberfroid 1996).
- the redox activity and the antioxidant protection based on a balanced, adequate supply of antioxidant (pro-vitamins) but also, undoubtedly, on non-vitaminic components such as the polyphenols and other antioxidants of plant origin for which precise pharmacokinetic data are still incomplete. The redox activity and the antioxidant protections, to some degree or another affect all the cells and tissues in the organism. Tissue and cellular imbalances are connected to many pathologies. There are many sound hypotheses on the mechanisms that cause food components to act; the greatest difficulty is demonstrating the beneficial effects of consuming them, - except for the complex case of fruits and vegetables.- modulating the metabolising macronutrients, carbohydrates, amino acids and lipids, no doubt via modulation of the equilibrium between insulin and glucagon and other hormones, to prevent Syndrome X-related and cardiovascular pathologies. This is where studies on the nutrition-gene expression interaction are the most advanced, both as concerns the direct role of glucose and certain polyunsaturated fatty acids and the more indirect role of certain non-digestible oligosaccharides on lipogenesis enzymes (Fiordaliso *et al.* 1995).
- the effects of the mother's diet on embryo development, and the effects of food in the first few months of life on the development of the infant and young child. These are very important subjects, especially as concerns the effects of certain polyunsaturated fatty acids on brain development early in life. Pediatricians are also very interested in the role of folic acid in the future mother's diet.
- the relation between the diet and psychological behaviour, and also diet and cognitive or physical performance. There are many questions to be asked about the effects of certain food components on these functions, and it is not easy to define the borderline between nutrition and pharmacology.

Methods of investigation are usually not rigorous enough, and it is not always easy to provide a quantitative interpretation of related data. One of the most promising subjects is controlling palatability; there are hypotheses on the role played by gastrointestinal hormones such as cholecystokinin.

HEALTH CLAIMS

The logical outcome of research on the science of functional foods is recognition of the fact that the results of such research can justify claims that the agro-food industry can convert into messages for the consumer. The research-development links presented above logically lead to two types of claims: functional or physiological claims and health claims.

A functional or physiological claim is based on evidence of a precise, well-defined effect on a clearly identified genomic, biochemical, cellular or physiological function of the organism. Examples: selective stimulation of intestinal bifidobacterial proliferation by the fructooligosaccharides, improvement of the bio-availability of certain ions by peptides in milk or certain non-digestible oligosaccharides, modulation of intestinal transit by certain insoluble, non-digestible polysaccharides, stimulation of specific functions in the immune system by certain probiotics. Research on these subjects has led to new concepts in nutrition, such as "colonic foods", "prebiotics", "synbiotics", and "bifidogenic factors" (Gibson and Roberfroid 1995).

A "health claim" subsumes demonstrations (even in human beings) that regular consumption of the functional food will prevent certain medical problems such as food allergies, defecation disorders (diarrhea or constipation), cardiovascular afflictions, cancer, osteoporosis, etc. How difficult it is to demonstrate these effects depends on the targeted disease. Studies always require a lot of time and in some cases lead to unexpected results. This was recently observed in studies on certain antioxidant vitamins that were thought to help prevent lung cancer (Hennekens *et al.* 1996).

Whether it be in support of a functional claim or a health claim, demonstrations of the effects on man will always be required as part of nutritional studies which use protocols and evaluation criteria that are not necessarily the same as clinical studies on drugs. Nutritional studies involve healthy people (or people presumed to be healthy), whose everyday diet is changed in order to cause a significant (biological?) variation in parameters established as markers of "good health". In most cases, these markers still have to be identified and validated.

PROCEDURES FOR AUTHENTICATING A CLAIM

On the one hand, functional foods aim to contribute to improving-maintaining-reinforcing people's health through better diets. On the other, the consumer and his "advisers" have every right to demand that the value of these claims and the quality of supporting scientific data be guaranteed. The public authorities, with this in mind, should introduce certification procedures to satisfy these well justified demands.

They have two possibilities. Claims can be substantiated by being on a positive list or after being reviewed and evaluated in a dossier of data demonstrating significant, specific, positive interaction with functions of the organism in order to maintain (or even improve) health and contribute to the prevention of certain diseases. Most of the data should be published in scientific journals. Claims on the positive list are substantiated by *a posteriori* scientific proof, while claims backed by a dossier are certified on the basis of a priori evidence. A multidisciplinary committee of experts could be made responsible for drawing up the positive list and analysing the dossiers. After confidential consideration, the committee could approve and suggest a priori the research protocol(s) that could provide the scientific proof required to demonstrate effects stipulated in the claim.

The claim authorisation should provide the consumer with a definition of the formulation (conditions and limitations) and a clear indication of the "effective dose". For each "functional" food covered by a claim or allegation, data will be required on the amount of "efficient dose" in each portion.

CONCLUSION

Functional foods provide a unique opportunity to improve food quality that contributes to human health. To guarantee the success of this new field of nutrition will require a rigorous scientific approach that includes constructive discussion between decision-makers, research scientists, industrial manufacturers and professional on nutrition. The science of functional foods must guarantee to the consumer that these new products are not merely new sources of profit, but that they are also, and above all, a source of progress that enables him to better control his health.

SELECTED REFERENCES

BLOCK, G., 1993. Micronutrients and cancer: time for action? *J.Natl.Cancer Inst.* 85, 846-848.

FIORDALISO M.F., KOK N., DESAGER J.P., GOETHALS F., DEBOYSERD., ROBERFROID M.B., DELZENNEN N., 1995. Dietary oligofructose lowers triglycerides, phospholipids and cholesterol in serum and very low density lipoproteins of rats. *Lipids*, 30, 163-167.

FRANCK- FRIPPIATA A., 1992. *Rafticreeming: the new process allowing to turn fat into dietary fiber.* FIE Conf Proceed. 193-197.

GIBSON G.R., ROBERFROID M.B., 1995. Dietary modulation of the human colonic microbiota - introducing the concept of prebiotics. *J. Nutr. 125,* 1401-1412.

HENNEKENS CH.H., BURING J.E., MANSON J.E., STAMPFER M., ROSNER B., COOK N.R., BELANGER CH., LAMOTTE F., GAZIANO J.M., RIDKER P.M., WILLETT W., PETO R., 1996. Lack of effect of long-term supplementation with beta carotene on the incidence of malignant neoplasms and cardiovascular disease. *N.Engl.J.Med.* 334, 1145-1155.

ROBERFROID M.B., 1996. Functional effects of food components on the gastrointestinal system. *Nutr. Rev.* 54, 538-542.

FOOD SAFETY

JEAN-FRANÇOIS GIRARD
Direction générale de la Santé
8, avenue de Ségur
75350 Paris cedex 07, France

INTRODUCTION

I thank the organisers of INRA's 50th anniversary celebration, and at the same time want to delve into the heart of my subject by asking myself out loud why a senior civil servant such as I was asked to talk about food safety. By asking the question I think I might cast some light on the problem. It is a sign that the regalian character of this problem has been recognised in most developed countries. In a country such as ours where a public body is responsible for funding health care, the Minister of Health is more that the minister of diseases, he is also minister of the determining factors of health.

I remember a newspaper article a few years ago that headlined, "La Rochelle, a rice and milk dessert out in the sun for several hours: nine dead. What does the Minister of Health do?". Our fellow citizens have a quite different way of "reading" their health today, considering what attention they should give it, and what attention they expect from outside. They closely watch what the Ministry of Agriculture is doing.

National and economic interests have been attended to for decades, but the limits to these interests become clear as soon as human health is mentioned. Health has become a subject of public debate and, thus, a subject of political debate. Political action is the domain par excellence of contradiction, confrontation and choices, together with all the attendant risks. It is certainly not per chance that health laws are being submitted more often to the Parliament. In my opinion, this explains why the organisers felt that the Director General of Health was qualified to speak about food safety.

THE VARIOUS MEANINGS OF THE TERM "FOOD SAFETY"

The Minister of Agriculture recently signed an article in one of the big daily newspapers in which he dealt with the importance of food being available to ensure proper nourishment for everyone in the world and the fact that this responsibility was incumbent on governments. But "food safety" can also relate to the protection of an economic sector which, in France, accounts for over 10% of the GDP, and quality, a subject which affects both production and consumption. The fourth way of "reading" the term food safety is the one which interests us most particularly. It concerns the question of food safety for human health and the eventual pathological consequences on man. Dwelling exclusively on the last aspect, we should remember that food is only one of the determinants of health; don't forget the environment, air, water, medicines, etc.

To speak of food safety implies that foods have a certain degree of risks. It is important to look at what this notion embraces, because man is somewhat irrational in managing risk. Just consider the differences in behaviour when we deal with risks that have been freely chosen, such as smoking, drinking, driving, mountain climbing in sandals, etc. We do not accept any restrictions, regardless of the consequences. Conversely, we are incredibly intolerant of imposed risk. For blood transfusions or surgery, for instance, we expect the risk to be as small as possible, or even non-existent. We might note that smoking is a risk somewhere between these two extremes since it combines the chosen risk (by the smoker) with the inflicted risk (by his neighbours).

This in-between situation also applies to food. The reality of a risk can be assessed on an individual level, but it is difficult to identify the precise percentage of the population that is exposed to it. On the one hand, the risk related to a drug prescribed to a few thousand people to treat a disease can be assessed, but it is quite something else if we want to judge the consequences of the harmful effect of a vaccine used on millions of healthy people. I am trying to draw your attention to this differentiation in risk analysis.

A distinction also has to be made between risks that bear immediate consequences, like salmonellosis, and risks related to a virus or to prion. In the latter case, the risk is "postponed" for at least a decade or even decades; this is a new factor that we have to take into account.

The importance of risk relativity: a concept that we are not really able to measure. Let me illustrate this idea with an example taken from another sector. In 1996, INSERM published a report on asbestos, with a risk scale that ended years of uncertainty concerning the exprofessional-risk. This scale provided indispensable relativity, without which the only reference that could be used in risk management would be zero risk. Regardless of the field of activity in question, however, zero risk is an illusion. The organisation of any society based on the Grailien quest for zero risk is a dangerous delusion that can only be avoided by a culture that accepts to compare risks.

When it reaches adulthood, our society should be able to see food-related risks in relative terms. But maturity seems far off because we increasingly see that any risk that has not been controlled or foreseen leads to a criminal lawsuit. Some writers even feel that if the State is not able to come up with a satisfactory public healthy policy, the criminal court has every right, and should, stand in for the government or even for a society that is incapable of total risk avoidance.

Communication on this subject is urgently needed. Reticence to talk about the concept of risk is perhaps inherited from medical confidentiality, the privacy of disease and our prudish attitude towards the dysfunctions of the body. From a collective angle, risk needs to be discussed publicly, otherwise the maturity I mentioned earlier will never be reached. We should speak about it in order to rank risks and then assume them, or not, once we know what they are all about. This would make the Manichean approach between zero risk and selected risk obsolete.

THREE APPROACHES TO FOOD SAFETY

But, as a preliminary measure, we have to apply three indissociable functions to all sectors for which we want to measure risk: expertise, control, warning. This is a theoretical chain that has already been applied to medical drugs, a field that pushed this tripartite logic to the extreme. I will not go as far as the market authorisation notion when talking about yoghurt, for instance, especially since the notion of "variable" consumption is very important for foods. This expertise-control-warning chain, thus, should be applied in adapted form, but it can be readily transposed. We could talk about expertise when considering composition and procedure, since we know that when the composition or a procedure is registered it is important to asked about possible risk-borne consequences, and not only think of consequences in relations to accidents, which can be left to later. Society no longer accepts the idea that we can wait until risk become reality before taking appropriate steps. Expertise has to be given its due importance and context; it has to be controlled by a set of rules and regulations, with recognition of the fact that the aim is not to systematically put the enterprise before an unscaleable wall.

As far as controls are concerned, I think that a lot remains to be done, both in France and in the rest of Europe. It is all well and fine to prepare standards and have ministers sign orders, but these orders have to be applied, and our systems have failed in this brief because of lack of adequate control measures. Giving warnings means being aware of unexpected consequences. Warning systems exist in most of our countries, but, at the European level, there are alarming shortcomings. It is incredible that the last Council of Ministers of Health again hesitated about setting up a European CDC.

Second, I feel very strongly that responsibility for economic security, production sectors, and food safety should not be left up to one authority alone. Since you cannot be on both sides at the same time, a technical and administrative solution will have to be found that ensures a separation of powers, otherwise the sector's creditworthiness will be lost.

Third, questions of organisation need to be answered. Should France, and Europe create agencies based on the American FDA model? The government is working on this, but, more than that, we need independent technical tools which will enable the public authorities to assume their responsibility, to produce rules and regulation, and not to interfere directly in the expertise of these tools. I am thinking of advisory councils, and the operational tools used by authorities to go beyond the expertise stage, by providing control and warning. The subject of this discussion should cross our national borders because this is a Europe-wide problem.

CONCLUSION

I believe that food safety is a theme with many, strong biological and social requirements. We cannot reduce it to nothing more than an economic constraint. On the contrary, if food safety is well controlled, the food line can meet the triple - physiological, cultural, and economic - stakes. And the public authorities are responsible for all three.

CHAIRMAN: STEIN W. BIE, *FAO, ROME, ITALY*

VICE CHAIRMAN: CHARLES RIOU, *INRA, BORDEAUX, FRANCE*

THEME 2

AGRICULTURE, ENVIRONMENT AND RURAL AREAS

Workshop 1a

Agriculture, natural resources and environmental constraints

WATER MANAGEMENT FOR AGRICULTURE

CARLOS TIO
Polytechnic University of Madrid
28040 Madrid, Spain

INTRODUCTION

The problem of water management in agriculture does not have the same economic and social importance in all regions and countries. It is a subject that especially concerns the arid and semi-arid zones and zones subjected to extended periods of drought because of strong interannual differences in rainfall levels.

Because of possible future climatic changes, regions that at present seldom experience a water shortage, may be more concerned with this problem in the future.

That is why water is increasingly considered as a strategic element that could play a decisive role in socio-political conflicts.

In regions that experience more or less long drought spells, agriculture has traditionally been the number one water consumer. In California the demand for water for agriculture is estimated to be 87% of total consumption, and in Spain the figure is 80%.

Considering the water shortage and the socio-economic changes facing agriculture, it is logical to wonder about water use efficiency in agriculture. This explains why water economics and technology are turning more to agricultural research and will do even more so in the future.

Any economic analysis must view water as both a basic natural resource and an "input" for economic production. But, on the other hand, variable physical and climatological as well as historical and cultural circumstances connected to the legal procedures applied to water rights and to all the related institutions give rise to a most singular situation because of which the problem of water must be very carefully studied, even at the local and regional levels.

On the other hand, since the water cycle in Nature is unique, continual and closed, we cannot ignore the importance of a global analysis nor of integrated hydraulic plans which, in some cases, involve two or more countries sharing a single basin, as is the case for the River Jordan in the Middle East, and the basins shared by Spain and Portugal.

Furthermore there is a possibility that water will be siphoned off from a basin or from inbetween basins (*e.g.* Rhone-Catalonia) which makes the problem of water utilisation even more complicated.

In such a complex situation, all aspects have to be considered globally through systematic analyses which might be structured as follows.

THE INSTITUTIONAL FRAMEWORK

- A global perspective is needed to analysis of the role the water cycle in Nature and the conservation of ecosystems, and the compatibility of the water cycle with the way in which water is used by man and society (domestic use, industry, agriculture, energy production). This analysis should lead to a basic conceptual framework that includes the public/private character of water, priorities, and limitations/restrictions in water utilisation.
- A system should be devised for allocating water resources to consumers, including regulations on individual water utilisation rights.
- The CAP and its orientations are part of this frame work.

WATER ECONOMICS AND TECHNOLOGY

This include an analysis and determinants of demand, a calculation of **water supply** and compensation for **water shortages**. Consideration must be given to the water supply (increasing the capicity of the hydraulic system; improving and modernising the infrastructures, reutilisation and purification, desalinisation, use of new groudwaters, etc), **the water demand** (water saving techniques and programmes), and the external inputs to the hydraulics system, or **siphoning water,** (water quality control and improvement systems, economic and financial regime for water).

Not all of these aspects are of universal importance. Many of them, which are meaningless for farmers and agro-foods entrepreneurs, are undoubtedly of keen interest to the ecology groups).

In my opinion, agricultural research should consider all these aspect before finetuning the subject of their particular study. This would provide a global perspective and avoid certain difficulties with the rest of their work.

Water management in agriculture includes a few additional considerations. Hydraulic and irrigation infrastructure have been used for hundreds, even thousands of years in certain regions, which explains why their institutional systems are often not adapted to the needs of modern society and agricultural economies.

In industrialised societies, where the economic dynamism comes from privatised economies and a "market logic", it is difficult to conceive of water use management being largely centralised or even that water is part of the public domain.

Water management in agriculture is often controlled by public or semi-public institutions or corporations like the "irrigation communities" in Spain that distribute water according to ancient water utilisation rights derived from century-old administrative entitlements based on a traditional method for drawing water, a retired legal structure, no personal or tangible line of responsibility, and economico-financial systems that seem disconnected from the dynamics of a market economy.

It is clear that these characteristics are not found in all the regions and countries because there are great differences from one to the next, but all the countries have met with some of these problems.

Traditions and century-old habits constitute the basis of institutional systems. In the United States, for instance, water rights differ from state to state with a water-sharing system in the east, a straightforward water ownership system in the west, and combined systems elsewhere. In the case of Spain, the situation is marked by graduated differences, depending on whether the issue is surface or underground waters, or else traditionally or newly irrigated lands.

Another element to consider is whether the hydraulic infrastructure for agricultural irrigation was financed by the government, as a result of pressure from public opinion, for reasons of land development and in some cases for highly complex reasons that include military interests, such as in Israel. In this type of scenario, economic rationale and strict cost-benefit calculations have often been of lesser importance.

Nonetheless, this order of things may change considerably in the future since the 21st century will be particularly sensitive to ecological problems and scarce resources. It seems quite obvious that institutions and water use regimes will have to adapt very quickly to new socio-economic realities, and increase water management efficiency through economic stimulus and market-driven considerations.

In many areas a "water market" already exists; it operates at the local level with a very singular management system. And this seems logical, especially in places where water resources are part of the public domain and where water rights are granted through a government concession. When this is the case, notwithstanding water transactions and markets, the market is really controlled by the public authorities since the state owns the water.

Furthermore, considering the new framework for "Ecosystems and Nature Protection and Conservation", and the fact that water plays a decisive role in maintaining life, any attempt to submit water rights to a line of logic based solely on economic and commercial interests would certainly be useless.

CONCLUSION

To recapitulate, let us look at a few conclusions to our analysis:

First of all, in the future the institutional framework will continue to be governed by certain standards, or restrictions, that stem from the all-important natural equilibrium of our ecosystems, or from our cultural and historical roots, or else from the psychological sensitivity to a resource as emblematic as water.

Second, interventions and public regulations on water rights, as well as communal and corporative supervision of water users, and even strategic planning for the protection and distribution of water will always be complex and essentially politically governed.

Nonetheless, it is urgent for economic rationality and vanguard technology to be adopted in water management for agriculture. Aside from restrictive and tempering elements that could be derived from the institutional framework and which undoubtedly will be permanent, the agricultural entrepreneur together with his professional organisations, and even the public services should adapt their operating norms and practices to present-day and foreseeable economic dynamics.

SOIL MANAGEMENT AND AGRICULTURE

CHRISTIAN DE KIMPE
Agriculture and Agri-Food Canada, Research Branch
930, Carling Avenue
Ottawa, Ontario, Canada K1A 0C5

INTRODUCTION

In the context of sustainable agriculture, and considering the increasing pressure on natural resources and the high costs of soil degradation, scientists are re-evaluating the role of soils, and the research priorities in this area. A new ethic prompts a moderate exploitation of natural resources. People are being better informed about the interactions between society and soils, not only for the immediate benefits of the agrifood production systems, but also about the role of soils in the ecosystems. In countries where food availability is not any more a critical problem, less than 10% of the population is involved in agricultural production, and urbanites consider the rural space as the essential lung of the large cities.

Soil is one component of natural habitats that integrate human beings; therefore, concepts on soils are being extended to include social sciences (Warkentin 1995): understanding the role of soil in the ecosystems will provide a sound basis to evaluate land-use decisions, to plan for the conservation and the quality of the natural resources, and to analyze those elements of the quality of life that are soil-dependent.

ECOSYSTEMS AND SOIL FUNCTIONS

An ecosystem is the integration of a community (biotic resources) and its environment (abiotic resources) functioning as an entity, with its own characteristics. In an ecosystem, the soil provides a diversity of habitats for biota and the stability of these habitats. Soil also acts as an accumulator, a transformer and a transfer medium for water and other inputs.

In order to achieve its role, the soil must fill specific functions (Larson and Pierce 1991, Doran and Parkin 1994):

Facilitate biochemical and geochemical cycles

such as the recycling of dead biomass to produce a new biomass, the weathering of minerals to provide nutrients, and the decomposition of toxic materials brought to the soil.

Store nutrients temporarily to prevent leaching, and progressively *release* these nutrients to plants, *store* water for plants and soil organisms, and immobilize residual toxic materials.

Partition water by determining surface runoff and infiltration in the soil, for accumulation as soil moisture and transfer at depth to groundwater and aquifers.

Act as a buffer by moderating temperature fluctuations and water inputs, modulating excessive changes in the composition of soil solution; adsorbing toxic materials for their degradation by soil organisms, and purifying water.

Partition energy by regulating temperature in order to make the soil viable for biota, and by acting on the gas circulation and the atmosphere composition.

DEMOGRAPHY AND FOOD PRODUCTION

Relations between humans and soils may be seen from different points of view. Apart from their role in the ecosystem, soils are perceived by many as a renewable resource to be exploited as a *source of food, fibres, pharmaceuticals, and other materials.* This "predatory" approach neglects soil functions. Soils are also the *physical support for buildings and infrastructures*, and are used as well to *dispose of wastes* generated by human activities. Finally, soils play a role in the maintenance of civilizations, and are a *"memory", i.e. they preserve the cultural heritage of archaeological information about previous civilizations* (Blum 1990).

According to the United Nations Development Program, world population was 5.48 billion in 1992, and will reach approximately 8.47 billion in 2025. In the 1990s, the mean annual increase was 1.7%. In 1995, the average consumption of grains, directly or indirectly as meat, was slightly over 300 kg per person per year (Postel 1996). Providing enough nutritious and safe food under conditions of high pressure on natural resources will be one of the major challenges in the future. Over the period 1950 - 85, the green revolution contributed to reducing food shortage in Asia by the release of new high-yield, short-cycle varieties, having also a good resistance to diseases. However, to achieve the potential of these varieties, it was necessary to increase fertilizer inputs and to apply more pesticides; this was the source of problems for the environment and the ecosystems.

GEOGRAPHICAL CONSTRAINTS AND SOIL DEGRADATION

Agriculture dominates largely in the debates on land use: approximately 11.3% of the earth's emerged land areas has no or few restrictions for crop production, and another 24.6% is occupied by permanent pastures. Population increase is at the origin of an increasing competition between agricultural and urbanized lands: between 1972 and 1992, agricultural lands decreased by 14% in the developed countries, and by 29% in the developing ones (Hopper 1995). In 1992, availability of land for food production was 0.76 ha per person in industrialized countries, and 0.17 ha in developing countries, *i.e.* less than a quarter of the land available in developed ones.

Land suitability for agricultural production is determined by soil quality and origin, climate, and the need to maintain lands under forests and as natural areas (King, 1990). Availability of arable land is another factor: for example, Canada is the second largest country in the world, but it has only 5% of the land that is adequate for agricultural production; in the European Union, the average figure is 58% (Table 1).

Table 1- Selected agricultural statistics for some industrialized countries, in temperate regions suitable for different crops (adapted from Wicherek, 1994)

Country	Population (million)	Total area (1000 km^2)	L.A.A.(*)/ total area (%)	Cropland/ L.A.A.(%)	Population in agriculture (%)
E.U.	346	2260	58	52.5	7.4
Eastern Europe	137	1246	69	65	18
former USSR	284	22 275	27	62	15
U.S.A.	255	9372	46	48	2.9
Japan	124	372	15	51	7.9
China	1166	9 596	33	11	61
Canada	27	9 959	5	85	6
France	57	551	59	56	3.6

(*): L.A.A. = land available for agriculture

The balance between biotic and abiotic components determines soil productivity and stability, and the quality of programs aiming at sound land management. Changes in the balance among soil functions occur when natural ecosystems are transformed into agroecosystems (King 1990); soil degradation results from a major unbalance between natural and human systems. The optimum functioning of the soil decreases along a chain reaction: for example, a loss of soil organic matter generates soil compaction, less gaseous exchange with the atmosphere, less water infiltration in the soil and more runoff, less retention capacity for nutrients and toxic material, less biological activity, less soil detoxification... On a global scale, various forms of soil degradation affect seriously some 305 million ha, and another 910 millions are moderately degraded (Oldeman *et al.* 1990).

Agricultural lands cannot be expanded at the rate of population's growth, and it has been proposed to aim at the maximum yield on the best agricultural lands, while setting aside marginal lands for wildlife and biodiversity (Avery et Avery 1996). Such a strategy cannot be successful unless it is clearly demonstrated that no major soil degradation will result from cropping intensification. Increasing yield has to be evaluated also from the viewpoint of the risks for the environment (Serageldin, 1995).

FUTURE PERSPECTIVES

If tomorrow's agriculture is to use more intensively the best soils in a sustainable way, it will be essential to maintain the right equilibrium among soil functions. This involves:

Characterizing and understanding the functioning of the soil ecosystem.

Knowledge on soil and ecosystem resilience is still limited. Much is known about soil's physical and chemical properties, much less about its biological properties and ecology, especially in the rhizosphere. Soil functions are measured at different levels (Table 2). A soil evaluation and monitoring program has been put in place in Canada, and a report on the health of the agricultural soils in Canada has been published (Acton et Gregorich 1995).

Table 2- Measurements of Soil Functions for Soil Quality (From B.P. Warkentin, personal communication)

Function	Level 1 Field evaluation	Level 2 Technical	Level 3 Research
Cycling	a) estimate of organic matter internal soil drainage	a) visual estimate of organic matter decomposition b) CO_2 evolution (lab) c) pH (field, lab)	a) macro and micro-organism species composition b) enzyme activity c) food webs in soil
Storage and Release	a) soil depth b) grain size by hand texturing c) estimate of organic matter d) field pH	a) root room b) water retention curve (lab) c) soil test for available nutrients (lab) d) pH (lab)	a) exchange capacities b) sorption isotherms c) kinetic constants for diffusion and exchange d) unsaturated hydraulic conductivity
Partitioning of Water	a) tilth b) observations of runoff	a) infiltration measurement in the field b) runoff measurements c) soil crusting	a) measurements with rainfall simulator b) modelling water balance
Energy Partitioning	a) soil color b) surface vegetation c) soil variability at km scale	a) soil and water temperature b) detail on soil variability at different scales	a) radiation measurements b) modelling
Buffering	a) organic matter b) clay content	a) organic matter b) partition coefficients	a) sorption isotherms b) modelling

Integrating better agriculture in the ecosystems.

Agriculture aims at producing selected species that are exported to a large extent out of the production area. Production systems based on less tillage, more crop rotations and pluriannual crops will maintain a higher level of soil organic matter, ensure a greater biological diversity, improve the balance among the soil functions and will contribute to improve soil quality (Stinner and Stinner, 1994).

Chemical inputs and organic amendments

A major percentage of the yield increase will continue to come from the use of fertilizers. Between 1972 and 1992, the worldwide use of fertilizers increased from 73.8 to 132.7 million tones of N, P_2O_5 and K_2O (Hopper, 1995). In Canada, the use of fertilizers increased from about one million tones in 1960 to approximately 4 million tones in 1985 (Government of Canada, 1991), while the percentage of land receiving fertilizers increased from 16% in 1970 to 50% in 1985. Fertilizers are becoming less efficient from the economical point of view. Large amounts are accumulating in the soil or are lost through leaching, creating important problems (Table 3a and 3b, Schreier, 1996). Biotechnology will help develop cropping systems with plants having different rooting depths to improve fertilizer efficiency. Soil management techniques with a beneficial impact on soil functions will require less fertilizer inputs: precision agriculture is making progress in Northern America. Remote sensing facilitates the monitoring of soil and crop conditions, and more accurate decision making about pesticide applications on a more timely and more localized manner.

Table 3a.- Annual rates of surplus nutrient applied to cropped agricultural lands in the Lower Fraser Valley in British Columbia, Canada in 1991 (based on large farms only) (from Schreier, 1996)

Surplus Applications per cropped farmland	Nitrogen Surplus in kg/ha/yr	Phosphorous Surplus in kg/ha/yr	Potassium Surplus in kg/ha/yr
Average	115	85	126
Range	-28 to 408	21 to 270	-31 to 340

Table 3b- Sources of nitrogen applied to the land and contributing to stream water pollution in the United States (from Schreier, 1996)

Sources of stream and lake pollution	non-point sources (agriculture)		atmospheric sources (%)	point sources (%)
	fertilizer (%)	manure (%)		
USA overall - N	50	30	14	6
Chesapeake Bay - N	51		26	23
Chesapeake Bay - P	61		5	34
State reports - N	76		12	11
San Joaquin River, CA - N	50	45	2	3
Red River, ND - N	84	16	3	1
White River, AR - N	25	56	16	3

* In the future, agricultural lands will most likely receive increasing amounts of organic matter (sludges, urban composts and from other sources), that will contribute to maintain soil structure under intensive cropping practices; damaged sites could also be restored. Long-term risks are associated with the accumulation of heavy metals, and their impact on soil organisms and the safety of food.

* Finally, as it may be expected that water available to agricultural production will be increasingly limited, soil management for improved water retention/utilization will complement irrigation strategies (Postel, 1996).

Agri-environmental indicators

will be used as management tools, by providing a measure of the changes or the risks of change of the resources used by the agri-food sector. Six indicators are under development in Canada (Table 4). Risk analysis will help identify geographical areas and production technologies that present a significant risk for the environment. Risk is measured by the relation: *risk = toxicity x exposure*, where toxicity relates to the nature and properties of the product or the practice (pesticides, fertilizers, soil tillage,...), and exposure refers to the extent and time of exposure. Measurements will help to determine whether those risks are acceptable socially and environmentally

Table 4- Agri-environmental indicators developed in Canada

1. *Farm resource management:* uptake by farmers of environmentally-sustainable management practices, by measuring soil residue cover and management of agricultural land, fertilizers, pesticides and manure.
2. *Soil degradation risk:* measure progress in reducing the vulnerability of agricultural soils to degradation processes and identify soils remaining at high risk from erosion, salinization or loss of organic matter.
3. *Water contamination risk:* assess progress in reducing the risk of water contamination from pesticides and nutrients used in agriculture and identify areas at risk of contamination.
4. *Agroecosystem greenhouse gas balance:* estimate trends in the net balance between accumulation and release of the three major greenhouse gases emitted by agriculture - carbon dioxide nitrous oxides and methane.
5. *Agroecosystem biodiversity change:* monitor biodiversity in agricultural ecosystems by measuring changes in habitat availability, and species diversity and abundance.
6. *Input use efficiency:* track the amounts of environmentally-sensitive inputs required to produce agricultural products by measuring the efficiency of fertilizers, pesticides, energy and irrigation water being used by farmers.

Considering rural development and integration of human habitats in agroecosystems implies various social aspects of soil management, both at the local and regional or national levels. Soil quality is an integrative concept linked closely to human needs. At the local level, the risks associated with the utilization of urban and suburban soils for food production and recreation are often underestimated and will require more attention in the future. At the regional scale, the vast plains in the temperate regions that were organized to sustain industrialized agriculture, were often subject to major floodings and droughts (Goudie, 1994; Wicherek, 1994, 1995). In addition to the economical factors that are commonly used (*e.g.* land-use decision towards agriculture or infrastructure), soil management must include social parameters such as integrated rural development including also wildlife habitats and wetlands. These parameters are more difficult to evaluate from an economic point of view, but they are important for the *natural resources* capital (Fig1, Serageldin, 1995) and in the context of sustainable agriculture. Re-evaluating rural development in order to reduce the risks linked with climate change and climate variability will require a major shift in the land management approach. Progress has been made in some countries such as France, but not yet in several other countries.

FIG. 1. Sustainable development framework

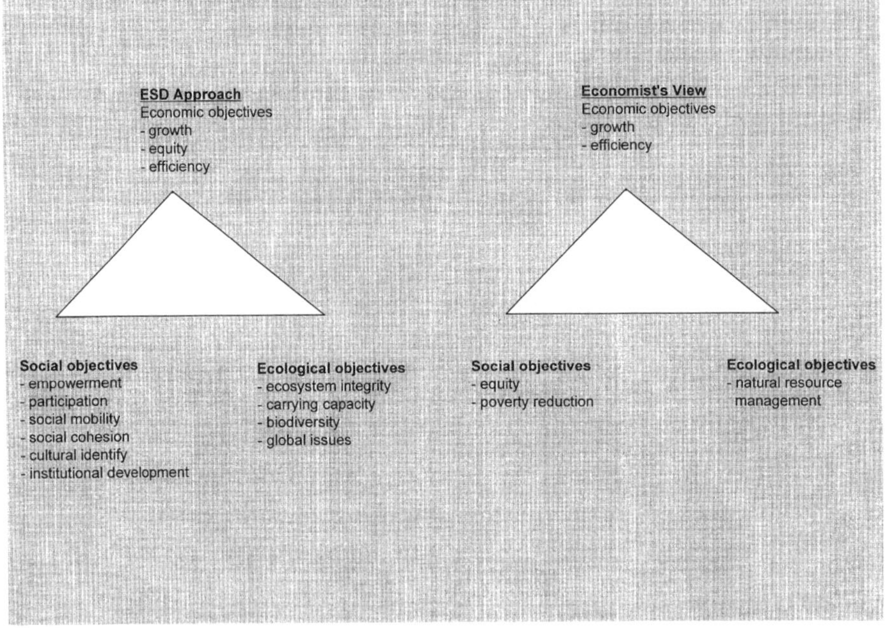

A decrease in soil functions that prevents the soil from filling its role in the ecosystem and that reduces its productivity, creates also a decrease in social and economical benefits obtained from the soil: soil erosion and other types of soil degradation are a socio-economical as well as a biophysical problem. Soils have to accommodate the impacts from the industrial and agrochemical agriculture. Next will come the effects of biotechnology. Managing the soils to ensure the sustainability of this resource from the environmental, social and economical points of view will be a permanent and major challenge.

SELECTED REFERENCES

ACTON D.F., GREGORICH L.J. (eds.), 1995. *The health of our soils: toward sustainable agriculture in Canada.* Agriculture and Agri-Food Canada, Ottawa, Ontario, 138 pp.

AVERY D.T., AVERY A., 1996. Farming to sustain the environment. *Hudson Briefing Paper* 190, 20 pp.

BLUM W.E.H., 1990. The challenge of soil protection in Europe. *Environmental Conservation*, 17.

DORAN J.W., PARKIN T.B., 1994. Defining and assessing soil quality. pp. 3-21. In: Doran J.W. *et al.* (eds): *Defining Soil Quality for a Sustainable Environment.* Soil Sci. Soc. Amer., Madison, WI, Spec. Pub. 35.

GOUDIE A., 1994. *The human impact on the natural environment.* The MIT Press, Cambridge, Mass., 454 pp.

GOVERNMENT of CANADA, 1991. Agriculture: searching for a sustainable system. Chap. 9. In: *The State of Canada's Environment.* Minister of Supplies and Services, Ottawa, Ontario.

HOPPER W.D., 1995. *An assessment of realities: an assessment of the expectations that follow.* Presentation to the Agricultural Institute of Canada's annual meeting, Ottawa, 13 pp.

KING L.D., 1990. Sustainable soil fertility practices. Chap. 5. In: Francis C.A. *et al.* (eds): *Sustainable Agriculture in Temperate Zones.* J. Wiley and Sons, New-York.

LARSON W.E., PIERCE J.F., 1991. Conservation and enhancement of soil quality. In: *Evaluation for Sustainable Land Management in the Developing World.* IBSRAM, Bangkok, Thailand.

OLDEMAN L.R., HAKKELING R.T.A., SOMBROEK W.G., 1990. *World map of the status of human induced soil degradation: An explanatory note.* Rev. 2nd Ed. ISRIC, Wageningen, The Netherlands, 14 pp.

POSTEL S., 1996. Forging a sustainable water strategy. In: *State of the World 1996.* The Worldwatch Institute, New-York.

SCHREIER H., 1996. *Agriculture and water on a collision course: linking water pollution with agriculture using Hypertext and GIS.* Presentation to the Agricultural Institute of Canada's annual meeting, Lethbridge, Alberta, Canada, 22 pp.

SERAGELDIN I., 1995. Sustainability and the wealth of nations: first steps in an ongoing journey. Third annual World Bank Conference on Environmentally Sustainable Development. ESD. World Bank, Washington, 19 pp.

STINNER B.R., STINNER D.H., 1994. *Ecosystem approaches to agriculture: towards a process of sustainability,* pp. 95-97, Abstr. 1st Int. Symp. on Ecosystem Health and Medecine, Ottawa, Canada.

WARKENTIN B.P., 1995. *The changing concept of soil quality.* J. Soil Water Cons., 50, pp. 226-228.

WICHEREK S., 1994. L'érosion des grandes plaines agricoles. *La Recherche*, 25, pp. 880-888.

WICHEREK S., 1995. Inondations: histoire à fleur de peau. *La Recherche*, 26, pp. 692-693.

MAN AND BIODIVERSITY

ROBERT BARBAULT
The Federative Institute of Fundamental and Applied Ecology (CNRS-FR3)
Université Pierre et Marie Curie
7, quai Saint Bernard
75252 Paris Cedex 05 , France

ABSTRACT

Although the word was coined in 1985, the concept of biodiversity really emerged in Rio de Janeiro in June 1992 with the global media coverage it received during the United Nations Conference on Environment and Development. It is therefore, not a uniquely biological concept. Over and above simple considerations on the diversity of the living world - the state and dynamics of genetic variability, the richness of flora and fauna, the diversity of ecosystems - biodiversity really refers to mankind: mankind who threatens it, mankind who depends on it for his needs and well-being, mankind as a species, and mankind as the species to be held responsible for it.

Thus the current debates on the conservation and management of planetary diversity transcend the problem of protecting threatened species and focus on a real challenge to civilisation.

INTRODUCTION

The Earth Summit at Rio de Janeiro in June 1992, heralded the emergence of a whole range of issues regarding the diversity of the living world. A Convention on Biological Diversity was signed at the summit and today has been ratified by more than 150 countries.

The diversity of life has long been known and scientists, nature-lovers and managers of natural resources alike did not wait until the United Nations Conference on Development and Environment to address the issue.

So what is new?

In the great debate which has opened up, apart from the state and dynamics of biological diversity, it seems clear to me that it is man who is singled out: man as part of this biodiversity, a species directly and indirectly dependent on a multitude of others; man as a responsible being challenged to assume overall accountability with regard to future generations.

FROM THE DIVERSITY OF SPECIES TO BIODIVERSITY

Our initial impression of biological diversity is embodied in the profusion of animal, plant and microbial species which populate our planet: 1.5 million known species, and several million still to be discovered (between 5 and 10 million according to the most plausible estimates). For a long time this fascinating prodigality hid two fundamental aspects of the diversity of the living world which were more difficult to make out: its genetic and ecological dimensions.

In fact, the several million living species which coexist with us on earth, as well as the hundreds of millions of species that are now extinct, are the evidence and fruit of 3 to 4 billion years of evolution.

At the origin of this profusion of species we find an omnipresent genetic variability, the prodigious capacity of living species to reproduce and the ruthless mechanics of natural selection.

Two clues enable biologists to detect order in a world teaming with life: a taxonomic strand to analyse the phylogenetic organisation of biodiversity and an ecological strand to express the functional organisation underlying the former.

Figure 1: a phylogenetic tree placing man among the other primates. The divergence between the species, characterised by their DNA percentage differences, yields an approximate dating.

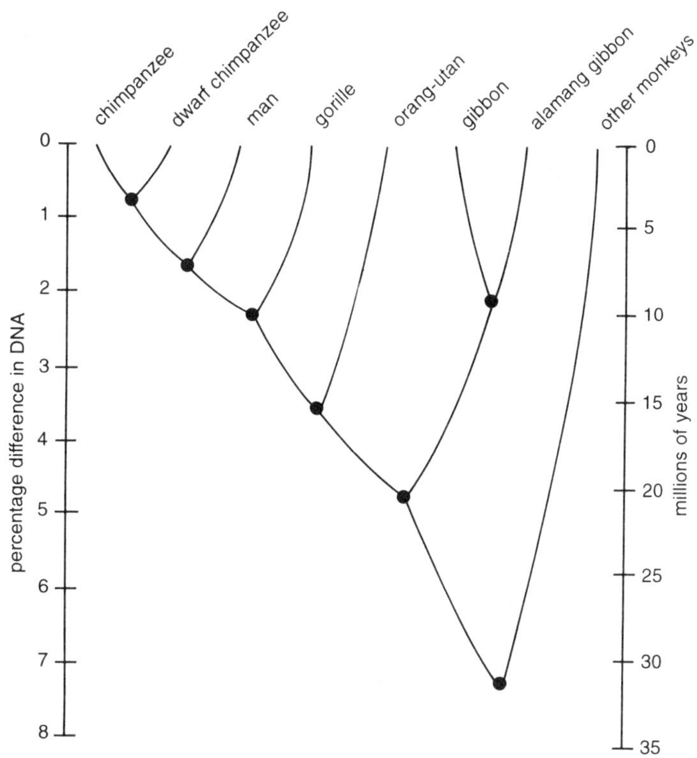

The taxonomic order is a consequence of the speciation process: new species that appear derive from a parent species. In other words there are relationships between the species: the organised image of the phyletic tree has replaced the first impression of some kind of virgin forest of species totally unrelated to each other. This ordered representation of the diversity of life, which modern biosystematics has built for us, also has the great merit of relocating us, as members of the species Homo sapiens, very close to our brothers and cousins, the chimpanzees and gorillas (Fig.1).

Ecological order is the outcome of the dynamics of interactions taking place within systems made up of natural populations and their physical environment.

To highlight this idea, reference is made to population-environment systems (Fig.2). Thanks to the interaction between competition, predators, parasitism and cooperation, and under the constraints of the physico-chemical and climatic setting, the various species adjust their ecological niches, evolve and adapt, move on or disappear.

Figure 2: a simplified representation of a population-environment system.

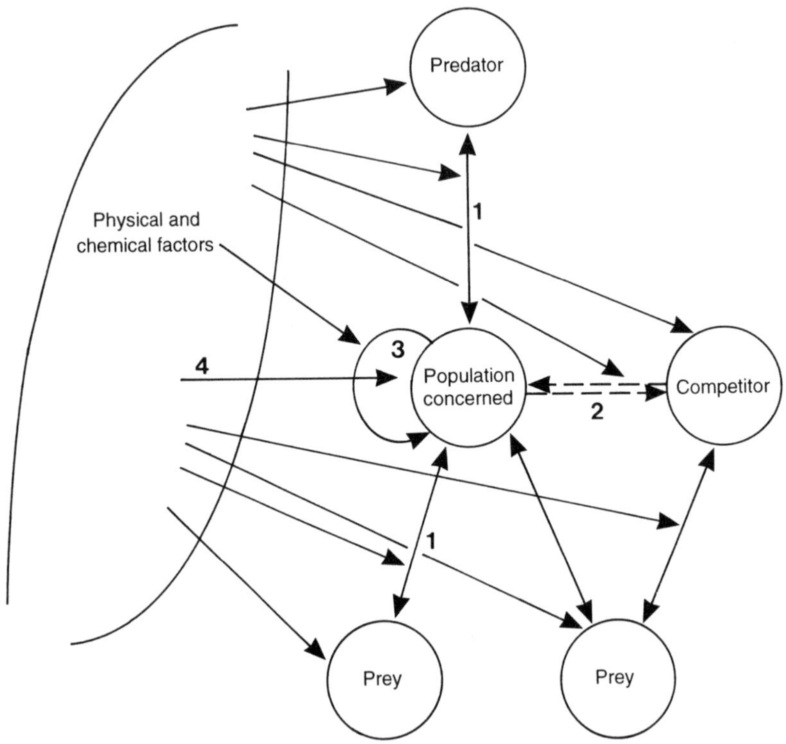

Each circle represents a population (all individuals of the same species). Each population is a part of a network of interactions which determine its dynamics and its evolution: predator/prey interactions (1 predator, parasite); interactions of interspecific competition (2) or intraspecific (3);

effects of environmental factors (4). Biodiversity is structured and renewed by these complex ecological dynamics.

In other words, this ecological diversity, diversity of species and diversity of ecological functions (= niches), is related to the evolutionary scale and organised spatially from the local to the global level.

Moreover it can be appreciated, from the point of view of the functioning of living systems, that genetic diversity, diversity of species and ecological diversity cannot be dissociated. The concept of biodiversity, which in itself appears innovative compared with the simple observation of diversity in the living world brings this out (Fig.3).

Figure 3: The concept of biodiversity is applicable to the whole and is made up of genetic diversity, the diversity of species and ecological diversity as well as their interactions (Di Castri and Younes 1996).

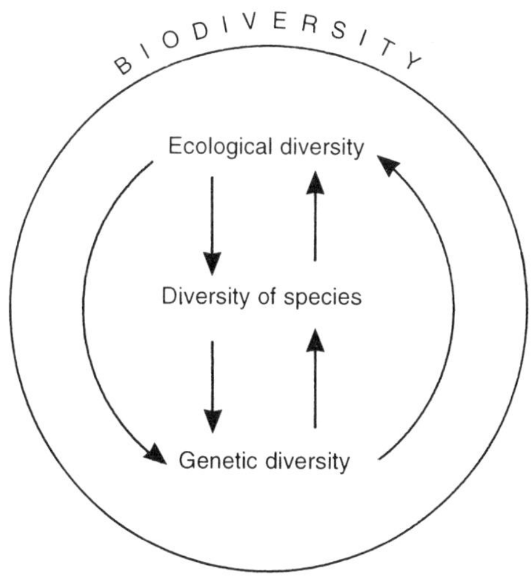

Furthermore, the ecological approach used here ties in with the socio-economic approach to biodiversity, where ecologists and economists rightly stress how clearly interwoven ecological space and economic space are (Fig.4).

But why should we be concerned about biodiversity? The answer is twofold: (1) because it is threatened; (2) because it affects us.

Before considering biodiversity from the viewpoint of our own interests, we should ask ourselves what importance it has and, what is its intrinsic significance.

Figure 4: Ecological and economic systems are clearly interwoven as regards biodiversity (Barier, Burgen and Folke 1994).

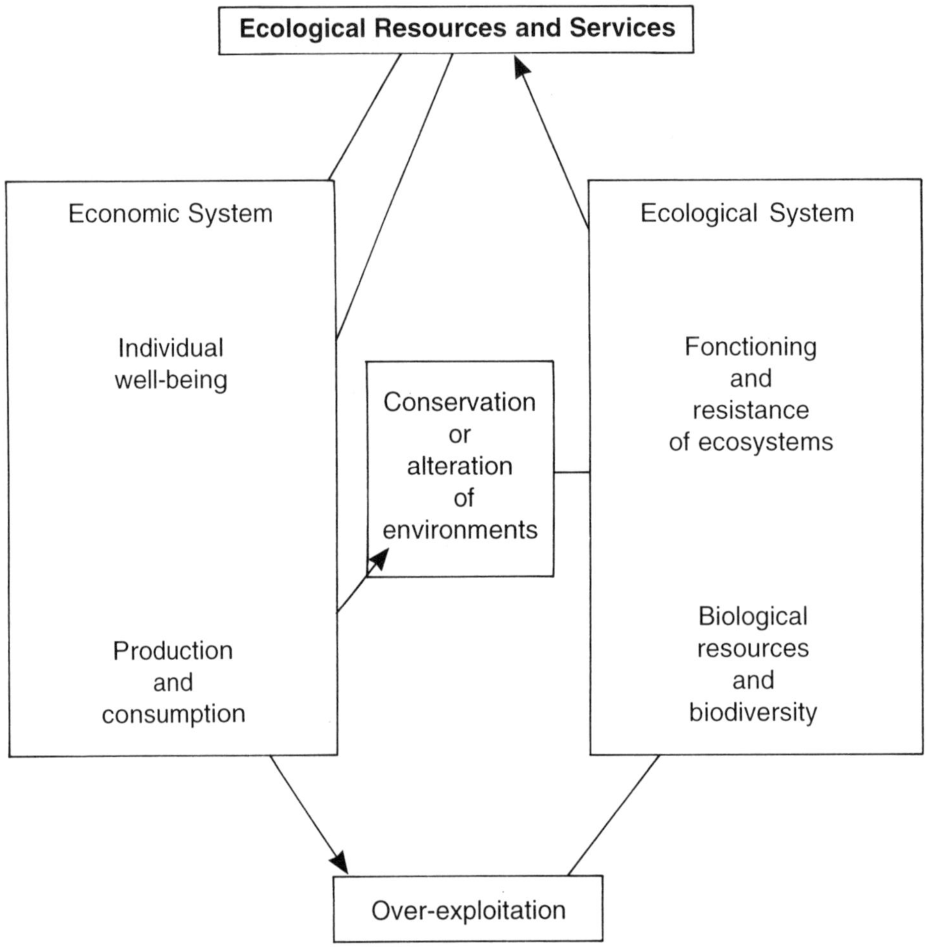

THE SIGNIFICANCE OF BIODIVERSITY

If diversity appears so omnipresent, constantly renewed and restored after each great wave of extinction, this is no doubt because it provides an essential function for the expression and the preservation of life. In fact, there is no life without diversity: biodiversity is an intrinsic characteristic of life.

We are fully aware today of the risks linked to the genetic impoverishment of animal and plant populations. There are three kinds of risks: (1) the increase of genetic homogeneity within populations confers on them a lesser degree of adaptability when faced with changes in the environment; (2) it favours the expression of harmful genes; (3) finally, the lack of variability in defence systems, immune or otherwise, exposes more individuals and populations to attack from pathogens.

Mankind has learnt at his own expense that by reducing the genetic diversity of the plant varieties produced and grown on an industrial scale he is exposing them to rapidly evolving pests, viruses, fungi and insects (Riba and Silvy 1989). Thus, in 1970, while the practices of hybridisation and selection had reduced 85% of corn grown in the United States to almost complete genetic homogeneity, the plant's resistance to helminthosporiosis, a cryptogamic disease, was overcome by the fungus, and the epidemic led to considerable crop damage. In 1980 for the same reason 90% of the Cuban tobacco harvest was destroyed by mildew.

In very general terms, genetic diversity of natural plant and animal populations appears to be a strategy used by natural selection, in response to constant pressure from a multitude of rapidly-evolving parasites. Thus, sexual reproduction, which involves genetic intermixing from generation to generation, would be one of the components of this strategy: in a world without pathogens or parasites asexual reproduction would easily prevail over the roulette of natural selection.

" If genetic variability is the way every species insures itself against the unknown, we can argue that the diversity of species, and consequently of ecosystems, should be considered in the same way by man, in terms of his own known or future needs. In fact, at a time when there is a great deal of talk about climatic or planetary change, and when the use of soils and environments is profoundly affected by the needs of man, there is no doubt that environmental conditions will change in the years and decades to come. In order to remedy or control these changes, and better manage the biological systems on which we depend in the long term, we need to possess the whole range of ecological facilities which exist in nature: genes, gene complexes, species, species complexes, ecosystems and landscapes. (Barbault 1994).

Beyond the obvious economic value of the genetic resources that species represent and the ethical reasons that militate in favour of their preservation, the erosion of biodiversity has ecological implications which it is important to underline, because this is perhaps whatholds the key to genuine sustainable development.

THE ECOLOGICAL SIGNIFICANCE OF BIODIVERSITY

Biodiversity cannot be reduced to specific genetic inheritance. Organised into ecosystems, it also provides impetus for ecological services: although our knowledge on this subject is scanty, there is absolutely no doubt that it is involved in many processes which drive or regulate essential cycles such as the water cycle and/or biogeochemical cycles (nitrogen, carbon, phosphorous) which stabilise climates, preserve the fertility of the soil and protect it from erosion.

Putting aside the disturbing erosion of genetic diversity, the real question is how and to what extent species ensure the sustainability of ecosystems through their diversity.

In an ecological perspective, that is to say in seeking to specify the functional significance of biodiversity, a distinction must be made between various types of species or functions. One can distinguish the species or groups of species on which the key processes of ecosystems are based - species which determine their patterns as well as other variables which set the rhythms and dynamics of these ecosystems. Various authors refer to structural species or keystone process species. Other species would appear to depend on niches created by them, but not to have any effect on the structure or the critical mechanisms involved in the functioning of ecosystems under normal circumstances. They might however play an important ecological role in showing resilience (= ability to return to their previous state) in periods of disturbance: reference is made to life insurance species, since they can be equated to an endowment which guarantees the ecological services of the future. It is known that the loss of species, the steady dwindling of natural populations and the simplification of habitats can reach a critical threshold and finally lead to a breakdown in the working and resilience of ecosystems and thus to their irreversible collapse: other species, pests or scourges, can become established and exacerbate the process.

Thus the concept of biodiversity leads inexorably to that of sustainable development.

MAN, A COMPONENT OF GLOBAL BIODIVERSITY AND A MAJOR ACTOR IN PRESENT DYNAMICS

Man, like every species is a product of evolution, fashioned through natural selection. A large part of his genetic inheritance comes from very distant ancestors and he shares it with his biological cousins, chimpanzees and gorillas: there is only 1% difference between human and chimpanzee DNA. Indeed, this small "detail" makes all the difference, but all the same it amounts to only 1%. It is a good example of the qualitative importance that is inherent in minute differences, sometimes just a single mutation. Thus the precious character of biodiversity does not only lie in the quantity of species, but also the singularity of each one: the erosion of biodiversity means that uniqueness is lost, it is not just a matter of running down the capital stock.

Modern man should give biodiversity serious thought first and foremost as he is both historically and biologically close to his animal cousins.

The second reason lies in modern man's role as a major player in the dynamics of biodiversity - a key species if ever there was one, much more an "ecological engineer" than the beaver or the elephant, quoted as examples to explain this new concept in the specialised literature. In other words, after a strictly biological approach to man we are invited to consider: an ecological approach to man as shaper of ecosystems and landscapes; man as a propagator of species (plants, animals, pathogens); man as a factor of extinction; man as creator of biodiversity (creating new varieties and breeds, transforming species by transgenesis) and man as dependent on his environment, that is to say on numerous other species for his well-being and long-term survival.

THE PRESENT WAVE OF EXTINCTION

The present wave of extinction differs from previous ones in that it is the result of man's actions and that it has not been scaled over millions of years but over centuries, even decades.

It is the result of four phenomena, of Jared Diamond's (1984) evil quartet: (1) degradation of the environment (pollution, fragmentation of habitats, deforestation); (2) the overexploitation of species (hunting, fishing, harvesting); (3) the introduction of exotic species (destroyers of the habitat such as sheep and goats or predators such as cats and dogs, effective competitors and carriers of diseases) and (4) knock-on extinctions (which follow for example, the extinction of a key species).

We know that it is the reduction in numbers and genetic diversity of populations that drives them to extinction; we also know that there is a close link between the area of a habitable environment, the size of the populations involved and specifically local resources. This emphasises that the key factor of sustainable preservation of biodiversity is the maintenance or restoration of diversified natural environments covering large areas: unfortunately our species also needs space! Therein lies the problem. In fact, beyond the immediate factors which lead to the extinction of species one must consider the primary causes, all linked to what might be called the ecological success of the species Homo sapiens: (1) the growth in the human population and its need of natural resources (Vitousek, 1994, reckons that we consume, divert or hold 39% of the planet's plant production); (2) the growing weight of an economic system which is unable to take into account the environment, the renewal of natural resources and the interests of future generations; (3) economic globalization and the progressive decline of products coming from agriculture, forestry and fishing; (4) the predominance of legislative and institutional systems favouring the non-sustainable exploitation of resources; (5) the lack of knowledge and failure to apply knowledge where it exists.

MAN IS DEPENDENT ON BIODIVERSITY

The prodigious diversity of species (including genetic diversity) is a resource that remains to be explored for agriculture, medecine and industry.

Agriculture constantly takes advantage of this biodiversity, not only regarding species which are efficiently grown on an industrial scale (rice, wheat, corn) but also by drawing on the many wild varieties and species constituting a gene pool with which to improve the varieties used (resistance genes for example).

The same applies to medicine. Just as they were an essential part of traditional medicine, so plants remain the basis of modern medicine. Thus Chinese medicine uses 5000 plants (a long way from the 250 000 known species). In the USA, 25% of prescriptions are for drugs whose active ingredients are taken or derived from plants. In 1990, the budget for these plant-based medicines amounted to 12.5 billion dollars.

Basically, for the biologist struck by the diversity and the sophistication of defence mechanisms developed over the millions of years of evolution by species such as ourselves, exposed to a multitude of pathogens, parasites and predators, what would be more natural than to try and turn these chemical weapons to our advantage. Everywhere natural selection has created pioneers, species capable of solving problems posed by a hostile environment - extremes of temperature (bacteria in hydrothermal springs), the struggle against pathogens or parasites, protection against over-avid consumers or dangerous competitors.

Thus many species should be seen more and more as agents at the service of man, not just the one or two entomophagous insect species used in biological control!

"If the extraordinary diversity of life is indeed the expression both of the results of natural selection and the issues at stake for species and ecological systems which exhibit diversity, then it is a treasure trove of solutions to many problems which confront our species. Like any other organism, man must in fact struggle against numerous other life forms, bacteria, viruses, fungi and parasites, which threaten his health and attack his resources: why not turn these biological weapons that evolution has created over billions of years in millions of species to our advantages?" (Barbault, 1994).

BIODIVERSITY REFLECTS DIFFERENT ISSUES

Biodiversity is fundamentally an asset for our species with which to adapt to a changing world, providing long-term security. But it is also a seedbed of conflict. When one realises that biodiversity is principally a resource to be found in tropical countries and that large industrial and pharmaceutical groups are international firms located in northern countries, one can better understand the polemics which arose before Rio and which persist in connection with the convention on biological diversity.

Thus man is also challenged in that realm which constitutes his very specificity, his humanity. A cultural being, man has to take up the challenge of preserving biodiversity to ensure sustainable development, here and elsewhere, today and tomorrow - a challenge for civilisation. A challenge for agriculture in the XXIst century!

SELECTED REFERENCES

BARBAULT. R, 1994. *Des Baleines, des Bacteries et des Hommes.* Odile Jacob, Paris, 327p.

BARBIER R.B., BURGEN J.C. and FOLKE C., 1994. *Paradise lost? The ecological economics of biodiversity.* Earthscan Publications Ltd, London, 268 p.

DI CASTRI F., and YOUNES T., 1996. Introduction: Biodiversity, the emergence of a New Scientific Field - its Perspectives and Constraints, 1-11, *In*: F.di Castri & T.Younes (eds), *Biodiversity, Science and Development. Towards a new Partnership.* CAB International, Oxford.

HUSTON M.A. 1994. *Biological Diversity: the coexistence of species in changing landscapes.* Cambridge University Press, 680 p.

RIBA G. and SILVY C. 1989. *Combattre les ravageurs des cultures. Enjeux et perspectives.* INRA, Paris, 230 p.

VITOUSEK P.M. 1994. Beyond global warming: ecology and global change. *Ecology*, 75: 1861-1876

WILSON E.O. 1992. *The diversity of life.* The Penguin Press, London, 424p.

CHAIRMAN: CHRIS **LEAVER**, *UNIVERSITY OF OXFORD, UNITED KINGDOM*

VICE CHAIRMAN: **J. BOYAZOGLOU**, *FAO, ROME, ITALY*

THEME 2

AGRICULTURE, ENVIRONMENT AND RURAL AREAS

Workshop 1B
Production improvement

THE GENETIC CONTROL OF PLANT PARASITES:

"UN PLAT DE RESISTANCE"

IAN R. CRUTE

Horticulture Research International
Wellesbourne
Warwick, CV35 9EF, United Kingdom

ABSTRACT

Protection of man's crops from parasites will undoubtedly be a primary preoccupation in the 21st century. This paper describes some important aspects of the phenomenon of parasite resistance in plants and suggests how recent and anticipated advances in knowledge of "plant immunology" may herald a change in the way genes and their products could be used in the coming decades to provide sustainable protection of crops from yield and quality loss.

INTRODUCTION AND BACKGROUND

I wish to thank the organisers of this colloquium for giving me the opportunity to present my personal perspective of how research in crop genetics may impact on agricultural production in the 21st century. In the last two decades there have been dramatic technological advances that have allowed the identification, isolation, molecular characterisation and manipulation of plant genes with enormous potential to enhance the productivity and value of crops. In the time available, I will not even make an attempt to itemise the many important agronomic traits that new genetic technology may provide the means to influence. I have chosen instead to talk only about the exciting prospects that are now being presented for the sustainable control of the innumerable parasites (invertebrates, microorganisms, viruses, viroids and even other plants) which can result in substantial loss of yield and quality, before and after harvest, of the crop species on which man depends for food, fodder, fuel and fibre.

In recent decades, the availability of antibiotics and other toxophores, have helped to keep in check the scourges of infectious disease and debilitating parasitism in man and his domestic animals. By analogy, the health of man's crops has also benefited enormously from synthetic chemicals (insecticides, acaricides, nematicides and fungicides). However, as with antibiotics, increasing problems are posed by the evolution of insensitivity to agrochemicals among target organisms and the associated loss of efficacy. In addition, the sustainability and acceptability of sole reliance on synthetic chemicals to maintain control of pests and diseases is being increasingly questioned.

Detailed knowledge of the mammalian immune system has made possible the second major influence on the control of animal and human diseases; the development and widespread exploitation of vaccines. In contrast, knowledge about the way plants recognise and respond to parasites ("plant immunology") is rudimentary. Although man has exploited natural genetic variation for resistance to pests and pathogens, within crop species and their near relatives, in programmes of crop improvement for almost 100 years, this has not been based on a mechanistic understanding of the resistance phenomenon. In my talk I will describe some important aspects of the phenomenon of parasite resistance in plants and suggest how recent and anticipated advances in knowledge of "plant immunology" may herald a change in the way genes and their products could be used in the coming decades to provide sustainable protection of crops from yield and quality loss.

The existence of plant genes providing resistance to parasites was demonstrated soon after the rediscovery of Mendel's seminal studies on inheritance. Biffen (1905) demonstrated that a single locus was responsible for the resistance of some wheat cultivars to yellow rust caused by *Puccinia striiformis*. Many hundreds of genes associated with resistance to a diversity of parasites have subsequently been identified in numerous plant species. At an early stage, it was discovered by McRostie (1919) that genes at different loci could be responsible for resistance to different parasite variants (races) but the full significance of this observation only became evident after the gene-for-gene relationship was elucidated by Flor (1971) in the course of 40 years of research on the interaction between flax (*Linum ultissimum*) and the rust, *Melampsora lini*. Following Flor's classical work, it became evident, that for many host-parasite relationships, matching gene pairs (resistance and avirulence genes respectively) controlled the outcome of interactions between different combinations of host and parasite genotypes (Fig.1).

Figure 1: Feautures of a hypothetical gene-for-gene relationship involving three interacting pairs

Host genotypes \ Parasite genotypes	*A1 A2 A3*	*a1 A2 A3*	*A1 a2 A3*	*A1 A2 a3*	*a1 a2 A3*	*a1 A2 a3*	*A1 a2 a3*	*a1 a2 a3*
R1 R2 R3	1	2	1	1	3	2	1	+
r1 R2 R3	2	2	3	2	3	2	+	+
R1 r2 R3	1	3	1	1	3	+	1	+
R1 R2 r3	1	2	1	1	+	2	1	+
r1 r2 R3	3	3	3	+	3	+	+	+
r1 R2 r3	2	2	+	2	+	2	+	+
R1 r2 r3	1	+	1	1	+	+	1	+
r1 r2 r3	+	+	+	+	+	+	+	+

THE GENE-FOR-GENE RELATIONSHIP

In interactions following a gene-for gene relationship, the expression of resistance or susceptibility of the host to a particular parasite is conditional on the pathogen genotype, and the degree of parasite virulence observed is conditional on the host genotype. Specifically matching gene pairs determine the outcome of any particular genotype - genotype interaction. Compatibility (extensive parasite development and reproduction in the absence of an effective host defence response) is the outcome of a host - pathogen combination unless an allele for resistance (*R*) at a particular host locus is specifically matched by an allele for avirulence (*Avr*) at a particular parasite locus; under these circumstances, degrees of incompatibility (reduced parasite development and reproduction associated with an effective host defence response) are expressed, depending on the particular matching gene-pair. *R-Avr* gene pairs resulting in incompatibility are epistatic (*i.e.* exhibit non-allelic dominance) over gene pairs that would otherwise result in compatibility. Gene pairs conditioning higher degrees of incompatibility are, in general, epistatic over gene pairs associated with lower degrees of incompatibility.

Figure 1 illustrates the features of a hypothetical gene-for-gene relationship involving three epistatic matching gene pairs. An interaction phenotype indicated by a + represents complete compatibility (susceptibility/virulence) and a score of 1 represents the highest level of incompatibility, observed whenever there is an interaction between *R1* and *Avr1*. Incompatibility results whenever there is a specific *R-Avr* combination *ie R1-Avr1*, *R2-Avr2* and *R3-Avr3* and incompatibility is epistatic over compatibility.

ORGANISATION AND STRUCTURE OF RESISTANCE LOCI: THE PLANT'S SURVEILLANCE SYSTEM

Detailed studies on the inheritance of specific parasite resistance typically reveal the existence of many genes each expressing unique recognition capability or specificity. For example, Flor and others identified the existence of genes expressing at least 32 different specificities for recognition of flax rust variants in flax. In wheat, at least 100 disease resistance specificities have been identified for rust and mildew pathogens. More recently, a few years study of *Arabidopsis* have revealed almost 40 genes with the capability to mediate specific resistance to parasites. Figure 2 shows a map of the five chromosomes of *Arabidopsis* indicating the location of these resistance genes.

It is evident that the number of genes in plant genomes capable of specific pathogen recognition is large and that these genes are commonly clustered in complex loci or less frequently as allelic series at a single locus. These loci sometimes comprise genes involved in the recognition of more than one taxonomically unrelated pathogen (for example, in *Arabidopsis*, genes mediating resistance to three species of bacteria, two viruses and two oomycete pathogens occur in a linkage group referred to as *MRC-J*). For some genes, alleles with a dual recognition capability have been identified (*RPM1* in *Arabidopsis* is such a gene; Grant *et al.*, 1995) while there is also evidence for the existence of genes expressing identical recognition capability present at different loci in the same species as well as in different plant species (Dangl *et al.*, 1992). There is now evidence that resistance genes are members of substantial multi-gene families. These large, potentially complex loci may, in the case of species not maintained by man as crops, provide several selective advantages. Such arrangements allow multiple specificities to be assembled and retained in a single haplotype. This preserves the potential for variation and the evolution of new gene combinations and alleles expressing novel recognition capability. Mispairing of repeat sequences resulting in unequal crossing-over, intra and inter-allelic recombination, gene duplication and gene conversion all have the potential to create novel variation (Hulbert and Bennetzen, 1991; Sudupak *et al.*, 1994; Hu and Hulbert, 1994). This is analogous to recombination characteristic of the mammalian immune system (expressed somatically). In plants, particularly those that have evolved to be predominantly inbreeding, recombination among recognition genes (expressed germinally) may be an essential prerequisite for survival in the face of rapidly evolving parasites. In natural plant populations, the progeny of an inbreeding species in the proximity of a parasitised mother plant are likely to be particularly vulnerable to parasitism themselves due to their genetic identity. The occurrence among progeny of rare recombinant variants, expressing novel parasite recognition capability, would allow survival of the potentially well-adapted parental genotype otherwise threatened by a matching parasite.

Figure 2: Chromosome map of *Arabidospsis* indicating the location and clustering of genes involved in parasite recognition and respons (see Holub, 1997)

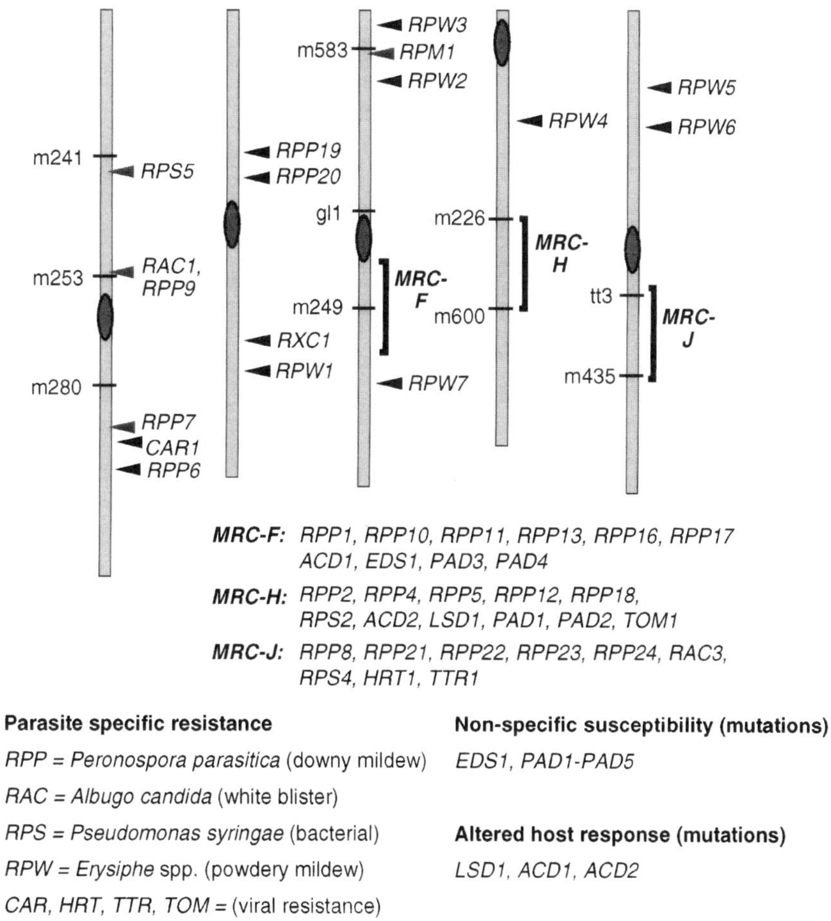

RESISTANCE GENE STRUCTURE AND THE DEFENCE RESPONSE

A number of genes conferring pathotype specific resistance to a diversity of pathogens have now been isolated from several different host species and sequenced; this represents one of the major achievements and break-throughs in plant science this decade (see Staskawicz *et al.*, 1995; Hammond-Kosack *et al.*, 1996 for reviews). Figure 3 illustrates some of the features of the four classes of genes identified.

Figure 3: Features of the inferred structure of plant genes involved in parasite recognition.

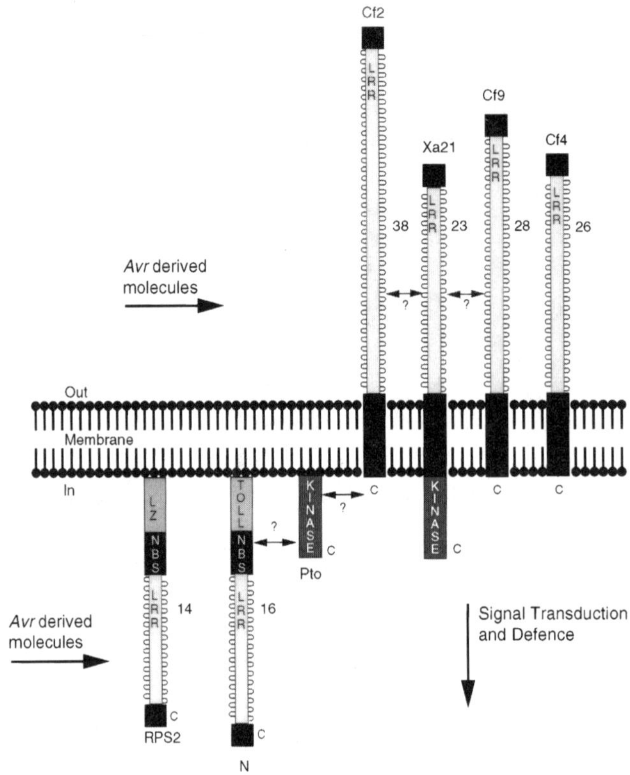

LZ = leucine "zipper"
LRR = leucine rich repeats (numbers of repeats to the right)
NBS = nucleotide binding site
? = possible molecular interactions initiating signalling

• Class 1: The *Cf* genes of tomato mediate resistance to variants of leaf mould (*Cladosporium fulvum);* the products of these genes have a membrane anchor and a domaine characterised by a variable number of imperfect leucine-rich repeats (LRRs)(Jones *et al.*, 1994)

• Class 2: *Xa21* mediates resistance of rice to bacterial blight (*Xanthomonas campestris* pv. *oryzae*); the product of this gene has a membrane spanning domain joining a cytoplasmic kinase and an extracellular LRR domain (Song *et al.,* 1995) .

• Class 3: *RPS2* and *RPM1* mediate resistance of *Arabidopsis* to pathovars of *Pseudomonas syringae*; *L6* is a gene for flax rust resistance and *N* is a gene from tobacco mediating resistance to tobacco mosaic virus. (Bent *et al.,* 1994; Mindros *et al.,* 1994; Ellis *et al.,* 1994; Lawrence *et al.,* 1995; Whitham *et al.,* 1995) All these genes have a nucleotide binding site (NBS) and a LRR domain.

• Class 4: *Pto* mediates resistance of tomato to bacterial speck disease (*Pseudomonas syringae* pv. *tomato*) and codes for a cytoplasmic serine-threonine protein kinase Martin *et al.,* 1993).

A striking feature of the molecular structure of the predicted products of most of the *R* genes that have been isolated and sequenced (with the exception of Class 4) is a domain comprising a variable number of leucine-rich repeats (LRRs). LRRs are often involved in protein-protein interactions and they may provide the basis for elaboration of the capability to recognise specifically molecules of parasite origin (analogous to the variable domains of antibodies). The variable number and amino acid sequence of the repeats may somehow provide specific ligand affinity. Hypothetically, inter and intra-cellular molecules with LRRs may interact with specific molecules of parasite origin. The LRRs of different molecules eg "*Cf9*- like" and "*Xa21*-like" may interact to transduce a perceived signal via activation of the protein kinase domain of the latter. Similarly, molecules with LRRs and an NBS, such as *N* and *RPP5* , may transduce the *Avr*-specific signal through a "*Pto*-like" kinase; and there are many other possibilities.

The working hypothesis is that the perception of specific *Avr*-derived signals by *R* genes, provokes a signalling cascade which results in the induction of a sequence of defence responses including: an oxidative burst; hypersensitive cell death; production of cell-wall appositions; synthesis of "pathogenesis-related" proteins including chitinases and glucanases; and synthesis of phytoalexins. As illustrated in Figure 4, there are likely to be a number of unique and shared steps in the networks that lead to any particular biochemical response. In addition to the *R* genes themselves, the genes coding for the enzymes catalysing these steps represent potentially important targets for manipulation and particularly those involved where pathways may converge.

Although gene-for-gene relationships are conceived as being essentially epistatic (as described previously; Figure 1) there is evidence for the occurrence of other forms of non- allelic interaction confirming the involvement in the phenomenon of defence of genes other than those responsible for primary parasite recognition. For example, resistance in bean (*Phaseolus vulgaris*) to bean common mosaic virus (BCMV) is dependent on complementary gene interactions; the gene *bc-u* (present in all but one cultivar examined) has no independent effect itself, but is required for the expression of specific resistance to BCMV that is determined by a set of *R* genes: *bc-1, 2* and *3* (Drijfhout, 1978). Speculatively, the genes that have been identified through their non-allelic interaction with genes determining specificity may be those involved in signal transduction events (Fig.4) leading to effective defence. Systematic attempts are now being made through mutational analyses to uncover more such loci essential for, or influential on, the expression of specific resistance. For example, in barley, the loci *Rar1* and *Rar2* (formerly *Nar1* and *Nar2*) have been revealed, by screening mutagenised plant populations, to be required for expression of Ml_{a12}- specified resistance to powdery mildew (Freialdenhoven *et al.,* 1994). Similarly, in tomato, two loci named *Rcr-1* and *Rcr-2*, have been identified which are required for the full expression of resistance to *Cladosporium fulvum* mediated by the gene *Cf-9* (Hammond-Kosack *et al.*, 1994). The *ndr*-1 mutation in *Arabidopsis* renders susceptible plants carrying different *R* genes to a bacterial (*Pseudomonas syringae* pv. *tomato*) and a fungal (*Peronospora parasitica*) pathogen. This indicates a common step in the pathways of resistance to a prokaryotic and eukaryotic pathogen (Century *et al.*, 1995).

Figure 4: Diagrammatic representation of how parasite recognition may provoke a signalling cascade with unique and shared steps identifiable genetically by mutation analysis.

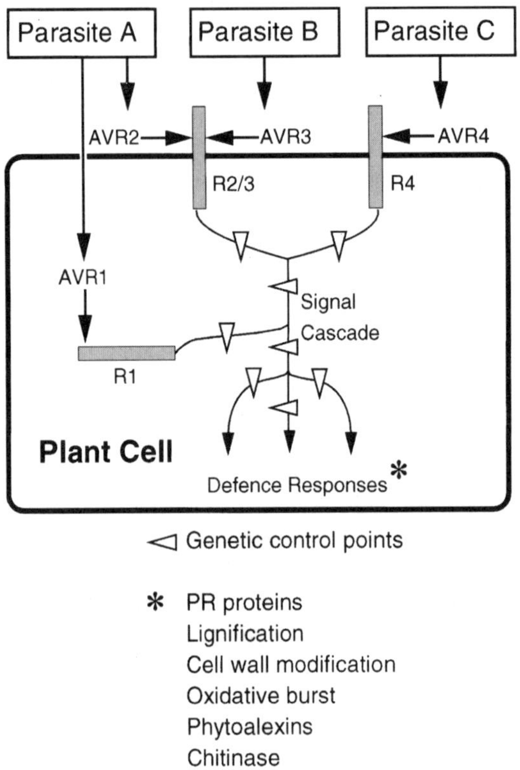

Direct evidence to support the notion that such genes required for resistance encode components of the transduction pathway or network leading from the perception of the signal provided by the *Avr* gene to the expression of resistance has been provided by Zhou *et al.* (1995) who identified a tomato gene, *Pti1* which encodes a serine/threonine kinase that is phosphorylated by the *Pto* resistance gene (also encoding a protein kinase). Interestingly, *Pti1* is not phosphorylated by *Fen*, another serine/kinase genetically linked and closely related to *Pto* (Loh and Martin, 1995) indicating that resistance specificity may also be a function of interactions between these "downstream" gene products.

ACHIEVING DURABLE RESISTANCE: A GOAL FOR THE 21ST CENTURY

Following the dramatic demonstration that disease resistance could be conferred by single major genes (early in this century), breeders of many crops initiated breeding programmes with the expectation that the resulting control of plant diseases would be permanent (Stakman *et al.*, 1918). Durable resistance based on the utilisation of one or more single dominant resistance genes has been achieved in some cases. More frequently, however, the rapid evolution of parasite variants virulent on previously resistant cultivars has forced breeders into a repetitive cycle of cultivar replacement, demanding the continual introgression of new resistance specificities.

Recent advances, elaborated here, suggest that a fuller understanding of the mechanisms by which plants perceive a potential parasite, discriminate among parasite genotypes and respond effectively to resist invasion cannot be far away. There is now the challenge, to utilise advances in knowledge and technology better to protect from loss due to disease those crops on which man depends. A major focus of the effort will be to provide more durable disease control systems than has commonly been the case hitherto.

Generating effective combinations of *R* genes by conventional hybridization and gene introgression by backcrossing, can be a lengthy process, often taking decades, particularly if a wild relative of the crop species in question provides the source of resistance. However, it is now reasonably straightforward to identify easily scored molecular markers that are linked to resistance genes. This facilitates marker-aided selection strategies that can improve the efficiency with which novel or particularly valuable resistance genes can be incorporated into well-adapted commercial cultivars. Moreover, by using markers that map to regions of the genome not associated with the target gene it is, in theory, possible greatly to speed up the selection process by ensuring that backcross progeny that carry predominantly the genome of the agronomically desired recurrent parent are preferentially selected.

The availability of sequence data from isolated resistance genes and the prospect of different classes of resistance genes being conserved between taxa provides the prospect of being able to isolate large numbers of "look-alike" genes which may or may not function in parasite recognition. Studies of collinearity between the genomes of different taxa may allow conclusions to be drawn about related genes residing in homeologous regions. However, a demonstration of functionality will demand substantial collections of variants of the target organism and plant populations derived from appropriate crosses to allow the necessary cosegregation studies. These latter two requirements may prove to be rate-limiting in being able to advance rapidly from the several genes isolated in the last few years to having available large numbers of isolated resistance genes of demonstrated function.

Although marker-aided selection will undoubtedly make the planned pyramiding of resistance genes more feasible than at present, the exploitation of transgenic technology will make such a strategy comparatively straightforward as it will not be constrained by the limitations imposed by the often encountered linkage between *R* genes. The availability of isolated *R* genes of known function will also make the production of multiline cultivars feasible. The response time for introduction of new cultivars in temporal gene deployment strategies may also be abbreviated. *R* gene deployment strategies based on the exploitation of temporal and spatial genetic heterogeneity are likely to feature more prominently in crop protection in future. For example, "mixed and matched" dynamic multilines could be produced which will have the agronomic uniformity required by farmers but a changing complement of *R* genes to prevent continuous selection pressure against matching *Avr* genes.

The often suggested prospect of using transgene technology to move resistance genes between sexually incompatible taxa can now be evaluated. The *Pto* gene from tomato which confers resistance to *P. syringae* pv. *tomato* has recently been transferred to tobacco (Rommens *et al.*, 1995) and the resulting transgenic plants proved resistant to *P. syringae* pv. *tabaci*. However, this example involves resistance to pathovars of the same bacterial pathogen, and it will be interesting to observe the interaction phenotype in response to infection by unrelated compatible pathogens when an *R* gene is transferred to a species which is a non-host of the pathogen originally recognised by that gene. There will undoubtedly be limits to the function of *R* genes when they are moved between taxa and this could relate to the specificity of signal transduction pathways in distantly related plant species.

Understanding why some resistance genes, or gene combinations, provide durable disease control while in other cases control is ephemeral will demand more information about the molecules of pathogen origin that are perceived by the plant. Durability is a function of the fitness deficit implicit in the allelic variation at an *Avr* locus or in particular *avr* (virulence) gene combinations. It is the selective advantage, in the absence of a matching *R* gene, that the *Avr* gene product confers on the parasite which determines the fitness deficit of the *avr* mutation (avoiding detection by the matching *R* gene). An understanding of the function of *Avr* genes should allow *R* genes, or *R* gene combinations to be identified which recognise those particular parasite *Avr* gene products imposing lethal or debilitating phenotypes when a mutation to virulence occurs. The organisation of *R* genes in plant genomes and such structural data as is available implies that considerable molecular variation may be possible among *R* gene products. When more is known about structure-function relationships, it should be feasible to create *R* genes whose products are designed to recognise parasite molecules unable to sustain genetic variation without a lethal effect. I consider this prospect to be one of the most exciting opportunities for the genetic improvement of crop plants in the next century.

In the search for durable resistance, it is not necessary to look further than so-called non-host resistance defined by the limits in the host range of a particular parasite. The gene-for-gene relationship provides the basis for the discrimination of variants by their virulence or avirulence on particular genotypes within an individual host species. However, in addition to this sub-specific variation among genotypes within a parasite species, variation is also frequently evident at the level of host species, genus or family. It is clear that such specificity represents the outcome of coevolution. There is some evidence that species or higher taxa specificity may also be a reflection of gene-for-gene recognition. It may be within the context of taxon specificity that gene combinations will be identified for which matching virulence is lethal; such genes thereby become genes for non-host resistance providing practical value by their durability as described in Figure 5.

Hypothetically, the progenitor host species carries two *R* genes (*R1* and *R2*) for which the progenitor compatible parasite lacks matching avirulence (*i.e.* carries virulence alleles *a1* and *a2*). Evolution of "new" host species 1 and 2 is associated with an alteration in the resistance specificity of *R1* to *R1** and *R2* to *R2** respectively such that *a1* and *a2* are specifically recognised. This results in incompatibility with the progenitor parasite. "New" parasites 1 and 2 coevolve compatibility with "new" host species 1 and 2 respectively by changes in *a1* and *a2* respectively (to *a1** and *a2**). Individually these mutations do not adversely affect fitness. Combinations of the mutations *a1** and *a2** are however lethal and *R1** and *R2** therefore become genes controlling taxon specificity or non-host resistance. This evolution of taxon specificity results from disruptive selection plus a severe (?lethal) fitness deficit. The resistance of "new" host species 1 to "new" parasite 2 and "new" host species 2 to "new" parasite 1 is durable. In other words, you can walk (a1) and chew gum (a2*) but you can't scuba-dive (a1*) and chew-gum (a2*) (lethal!)

Figure 5: Evolution of « durable » host range

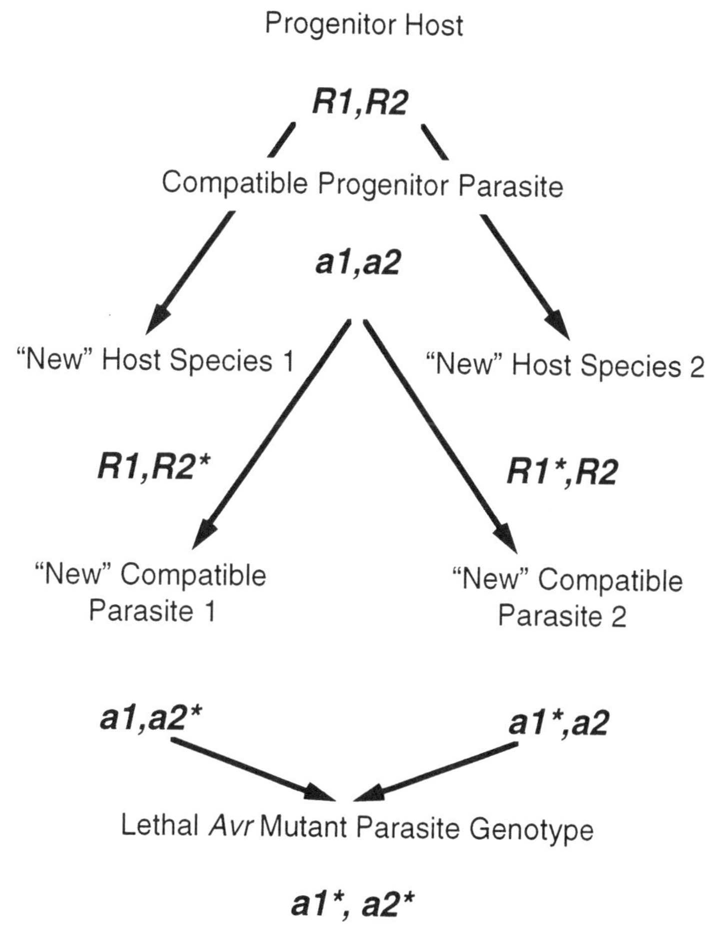

CONCLUSION

I hope that with this short overview, I have been able to provide you with a sense of optimism that through the application of advancing knowledge about the molecular and population genetics of the interaction between plants and their parasites, great opportunities await us in the next century to move from a generic mechanistic understanding of the phenomenon of resistance to the provision of stable and effective pest and disease control methods undoubtedly essential to man's future well-being.

SELECTED REFERENCES

BENT A.F., KUNKEL B.N., DAHLBEKD., BROWN K.L., SCHMIDT, R., GIRAUDAT J., LEUNG J., STASKAWICZ B.J., 1994. RPS2 of Arabidopsis thaliana: a leucine-rich repeat class of plant disease resistance genes. *Science* 265, 1856-1860.

BIFFEN R.H., 1905. Mendel's laws of inheritance and wheat breeding. *J. Ag. Sci.* 1, 4-48.

BISGROVE S.R., SIMONICH M.T. SMITH N.M., SATTLER A., INNES R.W., 1994. A disease resistance gene in Arabidopsis with specificity for two different pathogen avirulence genes. *Plant Cell* 6, 927-933.

CENTURY K.S., HOLUB E.B., STASKAWICZ B.J., 1995. NDR1, a locus of Arabidopsis thaliana that is required for disease resistance to both a bacterial and a fungal pathogen. *Proc. Natl. Acad. Sci.* USA 92, 6597-6601.

DANGL J.L., RITTER, C., GIBBON M.J. MUR L.A.J., WOOD, J.R., GOSS S., MANSFIELD J., TAYLOR J.D., VIVIAN A., 1992. Functional homologues of the Arabidopsis RPM1 disease resistance gene in bean and pea. *Plant Cell* 4, 1359-1369.

DRIJFHOUT E., 1978. *Genetic interaction between Phaseolus vulgaris and bean common mosaic virus with implications for strain identification and breeding for resistance.* Centre for Agricultural Publishing and Documentation, Wageningen, The Netherlands. 98 pp.

ELLIS J.G., LAWRENCE G.J., FINNEGAN E.J., ANDERSSON P.A., 1995. Contrasting complexity of two rust resistance loci in flax. *Proc. Natl. Acad. Sci. USA* 92, 4185-4188. FLOR H.H., 1971. The current status of the gene for gene concept. *Annu. Rev. Phytopathol.* 9, 275-296.

FREIALDENHOVEN A., SCHERAG B., HOLLRICHER K., COLLINGE D.B., THORDAT-CHRISTENSEN H., SCHULZE-LEFERT P., 1994. Nar-1 and Nar-2, two loci required for Mla12-specified race-specific resistance to powdery mildew in barley. *Plant Cell* 6, 983-994.

GRANT M.R., GODIARD L., STRAUBE E., ASHFIELD T., LEWALD J., SATTLER, A., INNES R.W., DANGL J.L., 1995. Structure of the Arabidopsis RPM1 gene enabling dual SPECIFICITY disease resistance. *Science* 269, 843-846.

HAMMOND-KOSACK K.E., JONES D.A., JONES J.D.G., 1994. Identification of two genes required in tomato for full Cf-9-dependent resistance to Cladosporium fulvum. *Plant Cell* 6, 361-374.

HAMMOND-KOSACK K.E., JONES D.A., JONES J.D.G., 1996. Ensnaring microbes: the components of plant disease resistance. *New Phytol.* 133, 11-24.

HOLUB E.B., 1997. Organisation of resistance genes in Arabidopsis. *In*: Crute I.R., Holub, E.B. and Burdon J.J. (Eds) *The gene-for-gene relationship in plant parasite interactions.* Walingford, UK, CABI publishers (in press).

HU G., HULBERT S., 1994. Evidence for the involvement of gene conversion in meiotic instability of the Rp1 rust resistance genes of maize. *Genome* 37, 742-746.

HULBERT S.H., BENNETZEN J.L., 1991. Recombination at the Rp1 locus of maize. *Mol. Gen. Genet.* 226, 377-382.

JONES D.A., THOMAS C.M., HAMMOND-KOSACK K.E., BALINT-KURTI P.J., JONES J.D.G., 1994. Isolation of the tomato Cf-9 gene for resistance to Cladosporium fulvum by transposon tagging. *Science* 266, 789-793.

LAWRENCE G.J., FINNEGAN E.J., AYLIFFE M.A., ELLIS J.G., 1995. The L6 gene for flax rust resistance is related to Arabidopsis bacterial resistance gene RPS2 and the tobacco viral resistance gene *N. Plant Cell* 7, 1195-1206.

LOH Y.-T., MARTIN G.B., 1995. The disease resistance gene Pto and the fenthion-sensitivity gene Fen encode closely related functional protein kinases. *Proc. Nat. Acad. Sci. USA* 92, 4181-4184.

MARTIN G.B., BROMMONSCHENKEL S.H., CHUNGWONSE J., FRARY A., GANAL M.W., SPIVEY R., WU T., EARLE E., TANKSLEYS.D., 1993. Map-based cloning of a protein kinase gene conferring disease resistance in tomato. *Science* 262, 1432-1436.

MCROSTIE G.P., 1919. Inheritance of anthracnose resistance as indicated by a cross between a resistant and a susceptible bean. *Phytopathol.* 9, 139-148.

MINDRINOS M., KATAGIRI F., YU G.-L., AUSUBEL F.M., 1994. The Arabidopsis thaliana disease resistance gene RPS2 encodes a protein containing a nucleotide-binding site and leucine-rich repeats. *Cell* 78, 1089-1099.

ROMMENS C.M.T., SALMERON J.M., OLDROYD G.E., STASCAWICZ B.J., 1995. Intergeneric transfer and functional expression of the tomato disease resistance gene Pto. *Plant Cell* 7, 1537-1544.

SONG W.Y., WANG G.L., CHEN L.L., KIM H.X., PI L.Y., HOLSTENT., GARDNER J., WANG B., ZHAI W.X., ZHU L.H., FAUQUET C., RONALD P. 1995. A receptor kinase-like protein encoded by the rice disease resistance gene Xa21. *Science* 270, 1804-1806.

STAKMAN E.C., PARKER J.H., PIEMEISEL F.J., 1918. Can biologic forms of stem rust on wheat change rapidly enough to interfere with breeding for rust resistance? *J. Agric. Res.* 14, 111-123.

STASCAWICZ B.J., AUSUBEL F.M., BAKER B.J., ELLIS J.G., J JONES J.D.G., 1995. Molecular genetics of plant disease resistance. *Science* 268, 661-667.

SUDUPAK M.A., BENNETZEN J.L., HULBERT S.H., 1993. Unequal exchange and meiotic instability of disease resistance genes in the Rp1 region of maize. *Genetics* 133, 119-125.

WHITHAM S., DINESH-KUMAR S.P., CHOI D., HEHL R., CORR C., BAKER B., 1994. The product of the tobacco mosaic resistance gene N: similarity to Toll and the interleukin-1 receptor. *Cell* 78, 1101-1115.

ZHOU J., LOH Y.T., BRESSAN R.A., MARTIN G.B., 1995. The tomato gene Pti1 encodes a serine/threonine kinase that is phosphorylated by Pto and is involved in the hypersensitive response. *Cell* 83, 925-935.

OBJECTIVES AND CHALLENGES IN IMPROVING BIOLOGICAL MATERIAL FOR ANIMAL BREEDING

PASCAL L. LEROY
Dept of Genetics, Faculty of Veterinary Medecine
Université de Liège
Liège, Belgium

INTRODUCTION

The improvement of animal productions is still one of the main objectives in our industrialised world.

The evolution of animal breeds is affected by various factors including consumer behaviour, the "price war", and product diversification, all of which contribute to a sort of inter-species competition. States are also worried about livestock production being developed in areas unsuited for intensification which means that production can never be very competitive.

Improvements of biological material should include these aspects of the matter, that sometimes seem contradictory. A good example can be seen in breeding to increase growth in broiler chickens vs. slower growth in labelled breeds.

Research has to fit into this context and, in particular, generate results that are of direct use to the whole animal production sector.

Regardless of species, and regardless of the agricultural intensity, it is important to maintain productivity, and guarantee jobs and, at the same time, increase production and quality, and decrease costs. Most efforts, therefore, have to focus on genetic improvement, improving food and reproductive efficiency, increasing disease resistance and longevity. It is obvious that regular progress will depend on concerted efforts by everyone concerned with breeding.

Production-related genetic improvement is changing quickly. It is based on the polygenic model and will increasingly take account of major genes and QTL identification (Quantitative Trait Loci). Furthermore, there are many gene-related physiological functions, such as prolificity, milk composition, and fat content in particular in poultry carcasses. Studying their expression should contribute to scientific knowledge.

GENETIC VARIATION OF POLYGENIC ORIGIN

The theory of quantitative genetics has led to remarkable results in improving characteristics that are linked to the economics of various sectors. Yet, as Pirchner(1990) said, progress in the growth rate of the broiler chicken is essentially due to mass selection, and progress in dairy cows is far below what was theoretically predicted (Van Vleck 1986).

The study of phenotypic and genetic parameters of characteristics connected to breeding goals starts by applying performance control systems to large well-studied populations and by considering data from breeding stations with fewer animals but for which there is more complete information and a complete genealogy.

Data are currently being analysed using precise mathematical models (such as REML) and with due consideration for the effects of breeding. This leads to an estimate of genetic parameters (genetic variation, genetic correlation among traits) and to an estimation of non genetic and genetic effects. Genetic effects, downstream, lead to evaluations and to the publication of estimates of the genetic value of reproducers. Genetic values are published and then used by scientists in breeding programmes, etc.

Improving prolificacy in the swine population in France, and the positive evolution of genetic trends expressed in kilos of milk in dairy cows in Belgium (Leroy and Detilleux 1996) (**Fig 1** and **2**) and elsewhere in Europe clearly illustrate repercussions from using the polygenic model.

Fig 1 : Genetic trends expressed in kilos of milk (305 d) in Belgian dairy breeds.
common base = milk (HF) + dual purpose)
(Source: Detilleux et Leroy, 1996).

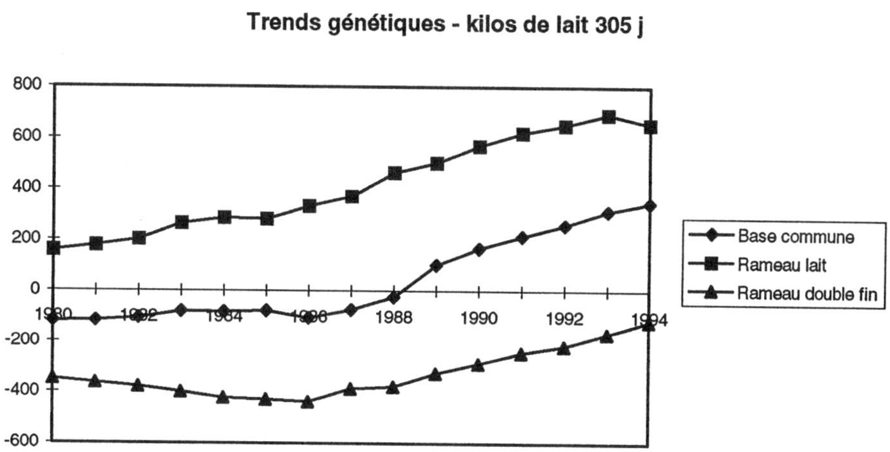

Fig 2 : Estimated trends for kilos of fat and kilos of protein (305d) in Belgian dairy breeds. (common base = milk (HF) + dual purpose)
(Source: Detilleux et Leroy, 1996).

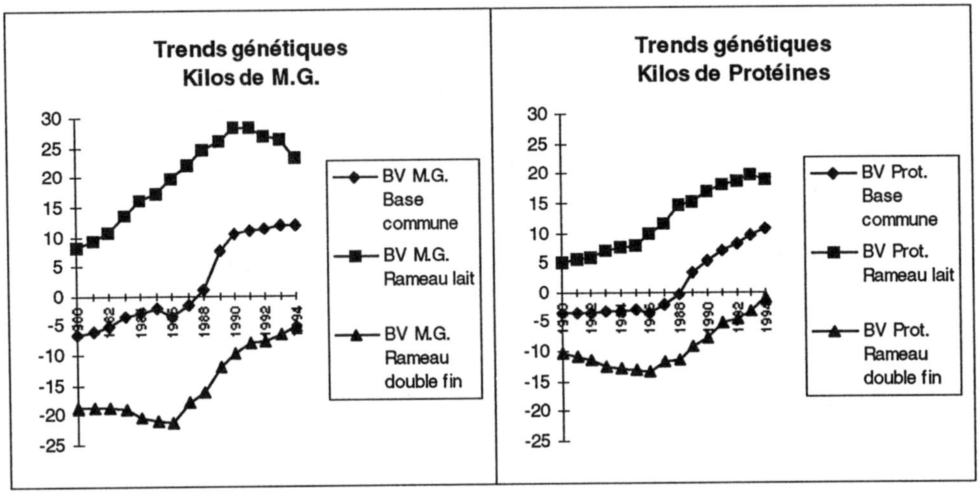

MAJOR GENES

Because of the nature and, in particular, the continuity of traits to be improved, it is not easy to define a major gene. For the geneticist, the concept of major gene applies when there is a least one phenotypic standard deviation between the average of genotypically extreme individuals (AA/aa). There are other definitions that draw more heavily on statistics, like the part of genetic variation that is explained by reference to a locus. They are more didactic and are easier to transfer to people working in breeding operations.

As Le Roy (1992) said, "the main advantage of a major gene is that a high genetic value that is immediate and stable can be obtained by fixing a favourable genotype in the major locus. It is always important to consider the effects of the major locus on the other traits".

Many major genes have been detected in domestic animals through the segregation method. The first step is to identify the location by using molecular markers. Introgression through repeated back-crossing is the most commonly used technique for transferring any major gene.

Pigs

In pigs, both research and industry have been especially interested in the Hal, Rn (ESR) and Mu major genes.

The creation of the stress resistant Pietrain ReHal© (University of Liège) clearly illustrates the value of introgression. Starting with the stress resistant homozygote Large White female produced on the INRA Station in Rennes, and the homozygotic sensitive Pietrain boar, Hanset *et al.* (1983) produced F1s, F2s and successive back-crosses and finally produced the BC5, 63/64 Pietrain. Pietrain were produced at each step; half of them were heterozygous for the Hal locus. In this example, introgression concerns the favourable allele in the Hal locus.

The analysis of the F2 and, in particular, the role of differential variation in the Large White Pietrain explained by the Hal locus and the polygenes is especially instructive. In other words, the differences in pH may be explained by looking at the Hal locus, but the percentage of carcass meat can only be partly explained by the Hal locus, which means that there are many QTLs that need to be identified (Figure 3).

Fig 3 : Part of variation of the Large white-Piétrain difference (F2 Large White x Piétrain cross) explained by Hal los and QTL.
(Source : IRSIA-IWONL Ulg project)

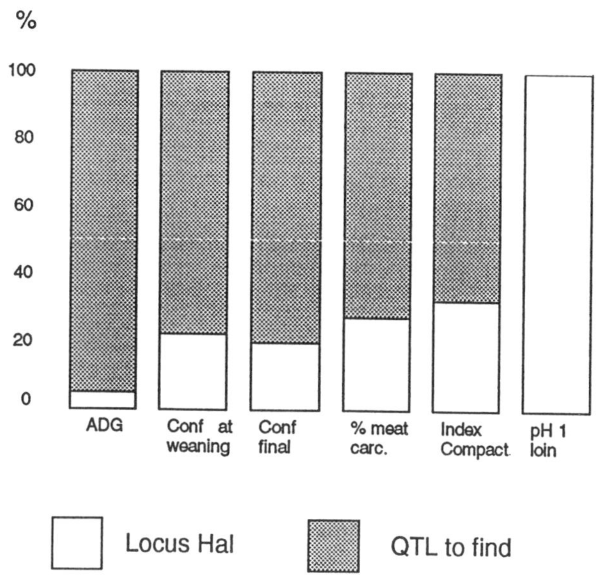

The Rn genes (Rn = rendement Napole - Napole yield) (Le Roy *et al.* 1994) and MU (effect on the percentage of muscle) (Naveau *et al.* 1993) have been brought out in France. In the case of the Rn locus, the RN- allele decreases the technological productivity of ham production by 8 points. These effects taken together lead to the notion of the "acid meat" syndrome. Milan *et al.* (1996) then identified the gene on chromosome 15.

The allele that was favourable to the MU locus affected the percentage of muscle in the carcass. It was brought out at the Pen Ar Lan company (France).

Goats

In goats, the hd locus has an important effect on milk the flow rate (Le Roy *et al.* (1995). In a sample of 1,139 Saanens, Alpines and crossbreds, born of 65 sires and 641 dams, to which the segregation analysis method was applied, the authors discovered a gene with a major effect on the variable "quantity of milk obtained during the first minute of milking". The difference between the averages for the hd hd and ++ individuals is close to three phenotypic standard deviations. Many QTL numbers still have to be identified since residual heritability is 0.25.

The major locus Casein αs1, together with the polygenes, control the protein level of goats milk. Manfredi *et al.* (1995) pointed out that estimates on heriditability of protein rates dropped from 0.66 to 0.34 when the locus was isolated. These results illustrate the importance of the effect of the major locus and the considerable role of polygenes in trait variation.

Studies are being made of the "strong" genotypes in the dairy goat. They are taken into consideration when mating is targetted to generating reproducers that then will carry favourable genotypes.

Cattle

In cattle, the MH gene, which was recently mapped for the Blanc-Bleu Belge breed (Charlier *et al.* 1995), had a considerable effect on the sales value of the animal and the carcass value. In animals slaughtered at age 12 months, the increase in carcass yield of double-muscled animals was about 4.8%, the increase in the percentage of muscle in the 7th rib was 20%, while the fat content dropped by 42% and the bone content by 13%. Charlier *et al.* (1995) presented a comparison of results obtained in Belgium and in France.

Hanset *et al.* (1989) showed that in the Blanc-Bleu Belge double-muscled breed, after the mh had been fixed, the variation coefficient for yield at slaughter and for percentage of muscle in the carcass was slight, while there was still large variation in growth and fertility traits. This gave an idea of the potential and the future for breeding work. (Table 1)

Sheep

In sheep, the Booroola gene which affects fecundity (Montgomery *et al.* 1994) and an entity called Callipyge, which is instrumental in muscular hypertrophy (Cockett *et al.* 1994, 1996), are the two main entities that were described.

Table 1 : Variation in main production and reproduction traits in the Blanc-Bleu Belge breed cattle after fixation of the mh gene. (R.Hanset, C. Michaux, P. Leroy, G.Detal, 1989)

Trait	μ	σ	CV	σ	Vmax	Ec Stad
Killing out %	65	1.5	2.31	1.1	69	2.67
% monocostal muscles	70	3	4.29	2	77	2.33
WT at 7 mo (Kg)	285	35	12.28	16	392	2.77
GQM 7-13 (gr)	1.42	0.17	11.97	0.11	1.967	2.46
Wt at 13 mo (Kg)	544	47	8.6	21	685	2.62
FB values	135.5	5.7	4.19	3.6	148	1.8
Culling of heifers (%)	11.9	30		5.11		
Age at Ist calving (days)	997	125.6		21.4		
%elimin infert > 1st clvg	14.8	39.1		3.71		
Age at 2nd calving (days)	1404	134.1		28.1		
Calving interval (days)	402	48.3		8.5		

GENOME ANALYSIS

The genome analysis has been added to complete the study on the genetic determinism of traits. It developed later in domestic animals because of the nature of the traits to be improved and the success of the polygenic model.

Furthermore, as Georges *et al.* (1995) showed, the structure of the bred animal populations contributed to reducing the degree of heterozygosity. They were analysed on the basis of the genotyping of micro-satellites. The analyses gave average values of 56% heterozygosity in dairy cows, and about 70% in man.

The first results of genome analyses came from monofactorial type traits. Starting with biochemical polymorphism in pigs, the Hal locus (malignant hyperthermia) was identified then molecular marking techniques were used to locate, and in some cases identify, many of the loci.

Georges and Andersson (1996) give a list of the main loci (Table 2) that control monogenic traits. In this list, next to the genes that govern the hereditary defects, there are genes for the coat colour and the horns (presence or absence). Certain genes have been eliminated, or especially appreciated, others, like the Hal gene in pigs, have been shown to contribute to a quantitative trait in certain selected populations.

Table 2 : List of main loci that govern monogenic traits in domestic animals (Source: Georges et Andersson, 1996)

Species	*Locus*	*Chromosome*	*Trait*	*Authors*
Pig	Hal	6	Malignant Hyperthermia	Fuji *et al.*, 1991
	I	8	White dominant	Johansson *et al.*, 1992;
	E	6	Extension	Mariani *et al.*, 1996
	Rn	15	Muscular glycogene	Milan *et al.*, 1995 & 1996; Mariani *et al.*, 1996.
	K88ab, ac	13	Resistance to colibacillus	Guerin *et al.*, 1993; Vogeli *ct al.*, 1994; Edfors-Lilja *et al.*, 1995.
Cattle	PDME	4	Weaver	Georges *et al.*, 1993a
	Polled	1	No horns	Georges *et al.*, 1993b
	Roan	5	Rouan coat	Charlier *et al.*, 1996a
	MH	2	Double-muscled	Charlier *et al.*, 1995
	E	18	Extension	Klungland *et al.*, 1995
	Sy	15	Syndactylia	Charlier *et al.*, 1996b
Sheep	FecB	6	Boorooala	Montgomery *et al.*, 1994
	CLPG	18	Callipyge	Cockett *et al.*, 1994 & 1996
Goat	Polled	1	No horns	Vaiman *et al.*, 1996
Poultry	DW	LG22	White dominant	Ruyter-Spira *et aJ.*, 1996
	SLD	Z	Sex-related dwarfism	Ning *et al.*, 1994

Research on the identification of the genes underlying intra- and inter-breed genetic various constitutes an important subject in genetic improvement.

There are two very distinct approaches to finding a gene that participates in expressing a trait. The first goes from the gene to the trait, the second starts with the trait and goes back up the chain to the DNA level.

The first approach involves the co-segregation of a marker (identifiable landmark) and a locus QTL. It gives the most important results in domestic animals, in particular in studies on major genes. Two important reports have been published on polygenic traits, one on dairy cows (Georges *et al.* 1995), the other on pig growth (Andersson 1995). In many other countries there are research teams working on species improvement programmes. The Consultative Group on International Agricultural Research (CGIAR) is also funding a programme on resistance to trypanosomiasis and to endoparasites.

The second approach has shown to be worthwhile. The most spectacular examples are unquestionably the BLAD locus in bovine species. Kherli *et al.* (1990) noted the resemblance between the affliction in cattle and in people (LAD) and the periodical paralysis in the Quarter Horse is an other example.

The second approach has shown to be worthwhile. The most spectacular examples are unquestionably the BLAD locus in bovine species. Kherli *et al.* (1990) noted the resemblance between the affliction in cattle and in people (LAD) and the periodical paralysis in the Quarter Horse is an other example.

The study of markers naturally leads to verification of filiation, research on descendants, conservation of genetic resources because of their diversity, the study of potential lines as part of heterosis, accelerating the introgression process and marker-assisted breeding.

The potential for marker-assisted breeding is considerable, because it is based on DNA, and thus avoids problems of phenotype measurement for traits expressed in only one sex (dairy animals) and the related costs. If the tests proposed are economically profitable, their widespread uptake will be guaranteed.

Most scientists are looking forward to the next step, the transfer of favourable genetic material within a species. Transgenesis, at least for laymen, is not unproblematic, but they forget that transgenesis is the accelerated version of the introgression carried out, for instance, through successive backcrossings. Public opinion fears potential excesses, and therefore has expressed certain reservations. The ban on importing "modified" cereals into the European Community clearly illustrates the present day constraints. But considering the extent of malnutrition in the developing countries, wouldn't it be logical to use all possible means to satisfy the world's demand for plant and animal productions?

Leaving aside genome analysis, as illustrated above, we see that there are many physiological functions that are connected to the genes. A study of the expression of these genes should increase our understanding of prolificacy, milk composition, and fat contents especially in poultry carcasses. In these examples, the main functions being studied are folluculogenesis, the mammary function, and lipogenesis.

Research connected to animal productions are connected to the agricultural economics of regions, countries, the European community and worldwide agricultural economics. Many constraints have already been identified.

Besides financial restrictions on under-subsidised research, there are other constraints linked to the application of the CAP and its reform, and to GATT. Since they are constantly changing and continually disrupting breeding programmes and targets, they are difficult to accommodate.

SELECTED REFERENCES

ANDERSSON L., HALEY C.S., ELLEGREN H., KNOTT S.A., JOHANSSON M., ANDERSSON K., ANDERSSON-KLUND L., EDFORS-LILJA I., FREDHOLM M., HANSSON I., HAKANSSON J., LUNDSTROM K., 1994. Genetic mapping of quantitative trait loci for growth and fatness in pigs. *Science* 263:1771-1774.

CHARLIER C., GEORGES M., LEROY P., MENISSIER F., MICHAUX C., *Hypertrophie musculaire bovine : cartographie du gene mh et perspectives pour améliorer la production de viande.* Renc. Rech. Ruminants, 2, 181-186

CHARLIER C., COPPIETERS W., FARNIR F., GROBET L., LEROY P., MICHAUX C., MNI M., SCHWERS A, VANMANSHOVEN P., HANSET R & GEORGES M., 1995. The mh gene causing double-muscling in cattle maps to bovine chromosome 2. *Mammalian Genome* 6: 788-792.

CHARLIER C., GEORGES M., LEROY P., MENISSIER F., MICHAUX C.,1995. *Hypertrophie musculaire bovine : cartographie du gène mh et perspectives pour améliorer la production de viande.* Renc. Rech. Ruminants, 2, 181 - 186.

GEORGES M., ANDERSSON L. Livestock genomics comes of age. Livestock Genomics .

GEORGES M., NIELSEN D., MACKINNON M., MISHRA A., OKIMOTO R., PASQUINO A.T., SARGEANT L.S., SORENSEN A., STEELE M.R., ZHAO X., WOMACK J.E., HOESCHELE I., 1995. Mapping quantitative trait loci controlling milk production by exploiting progeny testing. *Genetics* 139: 907-920.

HANSET R., LEROY P., MICHAUX CH., KINTABA K-N., 1983. The HAL locus in the Belgian Pietrain pig breed. Z. *Tierzuchtg. Zuchtbiol*, 100, 122-133.

HANSET R., LEROY P., MICHAUX CH., KINTABA K-N., oct.1983. *The HAL locus and its linkage with the Phi and 6-Pgd loci in the Belgian Pietrain and Landrace breeds.* Annual Meeting of the European Association for Animal Production, Madrid.

HANSET R., MICHAUX C., LEROY P., DERAL G., 1983. *Ann. Med. Vet.* 133:89-114.

LE ROY P. *Les méthodes de mise en évidence des gènes majeurs.* INRA Prod. Anim., 1992, hors série Eléments de génétique quantitative et application aux populations animales, 93-99.

LE ROY P., CARITEZ J.C., ELSEN J.M., SELLIER P., 1994. *Pigmeat quality: experimental study of the RN major locus.* 5th World Congress of Genetics Applied to Livestock Production, Guelph, Canada, 7-12 August 1994, vol. XIX, 473-476.

LE ROY P., ELSEN J.M., RICORDEAU G., BOUILLON J., MANFREDI E., CHASTIN P., MONOD F.,1995. *Mise en evidence d'un gène majeur influençant le débit de traite des chèvres.* Renc. Rech. Ruminants, 2, 177 - 180.

LEROY P.L. and DETILLEUX J., 1996. *Evolution de la production laitière en Belgique. Tendances des productions, importance économique des pathologies, évolution du prix du lait.*

MANFREDI E., PIACERE A., RICORDEAU G., ELSEN J.M., BIBE B., GROSCLAUDE F., 1995. *Polymorphisme de la caséine ml et selection caprine.* Renc. Rech. Ruminants, 2, 167 - 170

MILAN D., LE ROY P., WOLOSZYN N., CARITEZ J.C., ELSEN J.M., SELLIER P., GELLIN J., 1995. The RN locus for meat quality maps to pig chromosome 15. *Genet. Sel. Evol.*, 27, 195-199.

MONTGOMERY, G.W. *et al.* 1994. The ovine Booroola fecundity gene (FecB) is linked to markers from a region of human chromosome 4q. *Nature Genetics* 4: 410-414.

NAVEAU J., BAZIN C., LE ROY P., SELLIER P., ELSEN J.M., GELIN J., 1993. *Vérification expérimentale de l'existence d'un gène à effet majeur sur la qualité de la carcasse chez le porc et recherche d'un locus marqueur facilitant son utilisation.* Colloque bilan 1989-92, Agriculture Demain, Aliment 2002, 19-20 octobre 1993, Paris, 77.

PIRCHNER E., SCMNMUTH G., FLOCK D. *Management of animal material: adjustment to the economic evolution and to production constraints.* Proceedings of the joint symposium of all study commissions of the European Association for Animal Production, Toulouse, France, 11 July 1990, 72-78

VAN VLECK L.D., 1986. *Evaluation of dairy cattle breeding programs.* 3rd World Con. Gen. App Livest Prod. 9:153-164

INTEGRATED AGRICULTURE

JEAN THIAULT
Forum de l'agriculture raisonnée, respectueuse de l'environnement (FARRE)
10, avenue Charles-de-Gaulle
92100 Boulogne-Billancourt, France

ABSTRACT

The post-war food challenge was met, inter alia, through the complementarity of scientific progress made in the genetic, chemical, and mechanical fields. But since these new technologies were too complicated for a rural population composed mainly of people with elementary or empirical education, they had to be applied using simplified lessons based on a risk-free strategy.

Despite the movement that started in the early 1960s to decry the risks of disturbing environmental equilibria, cropping systems, (as the result of the farm price support policies designed to offset market imbalances), developed a uniformity that all too often did not respect the harmony of Nature.

Integrated Agriculture does the opposite. It starts with the observation of biological evolution and then develops a principle of diversification and adaptation. Ongoing thought is given to maintaining an equilibrium between ecological/toxicological requirements and the economic constraints and targeted product quality. Precision in agriculture demands exact analyses and cooperation among farmer groups that share a single target. The sine qua non condition is that their observations and knowledge not only be pooled, and also be constantly updated through dialogue with research.

The wager of European agriculture in the 21st century can only be won if production systems are integrated.

The world of agricultural professions has recognised this need. That is why associations, mainly interprofessional ones, were created to develop the integrated production mode in Europe and to let the public know that farmers are committed to living up to the expectations of society. Six national associations have joined together under EIF, the European Initiative for Integrated Farming, represented in France by FARRE (a forum for environment-respecting rational agriculture).

INTRODUCTION

The exceptional development in agricultural production during the last half century is the result of great production systems, but these systems do not respond satisfactorily to the expectations of the present day society and will respond even less to those of the 21st century.

These systems were born of World War II upheavals and food shortages, and benefited from considerable scientific and technological progress that had either been delayed or accelerated by the War. These "innovations" were complementary: genetics with its varietal selection and hybrid maize; chemistry with the first synthetic fungicides and insecticides, and weed-killing hormones; technology with mechanisation for traction, spreading and harvesting. Together they created development dynamics that fit in fully with the productivity doctrine, the economic credo of the times.

The agricultural world, with its essentially elementary, empirical training, was ill prepared to use these increasingly complex, interfering techniques.

Technical sequences scheduled according to a calculated timetable were popularised through simplified, schematic extension efforts. They were based on a principle of guarantees, and led to production systems that were farther and farther away from the basic principles of agronomy laid down during the previous century. But the food supply challenge had to be met. The context was very pessimistic, as had been forecast by the newly founded FAO which predicted a growing deficit in the world food supply.

As the 1950s met the 1960s, the risks of declining product quality and biological imbalances in the agricultural ecosystems were strongly voiced either by professionals who came up with a very radical proposal that favoured organic farming and rejected the use of any synthetic chemicals or by scientists who, as they created the International Organization for Biological and Integrated Control of Noxious Animals and Plants (IOBC) developed the concept of integrated control which, ten years later, led to the concept of integrated production.

But it was the market imbalances resulting from the sharp increase in output that was to guide agricultural policy. Agricultural support systems, mainly through the pricing policies that were gradually established, encouraged increased productivity and brought about a technical and production system uniformity that ignored the diversity of the natural environment.

21ST CENTURY AGRICULTURE: WHAT IS AT STAKE?

The major themes on our agenda here clearly define the stakes facing agriculture in the coming century. They imply three key objectives (that must be in balance) for agriculture production systems:

Supplying food is still the main concern as the world population grows and feed requirements are not systematically satisfied.

But besides maintaining its position as an exporter on the developing world market, European agriculture should, in its home market, increasingly satisfy the very strong and very diversified quality demands, in particular as concerns the relation between food and health.

Natural resource protection, first of all water, which, as mankind knows, is of limited supply and, on the global stage, is becoming a strategic challenge.

And then there are all forms of biodiversity, be it plant or animal, be it highly evolved or primitive. We now know how strongly long-term maintenance of an ecosystem is linked to maximum biodiversity. But let's stop thinking of man as an intruder in a terrestrial biosphere, where agriculture plays an important role. Let us recognise man's right to a central position in this vast ecosystem, on the condition that he uses his intelligence when carrying out his natural function as a predator serving his own interests.

Territorial development and maintenance are functions that, as farm populations dwindle, are becoming heavier to bear.

Remember that in an increasingly urbanised, industrialised society, agriculture is not only expected to reduce and eliminate the risks of causing disturbance through its activities, but is expected to serve as a offsetting pendulum for nuisance caused by urban and industrial concentration. It is generally the best activity for maintaining and upgrading protected areas and dispersing runoff waters, for conserving and developing periurban zones for recreation and leisure, for recycling urban refuse and wastewaters, and even for purifying the air (IAURIF 1994).

INTEGRATED AGRICULTURE: A PRECONDITION TO WINNING THE BET

There are many definitions of integrated agricultural production. The IOBC definition of 1980 seems the clearest for the farmer:

"A production system involving a farming package that, simultaneously, satisfies ecological, economic, and toxicological requirements as it strives to obtain an optimal quality harvest" (Milaire 1995).

This clearly shows the need to include characteristics of the natural environment in the agricultural technology rationale without foregoing the normal market-related economic targets. In France, the concept of integration often makes the farmer think back gloomily to the painful experiences of vertical economic integration, in particular in poultry production. That is why we have preferred the term agriculture raisonnée (well-reasoned agriculture), but the goals are the same.

Some scientists make a distinction between the two, and consider agriculture raisonnée as an intermediary step on the path to full-fledged integrated production. The stakes are too important to spend time quarreling over words. Above all, we must avoid dogmatic rigidity because this approach by nature must be flexible and must always be adaptable to changing natural and economic conditions.

The important thing is for agriculture to develop in this direction, to integrate agriculture into nature without trying to dominate it (Bonny 1995).

OPTIMISING KNOWLEDGE

Integrated agriculture broke away from the predetermined technical sequences that were applied more or less blindly after the Second World War. It is based on a synthesis of knowledge of all types, so that crop choices can be well adapted to local aptitudes using farming techniques paced with the agricultural season and environmental fragility.

The particular features of the villagelands (*terroir*) can be optimised through the special quality of their products, by integrating market opportunities and production potentials thanks to the precision of knowledge (Valeschini and Nicolas 1995). Regular monitoring of crop evolution to detect inadequate plant nourishment or the onset of pest or disease attacks early enough requires both time and knowledge.

In many cases, a diagnosis has to be made to assess the degree of risk and to decide whether or not, and how to intervene. Knowing when to harvest the crop also requires considerable expertise; tests and analyses are being used more and more.

Integrated agriculture, thus, is precision agriculture. It is the result of well-adapted early training, and the farmer's desire to learn not only through observation and reasoning, but also through field experience. Convinced farmers find it to be very profitable. Understanding the biological phenomena occurring in each of their plots, and knowing that whatever happens, be it good or bad, can be analysed, better equips them for the next agricultural seasons, and thereby gives them the feeling that they are no longer the victims of obscure forces and can better control their production conditions.

INDIVIDUAL ACTIONS: RESPONSIBILITIES AND LIMITS

Integrated agriculture has the farmer constantly making choices that condition the final outcome. It is a single farm responsibility that only the farmer can exercise. But it also makes him realise that he is responsible for the quality and integrity of the natural resources he uses in farming or maintains on his premises. Air, water, wildlife and flora, long-term soil fertility, even the environment surrounding his holdings are considered to belong to the society to which the farmer is accountable.

But agriculture impacts the environment as part of a complex ecosystem which covers all the farms in a given natural zone; this impact is the result of the relation between the agricultural techniques and the environmental fragility.

The individual responsibility is diluted into a group phenomenon whose sensitivity is especially difficult to appraise since the relationship between environmental fragility and agricultural techniques is hard to measure. The knowledge needed is complex and can only be acquired, and put to use, in an interdisciplinary, interprofessional context.

In a farm holding, collecting aggregate data, finding the right references, and preparing choices are beyond the capacity of one person alone. Belonging to a sufficiently homogeneous group is indispensable for pooling the observations, information and experimental data needed to prepare a line of reasoning - and choices, although the latter have to be made at the individual level.

Participating in such groups gives the farmer the opportunity to speak with people working in other parts of the sector, such as suppliers and customers, in order to define qualitative objectives together, choose technical sequences that can remove uncertainties about product characteristics, and improve technical logic throughout the sector (Sylander 1994).

This is also where it can be efficient to call on research. For integrated agriculture, or knowledge-based agriculture to develop requires constant contact with research so that questions can be clearly defined and solutions can be built up step by step.

A EUROPE-WIDE MOVEMENT

The common agricultural policy, that did so well in achieving its initial productivity targets, has inhibited the innovative capacity and attempts at personalising product quality in European agriculture for too long. Until a few years ago, European agriculture seemed to have lost the human values that marked the peasant tradition. Despite warning signs, the power of financial constraints and the inertia caused by debt tended to reduce agriculture into some sort of manufacturing activity, as the INRA President, Guy Paillotin, recently pointed out.

Fifty years of agricultural research and education gradually and deeply affected the rural population which, because of the rural exodus, now accounts for less that 5% of the total population although agriculture consumes 60% of the national territory. European public opinion spoke harshly about methods used in agriculture, thereby quickly making the farming communities aware of their responsibilities to society.

This was the context that gave birth to the movements that developed in Europe to promote integrated agriculture, starting with a network of farms whose owners agreed to testify to their commitment and report their results. This testimony is intended as examples and references for farmers who still hesitate to give up their oldtime practices. It is also intended to open discussion with the public at large who, in turn, should strengthen the farmers' perception of what society expects from them. But going further, it should convince the public to replace its globally negative view of agriculture by appreciation for the commitment that the farmers have made to changing their way of working.

To give their action international standing, the six national associations adopted a declaration creating the European Initiative for Integrated Farming (EIF). The EIF member associations work on a very broad interprofessional basis by bringing together not only organisations that represent the various stages in the production-to-sales chain, but also interested scientific and environment-protection organisations.

CONCLUSION

Integrated agriculture balances the farmer's economic objectives with the consumer's expectations for quality and the natural resource protection requirements in an attempt to stand up to the stakes of sustainable agriculture in Europe. Integrated agriculture is work of precision requiring the farmers to have thorough knowledge and up-to-date information. This means that more than ever before they need the support of dynamic research, working close enough to them to analyse their difficulties and swiftly make scientific progress applicable.

This type of agriculture is not a linear chain of automatically embedded techniques. It has to accommodate all the interactions simultaneously through a global view of the farm and its environment. Research is essential to provide the elements needed to understand the complexity of the agricultural ecosystems and propose objective criteria (other than those based on principles of precaution) for evaluating the impact of agricultural activities on the environment.

Europe needs modern, dynamic agriculture capable of providing the right quantity and quality of food and playing a growing role in supplying renewable raw and industrial materials and sources of energy (Guellec 1995). This is a prerequisite to maintaining an active rural economic fabric that can function, as it should, in counterweight to the urban and industrial zones. It is also an obligation in the name of solidarity with peoples suffering from insufficient access to food that we prepare ourselves to export more know-how than food.

This know-how does not only cover production, it also covers natural resource preservation.

The ambition of the EIF member organisations is to share the conviction that nothing is dead sure in biology. What is important is the state of mind, improving our comprehension through observation, and having the audacity to act since we know that techniques for producing wheat and other crops will be changing over the next 10 or 20 years.

The basic laws of biology will not have changed, but we will be able to better understand them and use them in our human activity; without ruining the environment. And once we possess that knowledge, aren't we obliged to share it with people who still go hungry, and hope that we can spare them the errors that we, in all good faith, may have made along the way?

What a beautiful programme for research!

SELECTED REFERENCES

BONNY S., 1995. Les possibilités d'un développement durable en agriculture, le cas de la France. *Courrier de l'environnement de l'INRA*, 23, 5-15.

COLLOMB PH., 1995. *L'homme et l'eau.* Population et société, 298 p.

GUELLEC J., 1995. *Agriculture-Industrie, une chance pour demain.* Espaces pour demain, Paris.

IAURIF., 1994. *Plan vert régional d'Ile de France.* Institut d'aménagement et d'urbanisme de la région Ile de France, Paris, 257 p.

MILAIRE H.G., 1995. *A propos de quelques définitions.* Phytoma-La défense des végétaux, 475, p 7-9.

SYLVANDER B., 1994. *La qualité, du consommateur final au producteur. Qualité et systèmes agraires*, Etude. Rech. Syst. Agraires Dév., 28, p 11-15. INRA, Paris.

VALESCHINI E. AND NICOLAS F., 1995. La dynamique économique de la qualité alimentaire.*In*: Valeschini E. and Nicolas F.: *Agro-Alimentaire, une économie de la qualité*, INRA Economica, p 15-37.

IMPROVEMENTS IN PRODUCTION SYSTEMS

DIETER SMIDT
Federal Research Centre of Agriculture
Mariensee,
31535 Neustadt 1, Germany

ABSTRACT

Livestock production serves a growing human population with an appropriate supply of food, other products and services in the non food-sector. Livestock production systems in Europe are widely characterised by considerable levels of specialisation and concentration as well as by high animal performances. Recent developments in public attitudes and agricultural policies have resulted in more complex considerations towards future perspectives for the livestock sector. Consequently, the following goals have to be integrated into the design of sustainable production systems: sustainable economy, good animal health, and welfare, conformity with ecological needs, high level of food quality and safety, and humane working conditions.

The following measures may contribute efficiently to respective achievements: selection of suitable livestock, adjustment of management systems to animal needs and human requirements, effective management of animal health, utilisation of biotechnological progress and technical achievements, and diversification of production systems.

EVOLUTION OF LIVESTOCK PRODUCTION SYSTEMS

Animal production serves to supply a growing human population with appropriate high quality food and important non food products and services. Therefore, in a broad sense, animal production has to be regarded as leading area in the fulfilment of human requirements. Nevertheless, increasing awareness has been developed, that in animal production, besides economic considerations for producers and consumers, further necessities and demands have to be observed. According to respective developments in public attitudes and agricultural policies, the following indispensable goals and principles have to be integrated into the design of sustainable production systems (Smidt, 1993):

- sustainable economy, including long-term stability of demand for animal products, which is widely based on consumers acceptance of production systems;
- animal health as an essential prerequisite for economy, food safety and animal welfare;
- animal welfare, including the fulfilment of animal requirements as well as the improvement of health aspects and the observation of legal regulations;
- compatibility of production systems with ecological requirements, including the preservation of animal genetic resources;

- food quality and food safety in order to meet consumers demands;
- humane working conditions in livestock systems.

Agricultural policy, including the veterinary sector, is aiming at the integration of relevant demands into general concepts for further development of animal production. These efforts usually result in respective legislative activities. Legislation and codes of practice are, therefore, important elements and preconditions for modern concepts of animal production. Overregulation, however, usually is counterproductive and should, therefore, be avoided.

The national legislation and codes of practice correspond to supranational European regulations, *e.g.*, European conventions and recommendations, issued in the framework of activities of the Council of Europe. Moreover, directives and guidelines of the European Union have to be included in the national legislation of the EU member states (Blumenstock, 1994; Scholz, 1996).

On the basis of legislation, principles, influences and public acceptance, systems of animal husbandry are to be developed and constantly adapted to necessities, conditions and demands. The following measures may be regarded as efficient contributions to respective achievements:
- selection of suitable livestock populations;
- adjustment of management systems to animal needs and human requirements;
- utilisation of biotechnological progress and technical developments;
- diversification of production systems.

Selection of suitable livestock populations

Successful and sustainable livestock production has to start with considerations concerning the choice of suitable animal populations for different livestock systems. Selection of respective animal populations has to consider the region of production, with its environmental conditions and natural feedstuff resources, as well as the type of production, including the kind of products it is aiming at. Preservation and maintenance of genetic diversity in farm animals is, therefore, one of the major issues in future perspectives for the evolution of livestock production systems.

Adjustment of management systems to animal needs and human requirements

Livestock management systems have to be carefully adjusted to:
- animal needs and welfare requirements,
- demands for humane working conditions,
- necessities for environmental protection,
- consumers' expectations.

Any management system, regardless whether it is intensive or extensive, large scale or small scale, professional or amateur, has to meet the regulations for animal welfare (Baumgartner, 1990; Rojahn, 1993; Smidt, 1983; Unshelm, 1987). This means in the first place: consideration of ethological needs, as well as active promotion of good animal health. Only healthy animals are able to experience welfare, and welfare is an important prerequisite for animal health. Welfare means the fulfilment of biological requirements of farm animals. These requirements have to be identified, evaluated and their fulfilment has to be provided. Ethological needs are determined mainly by investigating the behavioural inventory of animals. Since it is virtually impossible to meet every single ethological requirement, they have to be evaluated with respect to their significance for animal welfare. Ethological needs, which are regarded to be important for animal welfare, must be considered properly in the design of animal husbandry concepts (Smidt, 1993).

Besides the needs of animals, however, demands concerning good working conditions for people employed in animal production must not be neglected. These conditions have to be in compliance with modern standards. This mainly refers to requirements for human health and is, therefore, predominently aiming at the avoidance of accidents as well as at the elimination of other health hazards in animal husbandry units (Aumann, 1996).

All areas of terrestrial ecosystems are affected by livestock production. For this reason, environmental needs have to be carefully considered in modern concepts of animal husbandry (Peters, 1996). This mainly comprises:

- the reduction of pollutants and nutrient surpluses,
- the preservation of natural resources and habitats, and
- the maintenance of genetic diversity in farm animals and fodder plants.

The reduction of pollutants aims at a minimum of airborne emissions of dust, harmful gases and microbes from stables (Swierstra, 1991; van den Weghe, 1996). Also animal wastes should be as free as possible of unwanted pollutants (Strauch, 1992).

Measures to avoid nutrient surpluses in soil and water may comprise:

- the use of fertilisers, including manure and other animal wastes, according to season and to nutrient requirements of soils and plants (Werner, 1996), and

- the application of special feeding strategies in order to reduce N- and P- inputs, and special supplements like enzymes or amino acids to improve the efficiency of nutrients in the rations (Verstegen, 1996).

The unavoidable emissions of methane from ruminants , as related to product units, are by no means higher in intensive production systems than in comparison to more extensive farming (Pfeffer, 1992). High regional concentration of farm animals, however, always demands particular efforts in minimising environmental pollutions.

In modern concepts of animal production environmental protection, including the handling and management of wastes and manure, have been improved considerably. Moreover, effective steps have been taken to reduce emissions from animal facilities, with special emphasis on odour, microbes and dust. The related efforts are integrated into development concepts for rural regions and communities.

Animal production must not exploit natural resources to an extent, that it could result in environmental hazards. This holds true for global resources, *e.g.* tropical forests, as well as for local resources, like soil, water, landscape and others. An important aspect in this respect is the balance between the area of land available and the concentration of farm animals, which reflects more or less the level of intensity of production. Probably there will be a development in two directions, namely towards highly specialised production systems on the basis of high performances on the one hand, and a more extensive use of less productive areas with cattle, sheep and goats on the other hand (Langholz, 1992). Even more diverse developments may be expected to take place.

The maintenance of genetic diversity in farm animals is an important target within the preservation of natural resources. It requires the establishment of gene reserves by means of integrated systems including living animal populations as well as cryoconserved gametes and embryos of endangered farm animal breeds, which have proved to be worth being preserved (Ehling *et al.*, 1994; Scherf, 1995).

Consumers expect cheap, but valuable, healthy and tasty food from animal products. For this reason, the assessment of product quality by means of contents, physiological criteria, technical and hygienic traits as well as sensoric tests is an important area of concern in modern animal husbandry. Increasingly, however, methods and conditions of production are integrated into considerations concerning the quality standards of animal products (Kallweit, 1992). These refer, for instance, to the observation of animal welfare regulations and to aspects of environmental protection. Consumers' demands in this respect are very strong motives for the establishment of special strategies and programmes in animal production and in the marketing of animal products, which are aiming at the introduction of special brands and proprietary articles. For this reason, these programmes include certain codes of practice, which have to be strictly observed by the producers when embarking on respective production organisations. This refers to High-Quality-Meat programmes as well as to so-called "Bio"- or "Eco" - products.

Effective systems of control and certification of animal product quality are already in operation or in progress.

Effective management of animal health

Management of farm animal health today means an integrated system of hygienic measures, health control, and preventive treatment. It requires close cooperation of farmers and veterinarians. Other professional groups and institutions, like pharmaceutical companies, the food processing sector, companies for animal management facilities, and breeding organisations, should join them in an alliance for good animal health (Smidt, 1996).

The improvement of animal health is predominantly a matter of careful observation and conduction of up-to-date programmes in animal hygiene. The prevention of so-called technopathies, which means injuries, physical damage and pain, caused by unsuitable technical equipment, is a very important issue in health management in farm animals.

It is well known that severe acute and chronic stress can be the etiological background for animal diseases. Such stress situations therefore, have to be avoided. On the other hand, mild stress is essential for training the defence mechanisms of animals, especially for the stimulation of immunological reactions within the psycho - neuro - endocrine - immunological system.

Therefore, completely stress-free systems of animal production are neither possible nor desirable (Smidt *et al.*, 1988) .

Utilisation of biotechnological progress and technical developments

Biotechnology and technical achievements, if properly used, can contribute significantly to the solution of problems for future perspectives in livestock production. Sometimes, biotechnology and high technology are blamed as conflicting with actual demands on livestock systems. This, however, need not be the case if these new techniques are employed in a reasonable way and for important purposes, *e.g.*, cost reduction, preservation of genetic resources, improvement of animal health and welfare. The latter may be elucidated by means of two examples.

First, to make extensive use of genetic diversity in livestock, requires worldwide movements of genetic material. For numerous reasons, *e.g.* hygienic requirements, welfare considerations, and efficiency, it seems no longer desirable to manage this by the transport of live animals. The replacement of live animal transportation by shipment of long-term stored genetic material, like semen, embryos, and, maybe in the future, even conserved DNA, could be a promising way to solve this problem. Thus, biotechnological progress is a convincing contribution to problems of practical and public concern.

Second, concerning technical developments, modern technologies can help to improve animal welfare component in livestock systems, without being forced to give up all the advantages of existing and prevailing systems. By using high technology, like computer-aided management and feeding, we can allow the animals a certain level of self determination in production procedures by using group housing systems, instead of facilities for single keeping.

Diversification of production systems

Demands of producers and consumers, as well as animal needs can be fulfilled in a whole variety of livestock systems. Meeting the requirements is by no means bound to a particular mode of production. Diversification, therefore, is an important keyword for future development in livestock management. Diversification of livestock systems nowadays may be ranging, *e.g.* ,

- from intensive to extensive systems,
- from large scale to small scale enterprises,
- from specialised units to multiple farming,
- from individual to cooperative management,
- from professional enterprises to hobby farming.

All these systems, if applied appropriately, allow the actual requirements in today's' animal husbandry to be met.

CONCLUSION

The evolution of livestock production systems will result in further diversification, ranging from intensive, highly-specialised enterprises to extensive forms of land use by means of farm animals.

This development requires well-adapted and specific management systems. All of these systems, however, have in common, that a number of principles have to be harmonised in the practice of animal production. The keywords are: sustainable economy, high standards of animal health and welfare, environmental preservation, and high quality products.

Agricultural policy has to set proper frames for good chances for prosperous developments in diverse livestock systems.

An important prerequisite for successful livestock farming under different conditions is the application of scientific achievements in agricultural and veterinary research for the benefit of farmers, consumers, animals and environment. Moreover, sustainable livestock production requires an effective alliance of all people and institutions involved in this important sector of human life.

SELECTED REFERENCES

AUMANN, R., 1996. *Gesundheitliche Risiken für Tierhalter (Health risks for people employed in animal production).* Proceedings: Hülsenberger Gespräche 1996, 68 - 84.

BAUGARTNERNER, G., 1990. Tierschutzgerechte Haltungs- und Transportbedingungen für landwirtschaftliche Nutztiere (Conditions for the keeping and transport of farm animals in conformity with animal welfare requirements). *Dtsch. tierärztl.Wschr.* 97, 139 - 143.

BLUMENSTOCK, S., 1994. *Tierschutzgesetzgebung in Europa (Animal welfare legislation in Europe).* Schriftenreihe des Bundesministeriums für Ernährung, Landwirtschaft und Forsten, Heft 431.

EHLING, C., FALGE, R., GROENEVELD, E., NIEMANN, H., SMIDT, D., WEIGEND, S., Konzept zur Erhaltung und Nutzung tiergenetischer Ressourcen in der Landwirtschaft der Bundesrepublik Deutschland (Concept for the preservation and utilization of animal genetic resources in the Federal Republic of German). *Landbauforschung Völkenrode, Sonderheft 151.*

KALLWEIT, E.,1992. Tierhaltung und Produktqualität (Animal management and product quality). *Züchtungskunde* 64, 283 - 291.

LANGHOLZ, H.-J., 1992. Extensive Tierhaltung in der Landschaftspflege und als produktionstechnische Alternative (Extensive animal production in the preservation of landscapes and as an alternative production procedure). *Züchtungskunde* 64, 271 - 282.

PETERS, K., 1996. *Weltweite Wechselwirkungen zwischen Tierhaltung und Umwelt. (Worldwide interactions between animal production and environment)* Proceedings: Hülsenberger Gespräche 1996, 14 - 27.

PFEFFER, E., 1992. Umweltbelastung durch die Tierhaltung und Möglichkeiten zu ihrer Minderung - Ernährungsphysiologische Aspekte (Environmental loads from animal production and possibilities for their reduction - aspects of nutrition physiology).*Züchtungskunde* 64, 254 - 261.

ROJAHN, A., 1993. Entstehung und Entwicklung des Tierschutzgesetzes (Origin and development of the animal welfare law). *Dtsch. Tierärztl. Wschr.*, 100, 43 -47.

SCHERF, B. D. (Ed.), 1995. *World Watch List for Domestic Animal Diversity.* FAO Rome.

SCHOLZ, H., 1996. *Umweltverträgliche und rentable Erzeugung von Lebensmitteln tierischerHerkunft unter Aspekten europäischer Agrarpolitik und Agrarentwicklung (Environmentally conform and economic production of food of animal origin in view of the European agricultural policy and development).* Proceedings: Hülsenberger Gespräche 1996, 212 - 219.

SMIDT, D. (Ed.), 1983. *Indicators relevant to farm animal welfare.* Proceedings: CEC - Seminar, Martinus Nijhoff Publishers.

SMIDT, D., 1993b. 20 Jahre neues Tierschutzgesetz (20 years of new animal welfare law). *Der Tierzüchter,* 6, 38 - 41.

SMIDT, D., 1993b. *Modern concepts of animal husbandry with respect to health and welfare, environmental needs and product quality.* Indo-German Conference on: Impact of modern Agriculture on Environment, Hisar, India.

SMIDT, D., 1996. Gesunde Tiere - Grundlage einer verantwortungsbewußten und wirtschaftlichen Tierhaltung (Healthy animals - basis of a responsible and economical animal production). *Tierärztl. Umschau* 51, 519 - 523.

SMIDT, D., KALLWEIT, E., LADEWIG, J., 1988. Stress, Gesundheit und Leistung beim Schwein. (Stress as related to health and performance in the pig). *Tierärztl. Praxis,* Suppl. 3, 1 - 10.

STRAUCH, D., 1992. Umweltbelastung durch Tierhaltung und Möglichkeiten zu ihrer Minderung - Hygienische Aspekte (Environmental loads from animal production and possibilities for their reduction - hygienic aspects). *Züchtungskunde* 64, 245 - 253.

SWIERSTRA, D., 1991. Emissionsarme Betriebsgebäude für Milchkühe (Minimum emission stables for dairy cows). Landbauforschung Völkenrode, *Sonderheft* 125, 51 - 58.

UNSHELM, J., 1987. Die Anwendung tierschutzrechtlicher Bestimmungen aus der Sicht der Tierhaltung und des Tierverhaltens (Application of animal welfare regulations in view of animal husbandry and ethology). *Dtsch. Tierärztl. Wschr.* 94, 89 - 90.

VAN DEN WEGHE, H., 1996. *Gasförmige Emissionen aus der Tierhaltung und biologisch - technische Möglichkeiten ihrer Vermeidung (Gaseous emissions from animal houses andmeasures for their avoidance).* Proceedings: Hülsenberger Gespräche 1996, 108 - 115.

VERSTEGEN, M. W. A.,1996. *Reduzierungsmöglichkeiten für Nährstoffüberschüsse in der Tierhaltung (Possibilities for the reduction of nutrient surpluses in animal production).* Proceedings: Hülsenberger Gespräche 1996, 139 - 148.

WERNER, W., 1996. *Tierische Ausscheidungen als Nährstoffträger: Bedeutung für Boden und Gewässer (Animal wastes as nutrient carriers, and their impact on soil and waters).* Proceedings: Hülsenberger Gespräche 1996, 124 - 138.

CHAIRMAN: JEAN-CLAUDE **SABIN**, *APCA, FRANCE*

VICE CHAIRMAN: YVES **BIROT**, *INRA, AVIGNON, FRANCE*

THEME 2

AGRICULTURE, ENVIRONMENT AND RURAL AREAS

Workshop 2
Product diversification

NON-FOOD PRODUCTION: ECONOMIC, ENVIRONMENTAL AND LAND USE ASPECTS

PHILIPPE CHARTIER

Scientific Division
Agence de l'Environnement et de la Maîtrise de l'Energie
27, rue Louis Vicat
75737 Paris cedex 15, France

ABSTRACT

Non-food production has always existed in agricultural activity: plants used as raw materials for textiles are evidence of this. In addition, species that yield new high-added-value molecules have uses in sectors such as pharmacology, cosmetics and food additives, but occupy very little land area. Large-scale non-food production involves primarily potential substitutes for petroleum feedstocks for basic chemicals, liquid and solid fuels. Within the scope of these applications, this paper addresses the following points:

- current and future resources, with particular emphasis on the question of available lands
- potential economic competitiveness
- issues and limitations, notably from an environmental standpoint.

In this outlook, this paper deals with current steps taken in terms of regional niche markets and obstacles hindering the establishment of product channels, but also with the driving social forces that enable society to follow the learning curve of new technologies. These topics are illustrated by the case of the AGRICE group and liquid biofuels in France.

INTRODUCTION

In the last ten years, as the saturation of solvent food markets has imposed changes in the common agricultural policy, new non-food crops have increasingly drawn attention as a way to offset losses in farm income and use fallow lands.

This paper will focus primarily on biomass crops grown for energy value in the form of biofuels, by reason of their major potential impact in terms of available lands and the environment, but also because of the author's greater familiarity with this aspect of the subject. In passing we will see that this utilisation confronts conflicting views of the potential position of energy crops in agricultural and forestry production: the past versus the future, food versus fuels, biology versus physics, etc.

Finally, the thoughts which follow owe a great deal to a document drawn up by ADEME for the European Commission in September 1996, under the title Perspectives for Biomass Energy in the European Union.

CURRENT RESOURCES AND POTENTIAL FOR DEVELOPMENT

At the dawn of the industrial revolution biomass constituted the main energy resource. It has been largely replaced since then, for the most part by fossil products (coal, oil, gas), and marginally, except in France, by nuclear power and renewable energy, essentially hydropower. Overshadowed by this familiar state of affairs, it is often little known that biomass is today still a non-negligible source of energy in Europe via technologies that can be modern and efficient. More than 6% of the energy produced in the European Union (15 countries) is derived from biomass, *i.e.* the equivalent of one-quarter of nuclear power or of natural gas production.

Most biomass energy comes from wood used by industry and households (see Table 1). Waste constitutes a second non-negligible source. And lastly liquid biofuels are rapidly emerging, spurred by decisions made, notably in France.

Table 1: Energy derived from biomass – Natural resources in millions of tonnes-oil-equivalent(*Source: EUROSTAT 1993 12 member state*)

Wood	24
Municipal waste	3.5
Other wastes	1
Biofuels	0.5
Total	29

On the basis of information on the equipment pool and surveys, consumption would appear to be closer to 50 Mtoe, i.e. 3-4% of total energy consumption in the European Union. The difference is due to non-commercial supply networks.

The important role of lignocellulosic products should be underscored here, because in all likelihood they will constitute a major feedstock in the future.

The above-mentioned study carried out for the European Commission uses previous work to estimate the development potential for biomass at 130 Mtoe in 2015, compared to 50 Mtoe today[1] Of this potential, 50 Mtoe would come from forestry resources, and 30 Mtoe from energy crops covering 10 million hectares. This is both a fairly small amount from the energy standpoint (between 5 and 10% of consumption) and a great deal in terms of the land used.

The outlook is fairly favourable with respect to new forestry resources. Energy is a co-product of paper pulp and lumber production. The situation in Finland, where the wood industry drives the use

1. Mtoe = millions of tonnes-oil-equivalent. 1 Mtoe = 42,000 TJ

of biomass for its energy value, is a case in point. Since the expansion of the European Union to include Austria, Finland and Sweden, forested lands now cover 50% of EU territory, and this proportion is growing. It would appear that only 70% of annual biological growth is captured. Even in the wood industry only 60% of wastes are processed for added value. In another domain industrial wood wastes (palettes etc.) are now converted to new uses to avoid landfilling them. In short, signs are favourable for the mobilisation of 50 more Mtoe of forestry resources, if energy prices move upwards between now and 2015, as will be seen below. The outlook for energy crops is less clear, and more directly related to agriculture.

Factors of uncertainty are: world food supply and demand, and the resulting availability of land, if any relative prices for food and energy products extensification of plant and animal production level of industrial demand for biomass used for paper pulp, starch, lubricants, hydraulic fluids, detergents and plastifying agents extension of zones devoted to environmental (biodiversity) and tourism uses.

UNCERTAINTY AS TO AVAILABLE LANDS

Production of 30 Mtoe of biomass energy crops would require the planting of at least 10 million hectares of land, whether for lignocellulosic crops, oilseed, sugar or starchy plants.

Even though there is currently a certain degree of pressure on markets which has a tendency to limit land diversion in Europe, it is assumed that the long-term trend is to the extension of fallow lands and other formulas with similar effects. Several factors argue in favour of this position, *viz* in the medium term:

- the reconstruction of an exporting agricultural sector in the countries of the former Soviet bloc (Figure 1),
- the diminution of set-aside land area in the United States,
- the advancement of production in developing countries,
- continuing productivity gains per unit of surface area,
- reduced meat consumption.

Other factors weigh in the opposition direction. In addition to the inverse of the above conjectures, should be taken into account:

- the inevitable food shortfall that will be experienced by southern Mediterranean countries (from 40 million tonnes of grain imported today, to 70 million tonnes in 2010 according to FAO and World Bank projections),
- food supplies for Asian countries with very high population density and a rising standard of living (China, South-East Asia), although Europe has not been very active in this region to date.

As suggested by Jacques Poly in other domains, when confronting these uncertainties research should be used as a tool to counter the haphazard progression of discoveries, and to pursue learning curves via niche markets (*e.g.* liquid biofuels in France).

Figure 1: Potentiel agricole dans le monde - Surface agricole utile en hectares par habitant

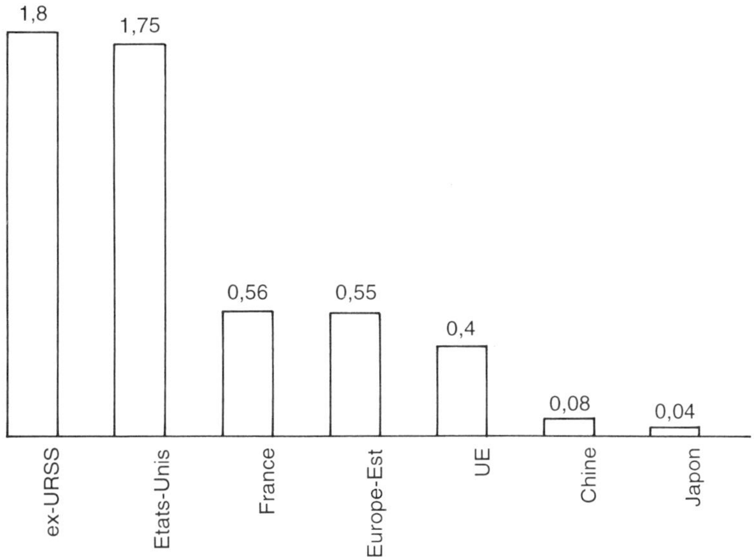

INDUSTRIAL DEMAND FOR BIOMASS

Currently 1.2 million hectares in Europe are devoted to starch production, notably for paper recycling, and to oilseed plants for lipid chemicals.

By 2015 twice as much land might be devoted to the same uses, and could produce up to one million tonnes of paper pulp from annual plants. The very high-added-value molecules that could be of considerable interest for industry require very little land. Today the outlook is better for the economic competitiveness of biomass for industrial applications than for energy uses, although market gluts might occur which would lower prices. In sum, the land area mobilised for industrial crops is likely to remain low compared to the lands required for energy, unless in the future massive feedstock substitution occurs in the refining of petrochemicals.

CONVERSION TECHNOLOGIES

Biomass can be used to produce motor fuels, electricity and heat. The feasible technologies can be divided into three categories, by degree of maturity.

- The first category covers mature technologies that may see incremental improvements such as clean combustion, alcohol fermentation, esterification of vegetable oils and anaerobic digestion.

- The second category involves developing technologies. These are at present primarily gasification and combined cycle production of power and heat that are the priority focus of the EU Joule and Thermie programmes, and of work in northern European countries.

- The third category groups breakthrough technologies that call for significant research and development efforts. Today the most prominent of these internationally is the production of a liquid biofuel derived by biological processes from lignocellulosic materials, after pre-treatment , or by thermochemical processes.

As a general rule, cost analysis shows that the limiting factor lies basically in the price of biomass delivered to the factory floor. This implies that improving competitiveness will require major efforts in agronomic and forestry research.

ENERGY AND ECONOMIC ISSUES

Currently, the main obstacle to the development of liquid and solid biofuels is found in low energy prices, and most particularly low oil prices which continue to lead the market. Niche markets have nonetheless been opened in this context, by policy decisions:

- exemption from taxation for motor biofuels in France,
- Non-Fossil Fuel Obligation for electricity production in the UK,
- CO_2 tax in Denmark,
- the consequences of the rejection of nuclear power in Austria.

In the medium term, factors of upward pressure on energy prices are:

- rising oil demand in newly industrialised countries, particularly in South-East Asia,
- the concentration of low-cost resources in the Middle East,
- the internalisation of the environmental costs of fossil fuels (greenhouse effect) and risk-related costs (nuclear energy),
- increasing recourse to financing for production facilities, due to privatisation and competition between electricity production and distribution.

By 2010 the price of oil will probably have risen past the $30 a barrel mark, and will stay there, compared to between $15 and $20 today. At that price level solid biofuels are sure to be competitive, and liquid biofuels should be competitive at around $40/barrel if research programmes such as AGRICE are successful (see below).

In an entirely different domain, it is worth noting that exploiting the energy value of biomass provides supplementary income for farmers and forestry operators, and thus justifies maintaining their activity. Extracting energy value also creates local jobs on a non-negligible scale. It is estimated that three jobs are created per 1,000 toe of forestry products, and nine jobs per 1,000 toe of liquid biofuels.

ENVIRONMENTAL ISSUES

The principal advantage of biomass is its contribution to mitigating climate change due to the greenhouse effect:

- by storing carbon in standing vegetation (via reforestation) and in durable goods (wood used in construction),

- by substituting liquid or solid biofuels for fossil fuels.

At the Conference of the Parties to be convened in Kyoto in late 1997, the OECD countries, including France, are slated to make quantitative and binding commitments for the reduction of CO_2 emissions for the periods up to 2005, 2010 and 2020. Given the narrow margin for the implementation of new measures in France, where 90% of electricity already comes from nuclear power and hydropower, the development of energy biomass will most likely be a major field of action.

However, it should be noted that the net balance of CO_2 and other greenhouse gases is positive for biomass only if advanced technologies are used throughout the entire chain of production, conversion and final use.

Biomass has other positive environmental impacts. These are the advantages of using high concentrations of rapeseed methyl ester in diesel fuel or incorporating oxygenates such as ETBE in gasoline to reduce emissions of benzene and aromatics. Using these biofuels helps improve urban air quality. Biomass also contributes to the upkeep of forested lands and has positive repercussions by maintaining economic activity in certain areas.

On the other hand, biomass energy crops, like all other crops, may be associated with the range of risks linked to poorly managed intensive single-crop cultivation: groundwater and soil pollution, diminution of biodiversity, uniformisation of landscapes. It is important to acquire better knowledge of the environmental impacts of different technical pathways and of the various ways to use regional lands. Agricultural producers are now pledging to follow good cultivation practices which are promising, and which could be further improved in their content.

The attitudes of environmental groups regarding biomass energy crops range from an extreme reticence, for the reasons mentioned above, to strong and solid support, depending on the primary focus of the group, nature conservation at one end of the spectrum, sustainable development at the other.

PROMOTING A STRATEGY OF REGIONAL NICHE MARKETS

In a context of uncertainty for the medium term (food supply and demand, energy prices, etc.), we must pursue the learning curves opened up by decisions favouring biomass that national, regional and local authorities are in a position to make. It is productive to identify and encourage initiatives taken by the driving forces that work in this direction, including:

- oilseed, grain and beet producers for liquid biofuels in France,

- electricity companies subject to the Non-Fossil Fuel Obligation in the United Kingdom, for electricity generation using lignocellulosic materials,

- industrial paper companies seeking to exploit their wood wastes and forest lands in Sweden and Finland,

- agro-food companies seeking added value from their wastes and effluents via energy recovery,

- local authorities in Austria, Denmark and Sweden who want to promote renewable energy.

TOWARDS THE PROMOTION OF RESEARCH FOCUSING ON ENERGY AND INDUSTRIAL USES FOR BIOMASS

Most research directions are not specific solely to energy and industrial applications for biomass, even if these orientations are inflected by these outcomes.

Current research in these fields in France is structured around the AGRICE group (Agriculture for Chemicals and Energy) which is made up of thirteen members, including INRA and ADEME, and located at the offices of the Agriculture and Bioenergy Division of ADEME.

The topics addressed by AGRICE involve liquid and solid biofuels, biomaterials and biomolecules. For liquid biofuels the aim is to narrow the cost gap between biofuels and fossils fuels, from two francs per litre to one franc per litre by 2005. In addition to the probable rise in oil prices, taking into account the positive externalities of liquid biofuel production (employment, greenhouse effect, trade balance) would make these products competitive.

Research programmes fall into three categories:

- energy topics, covering esters, ethanol and solid biofuels,
- chemicals, including surfactants, lubricants, various lipochemical products, and biomaterials,
- research focusing on the environmental and economic problems related to each product process.

In 1995 funding for these programmes amounted to 118 million francs, including 38 million francs in public funds, of which 18.6 million were provided by ADEME and the Agriculture Ministry in the form of incentive credits, and 47 million francs provided directly by business.

AGRICE is presided by Philippe Mangin, Chairman of EMC2, and directed by Philippe Mauguin, Division Director at ADEME.

CONCLUSION

High stakes are involved in the challenge of attaining sustainable development for the world's population in the course of the next century. Food supplies and energy are two major aspects of this challenge. Agriculture must give priority to meeting the first need; will there be room enough to supply some 10% of the second?

I have tried to address this question in this paper, with particular attention to the potential within the European Union.

SELECTED REFERENCES

CHARTIER PH., BEENACKERS A.A.C.M, GRASSI G., 1995. *Biomass for energy, environment, agriculture and industry.* Proceedings of the 8th European Biomass Conference, Vienna, Austria. Pergamon, Elsevier Science, UK. 2426 p.

CHARTIER PH., BATTAIS L., DEFAYE S., DE SILGUY C., 1996. *Prospectives de la biomasse pour l'énergie dans l'Union Européenne.* Rapport remis à la Commission Européenne. 83 p.

GOSSE G., MAUGUIN PH., 1996. *Prospective for biomass for energy in EU: land availability, land use and agricultural development. Environmental impact of biomass for energy*. The Netherlands (sous presse).

DE SILGUY C., CORTE. P., MERILLOT J.M., 1996. *Biomass research and development strategies.* Rapport remis à la Commission européenne, 123 p.

NON-FOOD PRODUCTION: TECHNOLOGICAL AND SCIENTIFIC BOTTLENECKS

THE ROLE OF BIOENERGY IN EUROPE'S ENERGY STRATEGIES

WOLFGANG PALZ
European Commission - DG XII
200, rue de la Loi
B - 1049 Brussels, Belgium

ABSTRACT

Bioenergy is an essential part of current and future markets for non-food commodities produced in Europe. These products will have a role to play in any global policy for sustainable development and Europe's contribution to it.

Today, the exploitation of forestry and non-food agriculture for bioenergy, at least as far as "Common Agricultural Policy CAP" is concerned, is limited to approximately 1 Mio hectares. But it could grow to 40 Mio hectares and a yearly production of at least 200 Mio toe, *i.e.* appro. 15% of Europe's energy consumption.

In the last few years much progress has been made to stimulate the non-food sector. The European Commission has provided support to non-food projects within its energy development programmes Joule and Thermie and within its "agroindustry" programmes Air and Fair. Consequently, a new range of high-efficiency energy crops and appropriate implementation concepts have been developed. It is important to emphasise the success of this European-wide effort: from the development point of view a great wealth of different biomass crops are now ready for planning and starting large-scale implementation.

As far as economics are concerned, there is a considerable difference between markets. Solid biomass such as forestry or cellulosic agricultural crops are in close-to-cost competitiveness with conventional energies such as coal, in particular when they are in used in modern cogeneration plants. Liquid biofuels such as oils or alcohols are less competitive but immediately available for fast implementation. There are also strategic considerations to push transport fuels in some countries such as France, for instance the need to support traditional agriculture and to fight excessive dependence on oil imports.

As far as incentives are concerned, the Commission has failed to reach agreement in the Council of Ministers for several wide reaching initiatives for fiscal support of biomass or renewable energies in general, *e.g.* the energy tax or the Scrivener initiative for petrol blends. But while currently the sector relies rather more on national support in many member countries (France, Sweden, Austria, etc), the Commission continues to review all possible options on the European Union level.

INTRODUCTION

The development of biomass - a large indigenous resource - relates to several important societal policies such as agriculture, rural development, job creation industry and energy supply. Hence it is part of the broad endeavour for sustainable development. Environment, and in particular global change, will become an ever increasing driving force for biomass utilisation in Europe and worldwide. Indeed, the attempts to mitigate greenhouse gas emissions, which are currently moving high up on the political agenda of the world community, have to include biomass development as a relevant option.

It is important to note that biomass is deeply integrated in agriculture and forestry. The production of food stuff and animal feed as well as timber and pulp is impossible without causing similar amounts of residues and waste *i.e.* biomass. However, in a sustainable system, all biological material produced must be recycled in a closed loop: eventually all this material created from CO_2 and water under the sun's action must decompose completely into CO_2 and water again. As far as food and feed stuff is concerned, the decomposition is operated in chains which rely by and large on chemical and bacterial action. When the bacterial decomposition takes place anaerobically instead of aerobically, *i.e.* in the absence of oxygen, methane is produced, a dangerous greenhouse gas. But besides methane, the digestion and recycling of all biological material involves many problems. Traces of minerals and other nutrients which are essential for plant growth (in general this includes also the pesticides), must be recycled. During growth, most plants accumulate heavy metals (some chlorine etc) and during digestion, such pollutants are also concentrated so that biological waste, sludge, etc. cannot (always) be recycled easily. In any case, decomposition of biological material is exothermal, and in a sustainable system a maximum of the energy becoming available should be employed as useful energy for man. This is the global concept of biomass.

The challenge is that in Europe only a minor part of biological by products and residues are properly recycled, leaving alone their use as crude energy material. The potential of non-food crops for energy and industrial raw material is exploited even less today. This potential is huge. A biomass policy for the future will be necessary which copes with the complexity of the different biomass routes, overcomes macro-economic, economic and institutional barriers and gives Europe's biomass a firm place, too, in international trade and the yet to-be-defined agreement on global change.

THE CASE FOR BIOMASS / MACRO ECONOMIC ASPECTS

Global change

In 1990, the European Union countries (EU 15) released 3.500 Mio t of CO_2 or 840 Mio t of carbon. This was approx. 13% of total anthropogenic emissions in that year. World emission per capita is approx. 1 t of carbon while for the EU it is 2.3 t of carbon, *i.e.* more than double the average.

When biomass is used in a sustainable way, CO_2 emission is zero.

Environment

The EU emitted in 1990 also 16 Mio t of SO_2, 15 Mio t of NOx, 5 Mio t of methane, 0.15 Mio t of N_2O, a very strong greenhouse gas and others. Unlike most fossil fuels, biomass contains virtually no sulphur which gives it an important benefit, also in terms of cost, as sulphur or SO_2 must be removed from the fossil fuel combustion process.

However biomass utilisation may involve various other emissions such as alkali metals, NOx, chlorine, CO, aldehydes, etc.

Agricultural policy

The Common Agricultural Policy (CAP) of the EU includes no specific element on biomass. Nevertheless, biomass growth is supported by the set-aside scheme. The area-related subsidy is fully paid if the set-aside land is used for non-food crops.

Geopolitics

As an indigenous resource, biomass utilisation displaces imports and reduces dependence, in particular from politically unstable regions. It increases reliability as it contributes to diversification of supply. Power production from biomass fits well with cogeneration, hence leads to larger decentralisation and smaller power plants; this in turn alleviates the siting problem for new plants which has increasingly become a major barrier for the large conventional plants in Europe. Also worth mentioning is the potential of biomass for international trade and export.

Employment policy

Currently the EU counts approx. 18 million unemployed people at a cost of more than 200 billion ECU, not mentioning the social misery involved. New jobs can be created by possibly covering the cost of new-job stimulation in a to-be-created biomass market from the social security budgets.

Under realistic scenarios for the EU more than 1 million new jobs could be created in the biomass sector by 2020 - if a sustainable development policy were implemented now.

Rural development and industrial policy

It is important to realise that 75% of the new biomass employments would be created in rural areas where traditional activities are continuously declining.

As a considerable part of the new biomass activities will be rooted in small- and medium-size enterprises (SME), such development is consistent with the drive for larger decentralisation and rural support.

For example, Austria has in the last few years displaced 800 Mio ECU/y worth of fuel imports and created 60 SMEs through biomass deployment for heating.

Macro-economic accounting, external costs

All the macro-economic benefits eventually boil down to potential financial benefits which in a sustainable and rational national economy must be reflected in actual market prices. Here are just two examples:

In France, the macro-economic benefits of rapeseed ester for diesel fuel in terms of national balance of payments and job creation have been estimated at US$ 0.11/litre.

In Spain, macroeconomic benefits have been calculated in the hypothesis of a 20 MW power plant employing the energy crop cynara. Table 1 gives an overview.

Table 1- Macro-economic benefits of a 20 MWe plant in Spain with local biomass crop Cynara

	ECU cents/kWh
Increase of GDP	1-2
Avoided emissions health costs	2-4
Avoided CO_2	0.7
Avoided soil erosion	0.7
Job creation	0.5
New tax income	0.2
	5.1 - 8.1

BIOMASS AND BIOENERGY: CURRENT RESOURCES

As mentioned before, all biological material will ultimately decompose into CO_2 and water: food and animal feed, saw wood and pulp will eventually be turned into biomass which represents the last step in the conversion chain. It is the integration of biomass in the many facets of biological synthesis and use of matter which gives a sense of the relevance, but also the complexity of the biomass concept. In the following, an overview is given of some important biological growth sectors of today.

Table 2 starts with the forestry sector of the European Union. The roundwood figure is for stems without bark. Half of this harvest is for pulp and paper for which the net production is 90 Mio t; hence 30 Mio t are residues from the processing and a lot of this is used today for internal energy supply of the pulping plants. The other half of the crude roundwood harvest is sawlogs; in terms of energy and environment the use of timber for construction, and in particular for private residences is actually very desirable because its energy content is much lower than that of concrete or steel. Hence there is scope for increase in this market sector.

Half of the harvest comes as bark, branches and roots and is left as residues. The Table 2 shows that a substantial amount of these residues as well as old furniture, demolition wood etc. is already used today as wood fuel. It is important to realise that the latter has to be used with precaution as it may be polluted from paint etc.

Table 2- Important biological productions and their link with biomass. EU 15 for 1992/93
(Source : Eurostat, Statistical Yearbook Agriculture, 1995, and other sources)

	Total production / year		Biomass / resp Non Food Use	
Forestry	242 Mio m^3			
Roundwood		75 Mio toe	X---------------	
Residues		75 Mio toe	X-----------	
not harvested		30 Mio toe		
Derived bioenergy				40 Mio toe
Agriculture				
Cereals	175 Mio t			
animal feed		84 Mio t		
human consumption		40 Mio t		
export		39 Mio t	X	
non food		12 Mio t	X	
straw residues	150 Mio t			
Oil seeds				
production		23 Mio t		
imports		20 Mio t		
animal feed				
biodiesel, esters			X	
Sugar	17 Mio t			
non food		0.5 Mio t	X	
alcohols		0.3 Mio m^3	X	1.5 Mio m^3
ETBE		0.075 Mio t	X	1.5 Mio m^3
agroindustrial residues		100 Mio t	X	
Wine	19 Mio m^3			
Potatoes, veget, fruits	117 Mio t		X---------	
non food (starch)		6 Mio t	X--------------	
Various animal feed (fodder, beets, green, maize, etc.)	165 Mio t			
Livestock				
Cows, pigs	200 Mio (manure crude 44 Mio toe)			
Sewage of human origine	(crude 15 Mio toe)			

Table 2- Important biological productions and their link with biomass. EU 15 for 1992/93
(Source : Eurostat, Statistical Yearbook Agriculture, 1995, and other sources)

	Total production / year		Biomass / resp Non Food Use	
Forestry	242 Mio m³			
Roundwood		75 Mio toe	X---------------	
Residues		75 Mio toe	X-----------	
not harvested		30 Mio toe		
Derived bioenergy				40 Mio toe
Agriculture				
Cereals	175 Mio t			
animal feed		84 Mio t		
human consumption		40 Mio t		
export		39 Mio t	X	
non food		12 Mio t	X	
straw residues	150 Mio t			
Oil seeds				
production		23 Mio t		
imports		20 Mio t		
animal feed				
biodiesel, esters			X	
Sugar	17 Mio t			
non food		0.5 Mio t	X	
alcohols		0.3 Mio m³	X	1.5 Mio m³
ETBE		0.075 Mio t	X	1.5 Mio m³
agroindustrial residues		100 Mio t	X	
Wine	19 Mio m³			
Potatoes, veget, fruits	117 Mio t		X---------	
non food (starch)		6 Mio t	X--------------	
Various animal feed (fodder, beets, green, maize, etc.)	165 Mio t			
Livestock				
Cows, pigs	200 Mio (manure crude 44 Mio toe)			
Sewage of human origine	(crude 15 Mio toe)			

Twenty to 30% of the yearly forestry growth in the EU 15 is not harvested. Hence we have a CO_2 sequestration effect in our national forestry. In the longer term, however, a sustainable forestry would mean that the whole yearly increment is actually recycled somehow.

As far as agriculture is concerned, the biomass issue becomes more complex as it implies three distinct levels. One is the non-food use of agricultural commodities such as starch, sugar or oil. A second one concerns the residues arising in the field during harvest and during processing in agroindustry. The last one is linked to the waste problems after human consumption of the food directly or as meat after the detour of agricultural products through the livestock chain.

On the first level the most important sector is the non-food use of starch. In 1994 6.2 Mio t of starch were used in industry: 60% for paper and paper boards, 30% for fermentation in chemistry. In a broader sense, when including also oil seeds etc, the chemical industry in Germany, for instance, is currently relying for 10% of its feedstock supply on agricultural products which are grown on more than 4% of the agricultural land in that country. The 6.2 Mio t of starch in the EU are produced from 8.5 Mio t of cereals and 6 Mio t of potatoes. This non-food sector represents a yearly business of 3.7 Billion ECU with an employment of 15,000 people.

Also in the non-food sector there is the use of alcohol. In 1992 the countries of EU 15 produced 14.7 Mio hl of pure alcohol for industrial use. Most of it was distilled from wine and potatoes as a regulating measure for the wine and potato market. However, these feedstocks are much too expensive for industrial alcohol production and this use is not sustainable. The cheapest route to produce bioalcohol is via starch and sugar beets. 3 Mio hl were produced from sugar beets, mostly in France, and part of it was blended into 0.75 Mio hl of ETBE, a transport fuel. The yearly cash flow in the bioalcohol sector is 150 Mio ECU.

Lastly there is non-food use of biooils. This sector has recently attracted much public interest although it is currently not the biggest non-food market. Industrial use of biooil is widespread, for instance for lubricating, but the largest market is for biodiesel. The European market is fluctuating strongly and is currently decreasing (+/- 1 Mio t in 1995). It is largely dependent on the set-aside regulation in the EU and the oil-seed trade with the US (Blairhouse agreement). The yearly turnover in the EU today is approximately 400 Mio ECU.

Below non-food, the second level is agricultural residues. Hundreds of millions of tonnes of residues are arising each year but there is no comprehensive accounting. Almost half of the total cereal harvest is straw. Generally, it is no longer allowed in the EU to burn the straw in the field. Some is used in husbandry and cattle breeding, but a lot remains as residue. In a sustainable system, all must be recycled: there are opportunities for industrial use and much attention is given to combustion in power plants. Denmark is a leader in this use. There are environmental hurdles for clean combustion of straw as it has a high chlorine content. Eventually residues in the cereal industry include, besides straw, chaff which comes in equal huge amounts as the ultimate product starch.

In the field of oilseeds the situation is somewhat simpler as much of the residues is used as by-products to serve as animal feed.

Large quantities of wet biomass residues are arising from sugar beets, potatoes, vegetables etc. They may degrade to compost which is a sustainable way of recycling. But when they degrade anaerobically they produce, naturally, methane which must be collected and burnt. Today much of it goes uncontrolled into the air as a big pollutant. The collection of land fill gas is making some progress and purpose-built digesters are coming on stream.

The other way of recycling is combustion of agro-industrial and municipal waste which also raises environmental concerns. The EU produces 370 kg of rubbish per inhabitant, much of it as biological waste stemming from the food consumption discussed here above.

In summary, the recycling of wet agricultural wastes and residues has to deal with huge amounts of matter and is currently not properly solved. It is a difficult environmental problem, not least because of the content of chlorine, trace metals and other pollutants.

Lastly, the bottom of the agricultural barrel are excreta. The EU counts 200 million cows and pigs and more than 3 Billion poultry. Most of their excreta can be collected. The crude energy content per kg of dry organic matter is 20 MJ. For cows and pigs the total energy content of excreta per year is 44 Mio toe; part of it could be collected as methane. The excreta from poultry have a smaller total volume but also represent a big environmental problem. The EU has also 110 Million sheep and goats, but in general they always graze outside and their excreta would be difficult to collect.

Ultimately, one has to count here the rest of the food consumption of 370 Million EU inhabitants. The amount of energy in the sewage to be possibly collected by methane or combustion is up to 15 Mio toe.

CURRENT MARKETS AND SUPPORT SCHEMES FOR BIOENERGY

Bioenergy markets

In the EU 15 biomass currently contributes 3% to gross energy consumption, *i.e.* approx. 44 Mio toe. Most of it is wood fuel and straw, which is used for heating. The dominant market is domestic heating but an increasing number of district heating plants are coming on stream: Sweden consumes 1.5 Mio toe that way, Austria has several hundred units, Denmark and Bavaria have a hundred units each. Electricity is produced in a total capacity of 1.5 GW by electric utilities mainly in co-combustion with coal. More than 20 of the plants use grate combustion and more than 15 plants employ modern fluidised-bed boilers. Combined heat and power plants employing biomass are not used on any significant level.

As mentioned before, fuel liquids *i.e.* bioalcohols and biooils are introduced on a modest scale of less than 1 Mio t per year in the transport market. Lead countries in the EU are France, Germany and Italy.

The three new member countries of the EU Austria, Finland and Sweden which already rely on biomass for approximately 13% of their energy supply have a big impact on the EU's biomass accounts.

When they joined, the biomass contribution in the EU jumped from 2% to 3% and the overall forest resource doubled. Table 3 provides a closer look at current biomass prices in some Member countries.

Table 3- Prices and Markets (examples)

Solid residues	S	Unrefined wood 35 ECU/t	Upgraded wood fuels 80 ECU/t market 1994 : 0.2 Mio toe market future: 2 Mio toe
	DK	Straw 30 ECU/t market 2000 : 1.2 Miot/year in utilities	
	Fi	Chips form logging residues 25 ECU/t transported 100 km	
Solid crops	S	Now 18.000 ha of willow	
	D	70 ECU/t hay, cereals	
	F	60 ECU/t pulpwood stems	
	UK	0.10 ECU/kWhe willow, poplar (NFFO3)	
Oils, Alcohols	F	\$0.5 litre rape seed oil ester 300.000 m^3 \$0.54 litre ethanol, ETBE 75.000 m^3	
Methane	DK	0.30 ECU/m^3 (6 ECU/t of feed-stock) in large plants	

For an estimated average price of 100 ECU/toe, solid biomass utilisation in the EU raises a turnover of 4 Billion ECU per year. To this one can add 0.55 Billion ECU for liquid fuels.

It is estimated that currently 110,000 people are directly employed in the EU bioenergy sector; 100,000 are involved in the production and use of solid fuels while 10,000 work in agriculture and agroindustry for bioalcohols and biooils. Additionally, an equal amount of indirect employments are linked to European biomass.

Biomass support schemes

In the frame of the **Common Agricultural Policy** CAP, as a matter of principle, non-food feedstocks are supported in such a way that their price in the internal market comes close to that of the world market. This may not always be achieved. In 1995 the EU production fund, for instance for maize, was approx. 35 ECU/t but came 20 ECU/t short of the level which would have been necessary to get down to the world market price of 105 ECU/t.

For wheat, markets were very volatile in 1996. Earlier in the year when a shortage of harvest was anticipated and world-market prices jumped up, the Commission imposed an export tax on wheat, but later on, when the harvest turned out to be very good, CAP introduced again an export refund.

Within the set-aside scheme in 1995/96, 6.4 Mio ha *i.e.* 8.5% of the arable land were supported. The financial support per ha depends on the region and can vary strongly from country to country. It is calculated as 68 ECU per tonne per average productivity of wheat. If the average productivity is, for example, 6.5 t/ha, the area support becomes 440 ECU/ha. This refund is also paid for example when non-food crops are grown on the land no matter whether it is wheat, rape seed or short-rotation forestry. In 1996/97 the set-aside scheme will be applicable to only 5% of the agricultural land. Half of the financial support is paid from the Commission's budget, the other half from national budgets.

For forestry, the financial support depends on whether it concerns short-rotation forestry or long-term plantations of 20 years or more. For the latter, all planting and growing costs of approximately 5000 ECU per ha for 5 years may be borne by the CAP not including the fences (which may actually double that cost). In Germany, the support given up to 20 years for new forestry has reached 700 ECU per hectare and year. For short-rotation forestry the support is different and depends also on the country. For instance, in the United Kingdom the total support for 5 years is currently 2,000 ECU/ha on top of which a one-time woodland grant of 500 ECU/ha can be obtained.

On the **market side** there is an additional portfolio of financial support. Here are some more examples: for biofuels, France has a regulation which waives the excise duty. In Denmark biogas plants get a grant of 20% to 40% of the investment plus a special kWh price per electricity produced. In the UK, "landfill gas" benefits considerably from the Non Fossile Fuel Obligation. Germany and other countries have grid buy tariffs of approx. 0.07 ECU/kWh for bioelectricity. Austria and Bavaria offer an investment subsidy of up to 50% and reduced discount rates for biomass-fed heating plants. In Finland, Sweden, Denmark and the Netherlands biomass benefits from national energy taxes. Biomass is excluded from that tax, which amounts for instance for coal in Sweden, to 150 ECU/t (energy tax + CO_2 tax + sulphur tax) and makes biomass a clear market winner.

In essence, biomass support - where it exists - is essential to give bioenergy a leading edge in the market: biooils from rapeseed and sunflowers are competitive in France because they are exempted from the excise duty: before taxes, diesel costs 0.13 ECU/l as biodiesel production cost is 0.4 ECU/l. Production costs for bioalcohol in France are even slightly higher (0.43 ECU/l). The set-aside restitution comes as an extra benefit. For a typical production of 1.2 t/ha of colza oil and a set- aside support of 500 ECU/ha the farmer receives approx. 0.4 ECU per litre. One must recognise that this is not a real extra income for the farmer since he also receives the support if he grows nothing at all.

Similarly, biomass utilisation for heating is competitive in Austria, Sweden etc.: In Sweden house owners save 1,000 ECU per year when they switch to wood pellets (which cost 80 to 100 ECU/t).

As a conclusion, it is important to note that many types of support schemes exist in the EU which differ widely from country to country and even between regions. However, we have no single European-wide support scheme (not considering set-aside which is also dependent on the region) such as an EU energy tax, grid tariffs for bioelectricity, NFFO, reduced discount rates, tax credits, tax exemptions for transport biofuels, grants, etc.

FUTURE SCENARIOS FOR BIOENERGY USE

The European Commission has developed a number of energy scenarios for the future in which bioenergy is firmly anchored ("European Energy to 2020, A Scenario Approach", published by European Commission's DG XVII in Brussels, Spring 1996). For various assumptions, 4 projections have been made for the energy consumption in the EU 15 by the year 2020. We shall, in the following, consider only the "reference scenario" (conventional wisdom) and the one which foresees the highest consumption rate of biomass ("forum").

Table 4 summarises the scenarios with an emphasis on the biomass issue. It is important to note that all projections agree on a massive renewal of the power generation capacity in the EU up to a total of 500 GigaWatt. Because there will be many new plant constructions there are also opportunities for new bioelectricity inclusion. Also of interest for biomass is the anticipation of a larger share of co-generation (CHP). Such plants are particularly suitable for the combustion of energy crops which can not be transported over long distances for reasons of logistics.

Table 4- Future Energy Scenarios for the EU 15
(Source: EC DG XVII, European Energy to 2020, Spring 1996)

	Conventional Wisdom*	**Forum****
Gross Inland Consumption (GIC) 1995	***1400 Mtoe***	
BIOMASS	44 M*toe*	
Projected increase of GIC by 2020	+ 24%	+ *21%*
BIOMASS by 2020	113 Mtoe (+ 180%)	160 Mtoe (+ 300%)
Electric Generation Capacity (EGC) 1995	**550 GWe**	
BIOMASS	*1.5 GWe*	
Projected increase of EGC by 2020	+ 24%	+ 33%
New plants to be built(cost without network by 2020)	450 Gwe (540 Billion ECU)	500 GWe
of which new CHP plants	48 GWe	56 GWe
Share of auto producers (Increase from)	10% to 21%	to 25%

* CO_2 up 14%
**CO_2 down 11%. New taxation for end user/power generator. Promotion of carbon-free fuels.

The possible share of biofuels in the transport sector appears in Table 5. A share of 20% of this market within the next 25 years would be tremendous: but it is a practical possibility. There is a very high potential for new employments associated with an increased utilisation of bioenergy. The following estimates are based on more detailed investigations by G. Grassi ("Potential Impact of Bioenergy Activity on Employment". Proceedings of the 9th European Bioenergy Conference, Copenhagen, 1996). Implementation of the Forum Scenario of Table 5 would lead to 1.1 Million direct and indirect jobs: approx. 200,000 for heating, approx. 200,000 for bioelectricity and more than 700,000 for liquid fuels, in particular for transport.

Table 5- Biomass Market Projections by 2020
(Source: EC DG XVII, European Energy to 2020, Spring 1996)

	Conv. Wisdom	Forum
Direct combustion	40 Mtoe	43 Mtoe
Electricity Generation	30 Mtoe	48 Mtoe
Biofuels Production	43 Mtoe (= 12% of transport fuels)	72 Mtoe (= 20% of transport fuels)

Table 6 goes one step further towards the upper limits of bioenergy which could eventually be mobilised. They are based on the author's own estimate of residues which were analysed here above and the review of energy crops and their potential shown in the following section. The Table includes also an example of an extreme or optimistic scenario. Obviously, many alternate scenarios are possible, and the only meaning of Table 6 is to give an impression of where the limits are.

Correspondingly a more speculative survey is given in Table 7. A maximum share of biomass is introduced in this projection which is much more long range. It might become relevant eventually when maximising the use of renewables and minimising energy consumption are a priority in view of mitigating global change. This might become an ever more burning issue in the next century.

Table 6- The Bioenergy Resource in the EU 15 (derived from Table 2 and Table 8)

Maximum Amounts of Residues and Wastes	
Wood, timber etc.	180 Mio toe
Straw	60 Mio toe
Res. from fruits, vegetables, etc.	50 Mio toe
Res. oils	10 Mio toe
Manure	65 Mio toe
Sewage	15 Mio toe
Total	380 Mio toe
Basic Scenario	
50% Energy Use of Res+Waste	190 Mio toe
400 Mio t* of dry biomass = equivalent	170 Mio toe
Total	360 Mio toe

Ultimate Resource Limit - a Possible Scenario		
	66% Energy Use of Res+Waste	260 Mio toe
	800 Mio t** of dry biomass = equivalent	340 Mio toe
	Total	600 Mio toe
*	produced on 40 Mio ha or 15% of total agr./pastures/woodland at average productivity of 10 t/ha	
**	produced on 53 Mio ha or 20% of total agr./pastures/woodland at average productivity of 15 t/ha	

Table 7- Optimistic Energy Scenario for 2050

Primary Energy Consumption Cut by 30%	1000 Mtoe
Renewable Energies Supply 80% of Consumption*	800 Mtoe
Bioenergy Supply 75% of Renewables	600 Mtoe
* in order of importance besides Biomass Building Application of Solar Energy Wind Hydro PV Ocean Energy, Solar Thermal Power, etc.	

SCOPE FOR TECHNOLOGICAL AND SCIENTIFIC IMPROVEMENTS THROUGH R+D

It must be emphasised that all products and schemes for biomass utilisation are available today. R+D should not be seen as an excuse not to enter into major new applications and initiatives for their implementation right now.

But like all established economic sectors -not forgetting that biomass is the oldest of them all- biomass, and in particular bioenergy, offer much scope for improvement through R+D. Priority targets for R+D are biomass crops (in terms of productivity, biodiversity, stress resistance, accumulation of pollution elements and general economics) and improvements of biomass conversion and utilisation (in terms of environmental effects and economics).

Biooils

To achieve better economics of biooils, the productivity of rape seed and sunflowers, the main European crops, which ranges currently between 1 and 2 tonnes per ha and year should be increased through plant breeding and genetic engineering. This is quite a challenge since Europe's climate is not optimal for the growth of oil seeds as it is more suitable for cereals. It is not without significance that the self sufficiency of the EU for vegetable oils is only approximately 50% and half of the oil consumption for foodstuffs is traditionally imported. In this situation, it is encouraging that INRA in France has recently announced a new hybrid rapeseed with a 10% higher productivity.

A second item of investigation is environmental properties. An important issue which also needs further clarification is pollution through fertilisers, nitrates, nitrogen protoxyde etc. in the field. Biooils are environmentally friendly as they have low toxicity and fast biodegradability; sewage treatment costs are reduced when collecting waste oils and fats at the point of households, restaurants and industry.

Recently some arguments have arisen from different interested quarters concerning the combustion properties of biooils as a transport fuel and in stationary engines. In France, Académie des Sciences has issued in 1996 a rather critical report on biooils and bioalcohols insisting on the emissions of aldehydes which are carcinogenic. The fact which is recognised however, is that biooil has no CO_2 and no acid emissions (since they contain no sulphur); they emit less particulates than conventional diesel oils while the CO_2 emission is the same and the emission of NOx and with it ozone are slightly higher. On the critical issue of formaldehyde, the defenders of biodiesel argue that the emissions are so low that they remain well inside current environmental regulations.

Bioalcohols

As mentioned before the productivity of cereals, one of the potential feedstocks, is already close to optimal. Nevertheless, bioalcohols in Europe produced from starch or sugar are slightly more expensive than biooils; in Brazil and the US they are cheaper. Hence the economic problem in Europe could partly rely on lack of implementation and experience.

The preferred feedstock in Europe is sugar beet which has a strongly regulated market in the CAP. The search for alternative crops is a promising new sector. Sweet sorghum has attracted most attention in Europe as trial plants have been established for many years all across Europe. Sugar productivity is similar to that of sugar beets. An important additional economic bonus is the associated bagass. Sugar from sorghum could also be introduced at least partly on the sugar market to improve economics. But more R+D is needed for the crystallisation of these fructose sugars (16 different types) which are different from conventional sucrose sugar produced from beet or cane.

As to combustion of bioalcohols attention in Europe has been focusing on ETBE, an ether produced through blending of ethanol with isobuthylene, a refinery product. Like ethanol, ETBE is an oxygenate for petrol but unlike the former it has good blending properties with petrol. ETBE also raises concerns about aldehydes emissions which should be mitigated through proper R+D. It seems anyway that the aldehyde emissions of ETBE combustion are lower than those of MTBE, the non-biological oxygenate which is preferred today by refiners.

Biogas

There are still many improvements under investigation. An important issue is the elimination of pathogenic bacteria and animal viruses. Thermophilic anaerobical processes are necessary to cope with the problem or special sanitation tanks. Residues must be qualified as to whether they are clean enough to be used as fertilisers in the soil. This also requires appropriate regulations.

Crops for cellulosic material

Table 8 is an overview of best productivities achieved in Europe today. For short-rotation forestry with eucalyptus, poplar and salix or willow there exists large-scale experience but for most of the other plants, field trials were relatively restricted and should be further developed urgently.

Important criteria are minimum need for water and fertilisers - for sorghum they are much smaller than for conventional C4 plants such as maize - and "natural" defence against diseases such as rust and other pests.

Productivity and agricultural practice *i.e.* the need for machinery, irrigation or pesticides and fertilisers eventually determine the cost of the feedstock. A group of leading European experts under the Agricultural University of Athens is currently developing an economic simulation including land use, man power and all the above mentioned factors (BEAVER programme). In an optimised system production costs of 50 ECU/t (dry matter) for willow and somewhat less for sweet sorghum (sugar included) can thus be demonstrated.

Furthermore, various plants have the property to accumulate heavy metals and other polluting agents. This is well known for algae but it is also the case for trees and most other crops. This offers the possibility to clean polluted soils and may even be applied to pollution by Caesium and other radioactive pollution (*e.g.* the soils polluted through Tchernobyl). To realise this interesting potential of energy plants, a lot more R+D is necessary to develop the concept further.

Table 8- Some Energy Crop Options

		Max productivity t/ha year demonstrated
C3 Plants	Eucalyptus	25 *(depending on field fertility)*
	Poplar	12
	Willow	13
	Robinia	13
	Reed, Canary Grass	8
	Kenaf	15 *(27 when irrigated)*
C4 Plants	Sweet sorghum	22 *(32 when irrigated)* 10 Sugar
	Fibre sorghum	38
	Arundo donax, giant reed	40
	Miscanthus	40 *(very variable)*
	Cynara	35*(need little rainfall : 500 mm/year)* 3 Grain

Combustion

In small wood furnaces for residences or small district heating, CO, NOx and carbohydrate emissions are higher than in oil furnaces, if particular care is not taken. Problems can be solved through a secondary combustion chamber for high-temperature post combustion. NOx can be further reduced through controlled air admission in a special reduction chamber. Flue gas condensation increases efficiency by up to one-third and particle condensation takes care of heavy metals. For larger combustion units, fluid beds have proven most convenient: moisture content is not critical and NOx emissions are reduced. The high chlorine content of straw is a problem however, even more so than in the large plants employing pulverised fuels. Corrosion and fouling problems can be solved by modifying the combustion process.

Gasification and pyrolysis

From a scientific point of view gasification is a sideline of combustion. Integrated-gasification combined-cycle combustion (IGCC) of solid biomass is one of the most popular items in international R+D. The promise of high efficiency in combined gas + steam cycles only holds true in relatively big units of at least 30 MW of electric power. Otherwise the high complexity of the system - pressurised biomass gasification and cleaning, gas cycle + steam cycle - turn the economics prohibitive. A long tradition exists in Europe for small wood gasifiers which were extensively used for transport in the 1930's and 1940's. For power production in the range of some 100 kW this "fixed-bed" technology, which employs solid wood pieces or new fluid-bed gasification, are of interest. Liquid effluents of these gasification systems - the process produces a lot of water - must be treated for environmental reasons.

A lot of progress has recently been made in Europe on pyrolysis. Flash pyrolysis of cellulosic material is being developed in Spain, Italy and Finland and results are encouraging. The solid material can be totally converted into carbohydrates, a heavy fuel oil, at an efficiency of approximately 70%. It could be demonstrated that this oil is suitable for diesel engines and hence offers important new prospects of cheap biomass liquids for transport.

IMPLEMENTATION ISSUES

Table 9 summarises some of the important issues to be taken into account for bioenergy implementation. The biomass chains are very complex and it is necessary to establish some clear implementation scenarios for the various routes.

Today, two sectors are the most advanced: small-scale heating and district-heating systems employing forest residues and various other wood residues and straw; power generation in large power plants employing residues in cocombustion with coal. Problems which have not yet been totally overcome concern, in particular, the pollutants contained in the residues. For the power production of the future some additional options are around the corner: CHP employing medium-size fluid bed combustion, typically in the 10 MW size, and gasification in the range below 1 MW; a wider use of energy crops.

Table 9- Implementation Issues

- Combustion or cocombustion of straw, etc. has to accommodate corrosion, fouling problems (Chlorine, low melting ashes, fly ashes)
- Pretreatment of biomass to achieve a standard product is not imperative but can be important to solve many problems *e.g.*
* Corrosion, fouling, ash, problems *e.g.* with straw
* Forestry and in particular some C4 crops accumulate lead and other heavy metals which must be fractionated
* MSW and namely wood from building demolitions is a dirty fuel
* Humidity must be controlled
- Emissions and effluents from combustion must be controlled for NOx, CO, aldehydes, heavy metals (spraying water in fuel gases, electric filters, etc)
- Ash should be recycled for nutrients in the fields but needs pretreatment to allow leaching. Its very high pH neutralizes nicely the grounds affected by acid rain.
- CHP is an attractive option for solid biomass utilisation in particular for energy crops. CHP plants next to the field avoid long transport. For a 10 MWe plant, transport is already 10 to 15 km.
- Rust and other diseases of energy crops can be alleviated by mixtures of clones and biodiversity.

Besides heat and power, transport is a target sector of high priority for biomass. Bioalcohols should be further developed, but newcomers such as cheap pyrolytic diesel oil offer interesting new perspectives.

STRATEGIES FOR THE FUTURE

Table 10 lists some possible initiatives which may be most appropriate for further increasing biomass utilisation in the European Union. Needless to say these measures could not be implemented without the leading players in the energy sector and in particular :

- oil companies, refineries, etc;
- power utilities, municipal works, CHP operators, district heating;
- industry, SME's, power engineering, pulp and paper, forestry, car industry;
- farming communities;
- responsible organisations for norms and standards;
- research institutes.

Table 10- Promotional Needs for the Future

- Network of promotional centers to initiate projects and provide support for implementation. Their importance was already shown in some examples for biogas, district heating etc. Local politics, industry, agriculture and energy groups may take part in these centers.
- European norms and certification for conversion plants, plant siting and construction, biomass feedstocks and emissions.
- Support of energy crops in the EU agricultural policy is useful and should be continued, but for a limited time.
- Countries with an energy tax have already an inherent system for biomass promotion which works well. Other systems which proved successful and may lead the way in other countries and regions are *e.g.*

* Investment grants. Fixed-amount grants are preferable to percentage support as it stimulates competition among suppliers. Grants should cover part of the extra cost of a standard and approved biomass system compared to a system employing fossil energy

* Part of the support should be given per kWh delivered to motivate proper O+M. Green electricity tariffs come in this category

* Subsidized discounting and soft loans, flexible depreciation, tax deductible investments, green funds, etc. are important instruments for promotion.

- New technological developments *e.g.*
  * Improvement of productivity of energy crops, including those for oils and alcohols
  * Gasification and pyrolysis of solid biomass
  * Dual use of grain and sugar crops for both alcohol and solid feedstock
  * Biomass use in fuel cells

CONCLUSION

Biomass is one of the most promising options among the renewable energies and industrial feedstocks. It will have a major role to play in future to mitigate greenhouse gas emissions and global change. As an indigenous European resource it increases energy security and independence. At the same time it looks most promising to create jobs in industry, forestry and agriculture and to contribute to rural development.

Biomass is also an interesting commodity for international trade and offers itself for international cooperation, for instance in the frame of joint implementation of measures to mitigate global change. In general terms, we can note the following:

- Conventional forestry is a sustainable resource: between 1980 and 1990 it increased (by 16%) in the EU.
- Biomass fits perfectly with the environmental need to recycle residues and waste.
- An important challenge for the future is the development of energy crops on a large scale
- Biomass is not yet fully economical - mostly because it has only a very short development record. Solid-waste combustion in condensing power plants is presently the most economical, bioalcohol and biodiesel are the least economical.
- Other problems concern securing of a stable supply of well defined feedstocks, handling of MSW and other dirty biomass, lack of district heating for CHP plants.
- Some countries have a low biomass potential (*e.g.* NL), others a very high biomass potential.

Besides international cooperation, there is room for intra-European trade.

UNITS

1 Mio toe (Million tones of oil equivalent) =
10 Mio G cal (10 Mio x 10^6 kcal) = 11.62 TWh or (11.62 x 10^9 kWh thermal) = 41.9 PetaJ (10^{15} Joule) = 0.0419 EJ (10^{18} Joule)
1 t dry wood = +/- 0.43 toe   1 straw = +/- 0.4 toe
1 t dry manure = +/- 0.2 toe

1000 m^3 biogas = 0.4-0.7 toe
1000 m^3 producer gas (gasifier) = +/- 0.1 toe
1 t ethanol = 0.65 toe
Roundwood 0.31 toe/m^3 Pellets 0.26 toe/m^3
Chips/grass 0.06 + toe/m^3

FOREST AND AGROFORESTRY

JAN-ERIK HÄLLGREN
SUAS, Dept. of Forest Genetics & Plant Physiology
S-901 83 Umeå, Sweden
and
Swedish Council for Forestry and Agricultural Research
Box 6488
S- 11382 Stockholm, Sweden

ABSTRACT

This paper discusses some of the potentials for a future surplus of wood fiber material. Continuing growth of European forests, and a system of combined forestry and agriculture will allow for both industrial expansion and environmentally friendly development. New techniques and ideas must be exploited in a much more systematic way for possibilities exist to optimise the use of raw material. Much greater diversification of the use of wood fibers and other chemical raw materials from trees and cultivated plants is possible. Forestry and agroforestry are sustainable systems which lend themselves to environmentally friendly production and the recirculation of waste products and material. Current developments show the importance of the production of energy in different forms. The production of new products that can substitute environmentally hazardous products constitute other possibilities. The development in, and competition from the American continent and Japan must be taken into consideration when future possibilities are discussed. The role of the research system in this process is discussed.

INTRODUCTION

Thank you for giving me the possibility to be here and discuss the next fifty years with you. My take home message is that there is a great potential in the future if wood and fibers and other resources from forest and agroforest systems are utilised in a creative way. The forestry sector can be a good example of a future green industry with environment-friendly operations that can supply us with a sustainable production of renewable material, and can even improve the environment. I claim this because I believe that the next fifty years will see the breakthrough for biological techniques, and because I believe that the recognition of environmental problems is going to be strong. It is today obvious for everybody that our common future will need a knowledge-based utilisation of natural and renewable resources. I will, not during the next fourteen minutes, have time to go into details, and I will not therefore address the whole subject properly. Agroforestry, for example will constitute more of the background when plant resources are mentioned.

NATURAL RESOURCES

We do not know what the future will bring. If you don't know what the future is going to look like, you can always take a historical viewpoint. From a relatively short historical viewpoint, you can analyse the basis of the sector in the Figure 1.

Figure 1

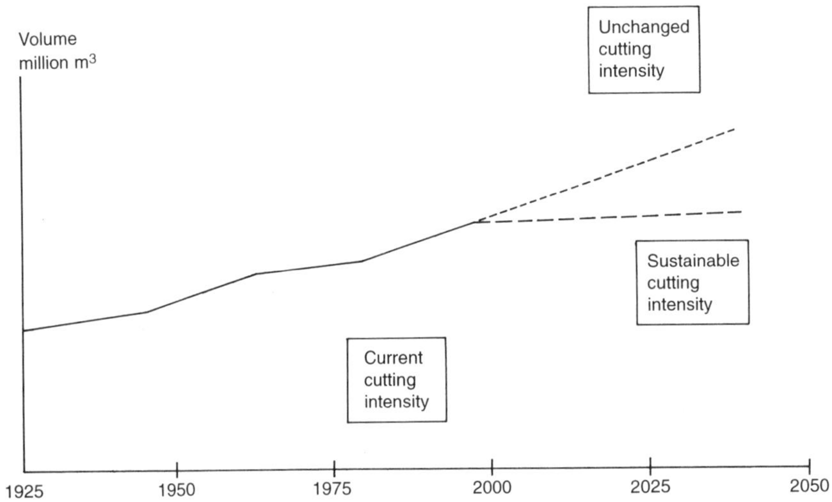

The volume of forest resources will be steadily increasing in the near future in our part of the world. The simplified graph show that if we continue with the cutting intensity that we have today, a much larger volume will be available tomorrow. This means that we will have new opportunities and can fulfil several important goals. We will probably not continue as we do today, but decisions taken today will influence the path that we take in the future. It is therefore important that we recognise the possibilities. With growing environmental awareness, the need to set aside some land for the wild life and for the benefit of biodiversity will be strong. At the same time recreation, tourism, and other interests will need some forest areas and undisturbed landscapes. But even if some land is set aside, the potential still exists for increased utilisation of the renewable materials. The gap between volume growth and current utilisation is so big that we can utilise much more of the logs without disturbing the sustainable growth of the ecosystem. Also the increasing use of recycled fibers will diminish the need for virgin fibers from forests. We also can take into account that reforestation of former farm land will creates a potential for the future. The economic possibilities depend on so many factors that it is impossible to speculate about the future here. We notice that the growing markets in, for example, South-East Asia, cannot be met by much increased utilisation of forest in that part of the world. A better economic future will of course increase the need for paper and wood products.

IN THE BEST OF WORLDS

In my dream about the future we can fulfil both an ecological approach and an industrial approach. These may be apparently contradictory, but in a dream contradictions can become complementary. In this world, forest and agroforest ecosystems are central in environmental management. In the landscape we see the link from forest ecosystems to clean air and drinking water. In the best of worlds, we live in a landscape with forestry and agriculture that increases biodiversity and gives enough refuges to endangered and rare species. In this world we understand how to use the raw material in a much more clever way than today. The farmlands and the forests are sources for production of medicine, pharmaceuticals and cleansing products with environmentally friendly techniques. In the best of worlds, lubricants are made of vegetable oils, and important information can be read on recyclable paper in books and journals produced in closed and energy saving factories. In the best of worlds, building materials are biodegradable and recyclable. In the best of worlds, technology will be used to improve the raw materials, to foster eco-efficiency, non-toxicity and biodegradability.

Unfortunately, we do not live in the best of worlds. We live in a changing environment, and we dream about sustainable development. We have to move step by step. Here is where the role of research has become increasingly important.

TARGETS

Research is one of the keys to enhanced competitiveness as we now experience globalization and internationalisation in this sector and as a whole. Where can research serve as a catalyst? If we look at the size of targeted markets and added value of the products we can visualise this from the following Figure 2.

Figure 2 (After J.Bergans and J.Comet, 1996)

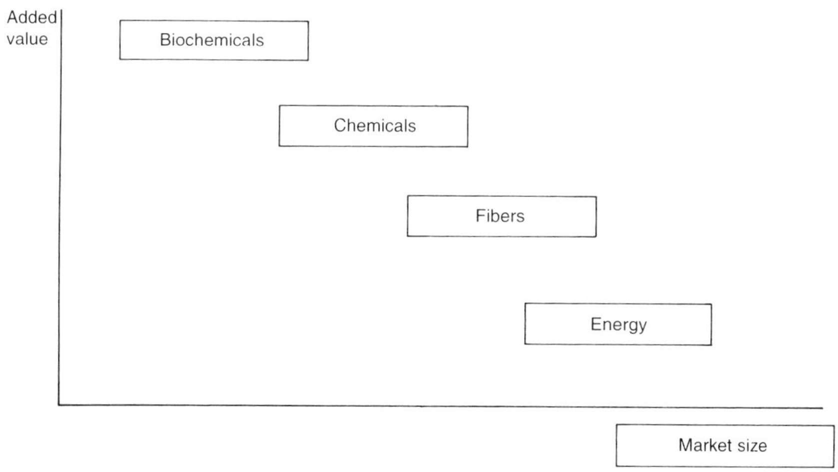

ENERGY

The energy sector is probably the most targeted market at a worldwide level. The decrease of the stock of fossil resources, the threats of planet warning, the pollution, especially from traffic in large cities, and the new markets for products coming from renewable resources are important driving forces. Pro-alcohol programs, clean-air acts, and increasing interest in using alternative fuel sources for cars will probably have a tremendous impact in coming years. So far fossil nuclear energies are still priviliged in political choices, but there are also opposite tendencies. However, a pro-alcohol program need political decisions. The advantages of the energy sector is its size, but there are some problems. You have already heard a lot about energy, so I will only take up a few supporting examples.

If we look at research in this area, it is obvious that there are great potentials. As I have understood it, we are close to a break-through in producing alcohol much cheaper than before. The raw material can be a mixture of wood chips and cheap waste from agriculture. Today the material can be fermented much more efficiently than in the past. Already by using commercial yeast, 30% of the energy in the raw material can be fermented. Increasing the temperature will speed up the process. The efficiency can further be improved by using traditional molecular biology methods. High temperature stable gene products from micro-organisms will further increase the speed of fermentation. Tomorrow we will have solved the problem of breaking down five carbon sugars, and the production of alcohol will be even more efficient and probably cheaper. Laboratory techniques today will emerge into demonstration factories and producing units tomorrow.

At the same time we can observe that there exist several initiatives to start alcohol production and biofuels in many places in Europe today. The only thing that is lacking is the political start signal. In this area we need political decisions, when this happens, and I am convinced that it will, we will see other actors on the scene. I am convinced that it will happen since politicians, as well as the car industry and other industries have to support this technological change to avoid losing credibility in the environmental area. The EU program already supports a lot of projects focused on energy *e.g.* the AIR and FAIR programs. These will eventually give results. You heard about them earlier today.

Wood has since the invention of fire been converted to heat. It is obvious that heat and electricity can be produced simultaneously and that combustion techniques will be more efficient in the near future. Innovative ideas such as bacterial celluloses will enable improvement of biogas generation from waste products in forestry and agriculture. Composting and biogasification have a future since their cost at the moment, is two to three times less than that of recycling the same materials.

TERMINOLOGY

There is a lot to be said about energy, but since this will also be dealt with by others, I will stop here. It should, finally, be noted that biomass-bioenergy are positive concepts, linked to renewable resources and environmental improvement. Bioenergy is a term that is easily understood and accepted by a large public and by a number of possible customers. At least I want to believe that this is of importance.

In many cases the use of the correct word is of importance. To my mind one should avoid terms like non-food and other negative concepts and, instead, try to find the most appropriate and, hopefully, also positive terminology for the products and processes that we want to foster. Competition should be initiated to find a replacement for the negative non-food concept when we mean new bio-products.

I will not have time to go into detail in the other sectors where the added value is higher, but will say a few words about fibers, chemicals and biochemicals.

FIBERS

Fibers have been discussed a lot lately, not only in the traditional sector of paper making. The traditional fiber industry will continue to polish their "diamond" which is to produce higher quality paper. If we look at other sectors, bio-fibers have generated a lot of interest, I will come back to this

CHEMICALS AND BIOCHEMICALS

The petrochemical industry has made extremely low-priced raw materials and semi-products available to the manufacturing industry in the past. Chemical products formally used to be large scale and manufactured through biomass. These products are today produced within the petrochemical industry at a much lower cost. However, forestry and agriculture play an important role as business partners and suppliers of raw material to the food industry. There is also a certain interest to develop renewable and/or recyclable products for food industry. Here the agricultural and forestry sector can provide some products of interest. One particular raw material has to be mentioned, and that is starch and starch derivatives. Also here biotechnological methods and new plant sources change the scene.

Vegetable oils with different qualities also constitute an area of considerable interest, outside the food-sector. The quality of fatty acids and oil can be further improved both directly in the plants and also chemically. The production of lubricants with different qualities has already taken its first steps into the market.

The pharmaceutical industries have the skills needed to use plant material for their own purposes. The same is true for producers of polymers, plastics, ink, and oil, etc.

Problems are important driving forces. The problem of resistance to antibiotics in animal and man is such a problem. The use of biological engineering facilitates the synthesis of a lot of new antibiotics and the extraction of very specific biomolecules used as active agents or as reactives in biological diagnosis. Many of these already come from plants and will continue to do so in the future, despite the new problems of patents. Progress in human and animal health moves towards less aggressive and non specific matters, and reduction of doses. This is achieved through plant-based adjuvants and selective resistance of crops to a few substances. Products like modified cellulose fibers to produce capsules is one example. Besides starches, sugars and fatty acids, the production of surfactant properties is of interest in the cosmetic sector.

A NEW CONTRACTOR-ENTREPRENEUR

These are perhaps examples of niche products but the high added value is still of interest and I believe that we will see more diversification in the future, and I believe that it is important for the forestry and agricultural sectors to take an interest, since they may become important players.

This reminds me of the classical statement by *Joseph Schumpeter* who reminded us the owner of the stagecoach is not expected to invent the railroad. The traditional forestry industries, or manufacturers of agricultural products may not be the appropriate partners in the future. They operate in markets where they focus on market shares for their special products and have very little competence or will too rapidly switch to other areas. New players in the field or new strategies for the game are perhaps what we need.

DOWNWARD AND UPWARD

I will continue to discuss some of the possibilities that we observe within the forestry sector. There are several directions in which this sector can develop. Figure 3 gives an example of four of them.

Figure 3

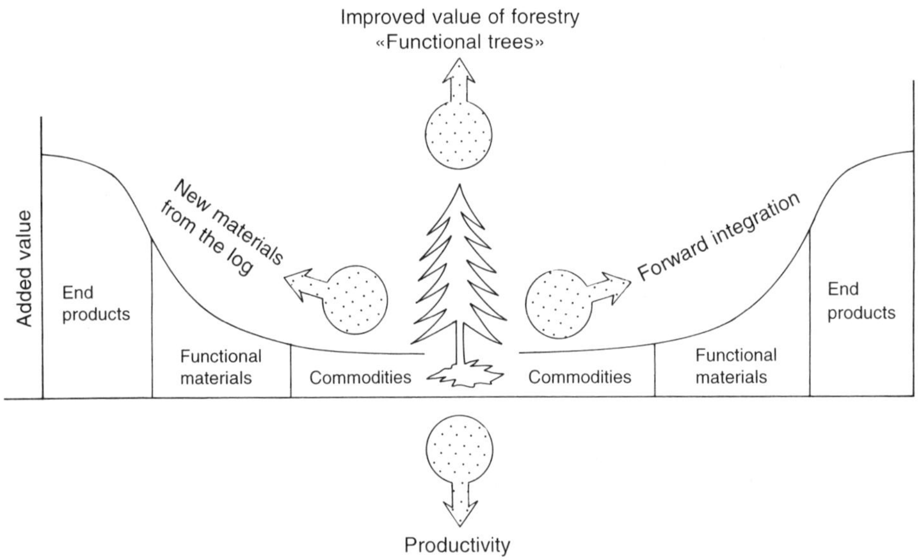

If we follow the arrow *downward* we are close to the present situation. We are improving the volume growth in the forest ecosystems by improving silvicullture. We try to make much better use of the raw material, and use the waste to produce energy. We are in the process of building faster paper machines. Paper is produced today at a speed of 14,000 m² per min. tomorrow we will see machines that run at a speed of 20,000 m² per min. The forest industry polish their diamond to make it shine brighter. The tradition is strong. The industry will continue to produce paper either from virgin fibers or from recycled paper. The amount of recycled paper will increase since the technique is now known. From time to time improved paper qualities will appear. We can also foresee better printing quality. The Environmental awareness exists, and soon we will see more closed operations with close to zero output of waste from the factories. This will probably enable the industry to move closer to the big market. A big paper industry can perhaps be seen in Paris and another in London.

The Olympic Games in Lillehammer, and the timber houses in Switzerland and Finland have changed the scene. Wood is already a competitive material for building three to five story houses. The obvious environmental and socio-economic benefits of wood utilisation favor wood as a substitute for non-renewable materials. The tendency toward market-oriented strategies brings a need to develop new products and to optimise the whole process chain, from the forest to the end-user. The mechanical wood processing industry produces more wood based panels and joinery products, prefabricated building components, furniture etc.

If we follow the arrow *upward* we can imagine a future with more functional trees. Several laboratories in Europe *e.g.* Belgium and France have already developed less ligneous trees. Tomorrow we will have new silvicultural techniques based on knowledge from ecological sciences. Silviculture and forestry will emulate natural systems, and the forestry and environmental groups, and the green movements will agree on principles on how to proceed. We will also see new types of plantations where fast-growing trees are grown in agricultural systems that use municipal wastewater for irrigation and receive optimal nutrients. Trees will be grown for special purposes, and the fiber quality will determine the final shape of the raw material. We know that all types of fibers that we currently are interested in exist in the forest. We will find out methods to collect them and distribute them to the right address. The quality-assessment will be important in the upward direction. Genetic improvement will further increase the possibilities.

The industry will use fibers to improve their products. In the paper industry new types of ink, paints and lubricants will be used. Deinking will be an enzymatic process. Biopulping will be improved. Energy will be saved by using fungi and other micro-organisms to degrade chips of both softwood and hardwood (*e.g. Populus*). Few chemicals will be used, and the waste products will be harmless. New methods to separate and modify lignin, cellulose, hemi-cellulose, etc are coming up. Paper will be a truly recyclable material. Today the paper contains too much clay and other non-organic products. These will be replaced by fibers from recycled paper, hardwoods or grasses. Protein, fat, starch and starch derivatives, and other natural plant substances are already, and will be increasingly used in the future. Vegetable oil from seeds of soya, rapeseed, linseed and sunflower can already be used to improve the technical result.

In the wood processing industry great interest is seen in modified wood products. The fastest growing market has been the so-called OTC board from USA. Composite materials from Japan are coming strongly onto the market. Modified wood products with more uniform engineering properties are needed. We see the results of thermal and chemical modification of wood, for example maleic acid and acetylation are used to produce dimensionally stable and biological resistant material. To produce non-degradable and fire-safe wood-material is a challenge for the future

FORWARD INTEGRATION

Forward integration indicates an interest in increasing the added value of today's products. In the food sector and also elsewhere, the emballage plays an important role. New fiber material can be used for food packages, and there is a need of replacing aluminium and plastics with renewable material. Plastic from plants, starch and starch derivatives are also of great interest in the paper-to-printing process. New types of ink and glues will increase the added value. All of this can be produced with the raw-material from the forest.

Fibber can also be used in other applications, such as fiber-mats and moulded interiors in cars and trains. Non-woven textiles are being used today in the construction sector, in the form of geomembranes, and also, to a lesser extent, in horticulture.

The arrow pointing *backward* is also an example of possibilities to produce new materials from wood. Non-destructive measuring techniques and intelligent assessment systems are needed to identify defects and to grade the raw material for optimal use. A computer aided product design need to be developed. We must realise that wood is an expensive material. The slow growth of forests, and the high cost of capital, and labour tell us that we should try to sell a product other than the mere log. New structures of plywood and wood composite structures generate higher added values.

Construction systems for the building industry are being developed and must be further improved. Today's board material must be replaced with green-polymers. Fire-safe wood construction, and new wood-based composite materials and structures must be available if the market is to compete with the building industry in USA and Japan.

RESEARCH AND DEVELOPMENT

These new products need more support from research. I will therefore finish my talk with a few words on R&D. The whole sector that I have been talking about today is generally regarded as low-tech. What I want to point out is that this is only true if you look at it with your coloured academic glasses on. It need not be the same tomorrow.

I have already noticed that the terminology in the framework program is misleading, but other things are also misleading. To my mind the EU system has concentrated too much effort on solving problems instead of trying to increase the knowledge-base for a competetive future. In the agricultural sector, the program is occupied with support for the CAP system, a system that probably needs to be changed in the near future. I therefore have my doubts that the EU program is organised to meet tomorrow's needs. I also fear that it has limited capacity to stimulate areas where European competitiveness can be improved tomorrow based on surplus raw materials and research. The forestry sector is an area where much more should be done. The potential is there, and one of the roles of R&D is to explore this potential. The area based on renewable material can be developed into a true science and technology area and will, if new products are developed, be a job creating area. Researchers must be allowed to explore their full potential, and the coupling of research and product development must be stronger in the future.

In order to achieve this, INRA and other organisations operating in this area, such as the one that I represent, must show the potential by stimulating innovative research. In some cases this will be basic research, in other cases applied research and development. I am convinced that the researchers will find an interest in searching for new techniques to produce environmentally friendly future products. The forestry and agroforestry sector will therefore be central in the next fifty years. I look forward to this development in the future, and I wish INRA all the best for the next fifty years.

TRENDS IN FISHERIES AND AQUACULTURE PRODUCTION IN EUROPE

ROBIN L. WELCOMME
FAO
Via delle Terme di Caracalla
00100 Rome, Italy

ABSTRACT

Fisheries in the marine and inland waters of Europe have been an important source of food and employment since the human occupation and development of the continent. Drastic changes have occurred in the last century which have altered the yield and the expectation from these resources. Rapid technological development created overcapacity in the marine fishery causing collapses in many of the more valuable groundfish stocks and their replacement by less valuable species. The total quantity of fish landed has also declined, and can only be restored through careful management. Environmental changes and pollution have damaged the fish communities in the inland waters with both resident and migratory species disappearing or declining in abundance. In any eventuality social and economic forces have changed the destination of the inland fisheries from food to recreation. Conservation lobbies are advocating even stricter regulations on the way such resources are managed particularly with regard to the stocking and introductions which are the keystone of present policies. Increases in food production can not, therefore, be anticipated from the capture fisheries where the main requisite is to manage to sustain present levels of production or even reduce them to allow stocks to recuperate. Shortfalls in production are largely filled by a growing and sophisticated aquaculture industry. Inland aquaculture concentrates mainly on salmonids in western and carps in eastern Europe. Marine aquaculture is heavily reliant on molluscs but recent developments in finfish, initially for salmon and later for bass, bream and turbot indicate the diversification and flexibility of this sector.

INTRODUCTION

Fishing has been a prominent food gathering activity from inland and coastal waters of Europe from the Palaeolithic onwards and remained central to many rural communities until the end of the last century. The Romans already had highly diverse and developed sea fisheries and also supplemented the capture fisheries with a well developed aquaculture, a practice that was intensified in some areas of Europe by the monasteries and land owners in the Middle Ages. Marine fisheries progressed in the North Sea and the Mediterranean during the Middle Ages when the Hansa league was based mainly on the capture and trade in herring. Fish from the sea were regarded as inexhaustible as late as the end of last century but rapid progress in fishing technology allowed for an increase of efficiency after the 1950s which rapidly depleted many of the stocks. The inland waters of Europe began to show signs of modification after the clearing of the forests for agriculture when siltation increased and discharge patterns were disrupted. The pace of modification accelerated during the industrial revolution with an increasing number of rivers becoming polluted and physically modified by revetments and damming. This inevitably led to the disappearance of many species and a decline in the general productivity of the waters. At about the same time the use of fish from freshwaters for food in western Europe began to decline in favour of recreational fishing. Aquaculture was maintained as a low level activity in certain countries until, in the 1960s, new technologies in rearing and marketing led to a rapid development of culture for salmonids in the western European countries. In eastern Europe, freshwater fish, principally carps, retained their importance for food and capture fisheries survived for a time to later be converted into recreational fisheries. The traditional fish culture for carps expanded to some degree through massive state owned farms. The present state of fishery resources arises from this past history. This paper mainly discusses the status and future of aquaculture in Europe with some mention of marine and inland capture fisheries

CAPTURE FISHERIES

Marine

The catch from the two sea areas surrounding Europe rose from 6 million tons in 1950 to a maximum of 13.8 million tons in 1977 and has since declined slightly to about 12 million tons (Fig1). In 1993, 87% of the production came from the Atlantic, North Sea and Baltic, 7% from the Mediterranean and 5% from long distance fleets. The rapid build up of the industry from 1950 onwards resulted in an overcapitalisation of the fishery which, coupled with high demand for seafood products, forced overexploitation particularly of valuable ground fishes such as haddock and cod in the North sea, and the groupers and sea breams in the Mediterranean. To a lesser degree there has been damage to the marine environment mainly caused by runoff of land based nutrients and pollutants. This is particularly severe in closed seas such as the Black Sea, the Baltic and the Adriatic. Catches from distant water fleets also declined after the introduction of exclusive economic zones by non-European countries and by the economic collapse of the distant water fleets of eastern European states such as Poland. The fall in absolute weight of catch has been paralleled by a decline in the quality of the fish caught with smaller and smaller species making up the bulk of landings.

Prospects for increased catches from marine fisheries are not good. The present philosophy would point to a need for retrenchment with the adoption of better management to restore the larger and more economically valuable ground fishes. Most of the production from the marine fishery is finfish (Fig.2) although the composition of this has changed in recent years with the decline in the percentage of valuable ground fishes and their replacement by less valuable pelagic forms.

Figure 1

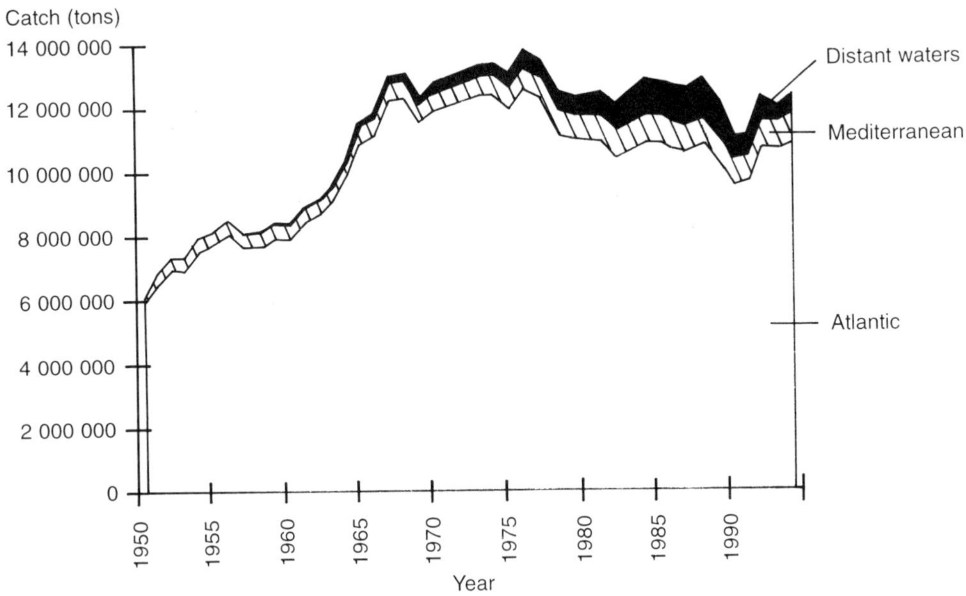

Figure 2

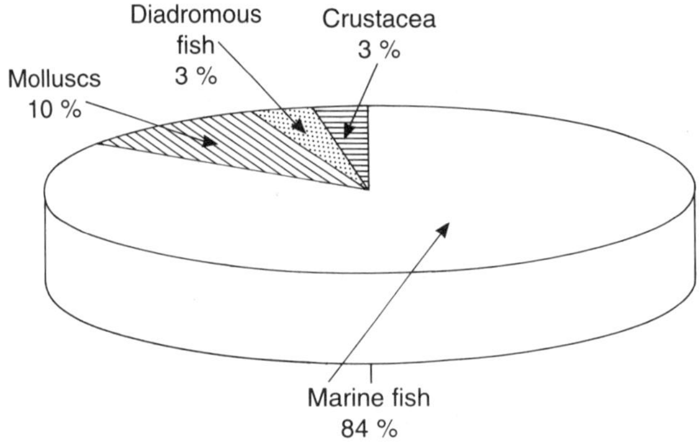

Inland

There are considerable difference in the orientation of inland fisheries and aquaculture in European inland waters. In part this originates from geographic differences with Eastern Europe consisting mainly of plains and mature river systems, and Western Europe containing mountains, torrential upper courses of rivers and large collections of lakes. In part it also originates from the different economic regimes.

The total production from inland waters from eastern and western Europe is shown in Figures 3 and 4 respectively and are differentiated between aquaculture and capture fisheries from 1984 onwards. Unfortunately the statistics are not available at present to extend the aquaculture time series backwards.

Figure 3

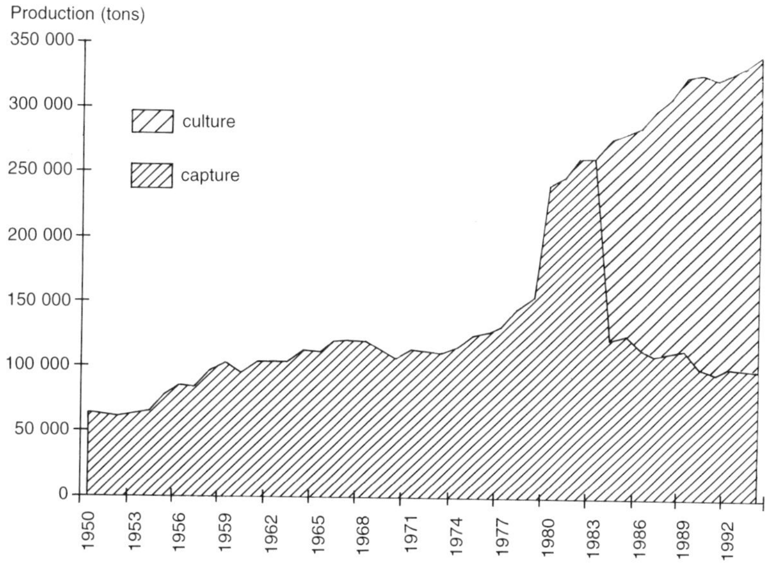

Figure 4

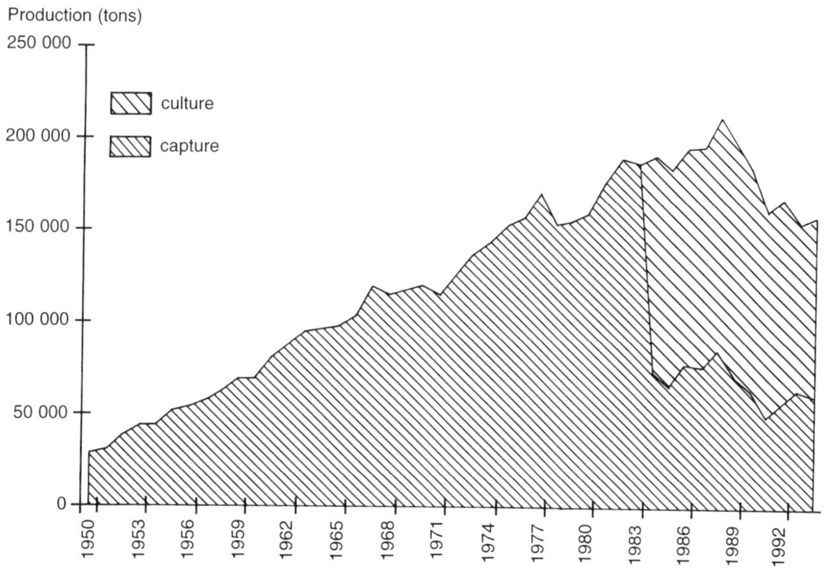

Commercial fisheries in eastern Europe

Total production from inland waters in eastern Europe increased from 28,100 tons in 1950 to 214,000 tons in 1988. There was a subsequent decline to 155,000 tons in 1993 which is attributed to the general collapse of administrative and production infrastructure during the period of realignment of the economies. Capture fisheries have contributed about 40% to this total since aquaculture and capture fishery records were separated in 1984. These fisheries have been based mainly on Lake Balaton, the Mazurian lakes, the Danube and some minor rivers. A large proportion of the production is cyprinids (40%) (Fig.5), mainly common carp, and much of the "other" category are also cyprinids not reported by species although a large range of other species including pike and *Silurus glanis* are caught. Many of these fisheries were maintained by stocking from state fish farms as a subsidy to the capture sector. Subsidies are no longer applied, and there has been a temporary hiatus in stocking until the private sector can organise to supply material to the now privatised fishery co-operatives. There has also been a steady deterioration in the quality of the aquatic environment which has unfavourably altered the species composition in many of the rivers and has taken many of the former floodable plains out of contact with the rivers. Although the capture fishery has recovered more rapidly than aquaculture and there has been a slight increment in catch since 1991 long term prospects for substantial gains in food supply from this source should be discounted.

Figure 5

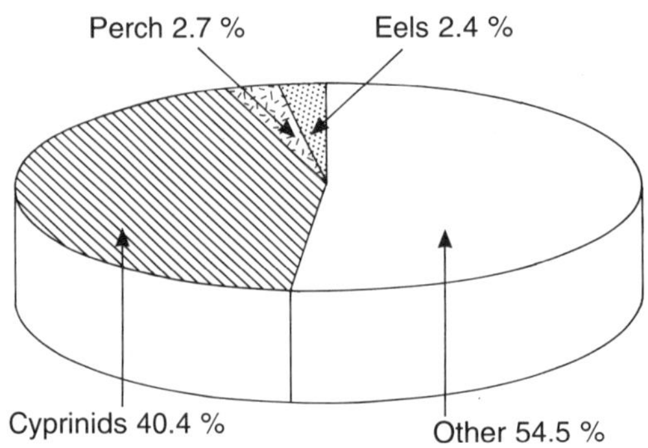

As the commercial fisheries declined so recreational fisheries have taken control of many of the fishing waters. Newly formed anglers associations have now assumed the responsibility for the management of smaller streams and rivers and some larger ones. At present the greatest need for the capture fisheries sector is to stabilise management particularly in view of potential conflicts between the recreational and commercial fisheries sectors. Stocking policies should be rationalised, and the infrastructure to produce fish for stocking purposes should be strengthened through support to the private sector. Damaged ecosystems should be rehabilitated in so far as possible.

Commercial fisheries in western Europe

Total production from the waters of western Europe has increased steadily from 1950 until the present. This increase has been entirely due to the expansion of aquaculture, at least since the records were separated on 1984, and the catch from the capture fishery has declined since that date. The decline is mainly due to the suppression of the commercial fishery in favour of the recreational sector, and many West European countries have no formal commercial fishermen. Such fisheries as remain are usually in alpine and Scandinavian lakes and catches are more varied than those reported from eastern Europe (Fig.6). Very little of the catch is exported, and most activities of this sort are closely associated with the tourist trade through supply to restaurants in the immediate vicinity of the fished water body.

Figure 6

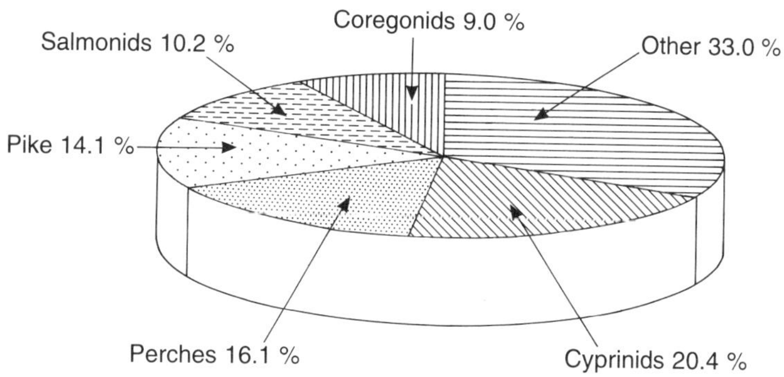

Recreational fisheries

Recreational fisheries now play a key role in inland water management as they represent one of the most intensive participant sports in Europe. About 20.2 million angling licences are issued per year in 20 of 31 European countries representing about 4.8% of the population. Although much of the angling is carried out on a catch and return basis, especially in western Europe, a substantial amount of the fish landed is eaten. Estimates of around 165,000 tons/year have been made much of which is not reported in the official statistics. This compares with the 163,000 tons reported as caught by the commercial fishery in 1994.

Recreational fisheries are coming into some conflict with animal rights and conservation movements throughout the continent. On the one hand these groups are questioning the ways in which the fish are caught and seek to suppress catch and return fisheries. On the other there is a move to reduce the stocking on which many of these fisheries are based. Ironically, as one of the most active of users of inland waters, recreational fishermen contribute substantially to the maintenance of water and habitat quality. This principal has been reinforced as Governments have sought to charge anglers associations increasingly with the management of the waters with which they are charged.

AQUACULTURE

With prospects of increases in future food supplies from marine and inland capture fisheries being extremely limited much hope is being placed in aquaculture as a possible means to make up the growing deficit between demand and supply. Examination of the progress of aquaculture over the last decade indicates that a more cautious approach to aquaculture potential may be advisable.

Inland

Eastern Europe (Table 1)

Aquaculture production in eastern Europe grew slightly from 118,000 tons in 1984 to 120,000 tons in 1987. This was in response to central governmental policies for large production units which were to a certain degree subsidised by the state. The reorganisation of the economies after that date led to a rapid decline in productive capacity as, in most countries, the private sector was unable to assimilate the large subsidised farms. The decline continued to a low of about 90,000 tons in 1993, representing a loss of about 30% of the total production, and showed some sign of increase in 1994.

Production systems are heavily biased toward extensive or semi-intensive farms for cyprinids which constituted 89% of culture fish in 1994 and particularly common carp (61.3%) which is a traditional food throughout the region. Until recently relatively few taxa (11) have been involved in culture although since 1992 some additional species are being farmed bringing the total number of taxa under culture to 16, indicating shifts in culture practice in response to market demand. Principal among these has been a move towards salmonids which comprised 8% of the production in 1994 against 3.5% in 1984.

It is clear that there is room for expansion of production in eastern Europe at least to previous levels or even beyond. However three factors are impeding recovery, lack of infrastructure, lack of capital and lack of markets. In the first instance administrative support to aquaculture is lacking as difficulties are being experienced in the transition to market economies. This is particularly serious as regards legislation, and at present the status of aquaculture is far from defined. It is hoped that these problems will soon be overcome as there is a concerted effort to assist the more disadvantaged countries in formulating legislation to deal with aquatic natural resource. Second, there is a chronic lack of the capital needed to modify the existing extensive co-operative farms into smaller and more efficient units suitable for individual operation. Third, and possibly most serious is the loss of markets. Many of the countries used to export part of their production to Russia, a market that is now barred. Alternative markets elsewhere are not viable due to the nature of the product as carp has a strictly limited popularity in western Europe. There is therefore a need to move towards a diversification of culture toward species that are more widely acceptable. The problem with this is that eastern European nations would then be competing more directly with the west not only for markets but for the feeds and other inputs that may well prove limiting to certain types of culture. Furthermore the extensive model of fish culture common in eastern Europe is generally regarded as relatively environment friendly, and intensification toward competitive systems may well introduce problems with pollution similar to those experienced in the West.

Western Europe (Table 2)

Western European production from inland aquaculture has shown a steady increase from 154,000 tons in 1984 to 241000 tons in 1994. This growth has been almost entirely through the expansion of intensive aquaculture for salmonids which comprised about 82% of the production in 1994, and about 79% came from one species, the rainbow trout (*Oncorhynchus mykiss*). Nevertheless there has been a continuous trend to diversification over the past decade with 16 taxa under culture in 1984 and 30 in 1994. As in eastern Europe a proportion of the production from aquaculture is used to stock natural waters in support of recreational or commercial fisheries. Of the fish produced for direct human consumption the majority of rainbow trout is marketed mainly as a frozen product through supermarket chains whereas many of the other species are cultured for local culinary tourism.

Aquaculture in western Europe is subject to three main constraints at present not experienced in the East. Firstly the heavy bias towards high food chain organisms such as trout implies a dependence on animal based feeds. At present these are drawn mainly from fish meal and are therefore sensitive to the fluctuations in supply of the oceanic pelagics. There is also growing competition for fish meal from other users, and although soy protein in some cases provides an alternative this does not apply to fish production. The question of feed supplies closely linked to the second constraint which is the market process. Western European aquaculture has traditionally aimed at high value species for sale to luxury markets. This implies low levels of production of an expensive product through systems that are relatively tolerant of inefficiencies. With the expansion of products such as the rainbow trout to a mass market the efficiency of rearing systems has had to be increased leading to more industrialised farms. The third constraint is that salmonid culture, especially at an industrial level, requires large quantities of water flow through facilities. This water tends to become polluted with the high levels of food and wastes generated. Awareness of the polluting potential of fish farms of this type has led to the formulation of water quality criteria which can limit the number of farms on any water course and can also add to the cost of production through the need to install water treatment facilities.

An interesting development in the last few years has been for the emergence of recirculating systems which control effluent emissions and limit the amount of water required. These are at present used mainly for exotic products such as tropical tilapias and catfishes but may point the way to a further industrialisation of production in the future.

Marine

Mediterranean and Black Sea (Table 3)

Marine aquaculture in the Mediterranean grew rapidly in the decade since 1984 from 77854 tons to 172980 t in 1994. Most of this production is molluscs (85.4%) of which mussels are by far the most important (87.6% of mollusc production). Mollusc culture saw its major expansion at the beginning of the decade, and growth now appears to have stabilised. The number of species under culture expanded rapidly at first but has since remained at 5 with clams rising rapidly in importance to overtake oysters as the second product. Fish now contribute 14.6% of the total production and the rapid expansion since 1989 still appears to be continuing. Early marine culture was based mainly on mullets (43% of fish in 1984) but has been overtaken by seabass (49.3% in 1994) and seabreams (43.8% in 1994). This sector is still active with a rise in the number of species under culture from 4 in 1984 to 10 in 1994). Culture of crustacea is unimportant and has shown considerable fluctuation. It should still be regarded as being in the experimental phase with only 3 species which together yielded about 100 t in 1994.

Atlantic, North Sea and Baltic (Table 4)

About 57% of aquaculture production from the Atlantic is molluscs although this category has fallen in importance since 1984 when it comprised 92% of production. It fell in terms of tonnage until 1993 with some upturn in production in 1994. The number of taxa under culture has risen steadily from 3 in 1984 to 13 in 1994 although mussels still make up the bulk of the yield (68.2%) with oysters contributing a further 30.8% in 1994. Crustacean culture is insignificant and has passed through several phases of expansion and contraction to its present 0.01% of the total yield. Fish culture on the other hand is one of the success stories of modern industrialised aquaculture expanding rapidly throughout the decade to its present 345,545 tons. The bulk of this (98%) is for salmonids and particularly Atlantic salmon at 87.2% of all fish produced. In contrast to the Mediterranean the number of species under culture has remained stable at 12 since 1988, and there are no provisions to open production to other species at present.

roblems

Aquaculture in both marine basins is subject to similar constraints to those in western European inland waters. As the major product is a high food chain fish the supply of fish feeds based on imported fish meal become critical not only to the sustainability of production but also to its further expansion. The expansion of cage culture for salmon, seabass and seabreams has caused considerable environmental concerns not only for the polluting effects of the intensive systems involved but also for their aesthetic impact. A further problem has been that of market saturation as world markets were unable to absorb the large volumes of previously luxury species. This resulted in drastic lowering of prices of some commodities such as salmon with the now familiar transition from a luxury to a mass product. This crisis seems to be temporary, and it is hoped that marine aquaculture will continue to supply an increasing quantity and variety of seafood products.

CONCLUSIONS

The prospect of increases in production of finfish and shellfish from conventional capture fisheries in Europe is limited. The fisheries of both the major marine areas are in some disarray from overcapitalisation and overfishing of major species. They consequently need a period of retrenchment if they are to recover and provide a basis for sustainable fisheries in the future. Inland capture fisheries are in decline as recreational and conservationistic objectives for management prevail both in eastern and western Europe.

Increases in the supply of fish and shellfish therefore rely on the expansion of aquaculture. This sector has made notable progress in the last decade both in terms of tonnage and in the variety of species on offer and will probably continue to grow, but at a reduced rate. A number of factors constrain possible future expansion. First, current trends in species reared are to switch from low food chain organisms such as carps or shellfish to higher food chain organisms such as salmon or seabass. This creates a situation where aquaculture is in competition with other activities for an unstable source of feed. This situation imposes penalties of price and security of supply on aquaculture products which can only be resolved in the long run by finding other sources of protein acceptable to fish. Second, intensive aquaculture carries with it certain risks for the environment which provoke concern in many European countries. Laying aside aesthetic considerations the potential polluting nature of the activity has provoked laws in many states which require plants for restoring water quality, improved expensive rearing systems, and more environmentally friendly feeds and additives, all of which increase costs. They also pose limits on the siting of aquaculture installations which will eventually limit capacity to expand. Third, aquaculture is sensitive to market pressures. In eastern Europe the problem lies mainly with a product that has a limited demand. In western Europe the transition from a limited luxury market product to a mass market one has imposed strains on several types of product. Rainbow trout, mussels, oysters and clams have already successfully made this transition, and salmon, turbot, seabass and seabream are in the process of doing so.

Table 1 Weight and species composition of aquaculture production from Eastern Europe

		1984	1985	1986	1987	1988	1989	1990	1991	1992	1993	1994	%1984	% 1994
Cyprinids		111083	110444	109165	112224	117733	118582	112388	103017	100046	80494	86513	94.06	89.67
	Chinese carp	1200	9201	8636	7204	6095	6290	5049	4853	5415	4749	4199	1.02	4.35
	common carp	72351	61851	62126	66596	71195	73807	71954	65834	68238	52281	58333	61.26	60.46
	other	37210	38910	38012	38005	40002	38012	35008	32004	26060	23154	23662	31.51	24.53
	tench	322	482	391	419	441	473	377	326	333	310	319	0.27	0.33
Eel		0	0	0	115	90	39	98	73	33	20	40	0.00	0.04
Other		2930	3504	3941	3178	3332	2930	2853	3195	4730	1925	2042	2.48	2.12
	catfish	10	80	85	69	98	97	90	97	169	198	242	0.01	0.25
	perch	254	195	203	201	249	200	176	164	228	246	278	0.22	0.29
	mullets	0	0	1	1	3	1	2	1	1	1	1	0.00	0.00
	non classified	2321	2920	3277	2550	2639	2290	2287	2680	4066	1253	1301	1.97	1.35
	pike	345	309	375	357	343	342	298	253	266	227	220	0.29	0.23
Salmonids		4073	4205	4553	4946	6979	6893	6397	5585	6633	7204	7710	3.45	7.99
	char	0	0	0	0	0	0	0	0	0	10	15	0.00	0.02
	trout	0	0	0	0	0	0	0	0	0	0	5	0.00	0.01
	non classified	1147	1255	1362	1343	1225	1256	1335	853	1502	456	437	0.97	0.45
	rainbow trout	4073	4205	4558	4946	6979	6893	6397	5585	6633	7204	7710	3.45	7.99
Tropical	tilapia	0	0	0	0	0	0	0	20	0	0	0	0.00	0.00
Whitefish		0	0	0	0	0	0	0	0	47	243	173	0.00	0.18
Crayfish		18	25	30	11	0	0	0	0	0	0	0	0.02	0.00
TOTAL		118104	118178	117689	120474	128134	128444	121736	111890	111489	89886	96478	100.00	100.00

Table 2 Weight and species composition of aquaculture production from Western Europe

		1984	1985	1986	1987	1988	1989	1990	1991	1992	1993	1994	%1984	%1994
White fishes		0	0	0	0	5	100	100	100	100	100	100	0.00	0.04
Cyprinids		23067	26478	29307	28177	30440	32159	28930	27192	23486	22810	22266	14.94	9.22
	Chinese carp	188	228	346	456	595	996	266	230	266	294	98	0.12	0.04
	other	0	370	1870	2000	2200	2500	2500	2500	2500	2501	2501	0.00	1.04
	tench	150	700	1490	1400	1710	652	539	586	472	470	520	0.10	0.22
	common carp	22729	25180	25601	24321	25935	28011	25625	23876	20248	19545	19147	14.73	7.92
Eel		1921	2217	3606	4224	5009	4561	5047	5499	4762	6101	6814	1.24	2.82
Other		3900	4330	3970	4994	5363	5688	6137	8425	8254	8144	8776	2.53	3.63
	catfish	1500	1500	1560	1505	1505	1810	1610	1850	1886	1900	2182	0.97	0.9
	non classified	300	310	310	549	878	1078	1527	3675	2599	3094	3246	0.19	1.34
	perch	0	0	0	0	0	0	0	0	400	400	400	0.00	0.17
	pike	0	180	390	400	440	450	450	450	500	500	500	0.00	0.21
	mullets	2100	2340	1710	2540	2540	2350	2550	2450	2869	2250	2448	1.36	1.01
Salmonids		125438	122895	133501	148715	152172	165201	181986	180982	187133	191291	199747	81.27	82.67
	American salmon	54	204	204	540	811	1019	512	33	20	30	74	0.03	0.03
	Atlantic salmon	0	0	0	0	0	0	0	350	350	350	400	0.00	0.17
	char	1	5	374	382	439	460	661	569	756	818	855	0.00	0.35
	European trout	4176	4102	4462	4701	5419	5993	5770	5705	6739	6377	6696	2.71	2.77
	rainbow trout	121207	118584	128461	143092	145503	157729	175043	174325	179268	183716	191722	78.53	79.35
Sturgeons		0	0	10	10	10	10	260	311	371	451	484	0.00	0.2
Tropical	catfish, tilapia	20	195	460	500	630	675	700	700	700	1100	1257	0.01	0.52
Crayfish		7	4	6	3116	2711	2709	2534	2246	2235	2044	2177	0.00	0.9
TOTAL		154353	156119	170860	189736	196340	211103	225694	225455	227041	232041	241621	100.00	100.00

Table 3 Weight and species composition of aquaculture production from the Mediterranean and Black sea

		1984	1985	1986	1987	1988	1989	1990	1991	1992	1993	1994	%1994	%1984	% Category
Crustacea															0.06
Crabs	Natantia	0	0	0	0	0	0	0	0	0	0	5	5.10		
Prawns		0	16	16	20	55	84	75	66	76	80	93	94.90		
	Penaeus japonicus	0	16	16	20	55	83	73	64	75	79	92	93.88		
	Penaeus monodon	0	0	0	0	0	1	2	2	1	1	1	1.02		
Weight		0	16	16	20	55	84	75	66	76	80	98			
No of Taxa		1	1	1	1	1	3	3	3	3	3	4			
Fish															14.57
Eel		844	852	854	853	1029	1264	1639	1590	1323	1028	930	3.69	33.09	
Mullets	Mugil cephalus	1100	1100	1700	1200	1200	1100	1200	1300	981	1000	750	2.98	43.12	
Other		0	58	79	125	66	262	389	56	56	59	59	0.23	0.00	
	Psetta maxima	0	0	0	0	0	0	0	0	0	0	1	0.00	0.00	
	Seriola dumerili	0	13	9	20	13	17	21	31	31	30	30	0.12	0.00	
	Solea vulgaris	0	7	10	6	6	8	7	9	9	9	8	0.03	0.00	
	Thunnus thynnus	0	38	60	99	47	237	358	16	16	20	20	0.08	0.00	
	Tilapia	0	0	0	0	0	0	3	0	0	0	0	0.00	0.00	
Seabasses		280	430	830	1010	1255	1850	3462	4732	8152	11249	12424	49.29	10.98	
	Dicentrarchus spp	0	20	100	50	70	200	160	100	383	174	182	0.72	0.00	
	Dicentrarchus labrax	280	410	730	960	1185	1650	3302	4632	7769	11075	12242	48.57	10.98	
Seabreams		327	380	470	660	1200	1460	2528	3154	5491	7958	11043	43.81	12.82	
	Pagrus pagrus	0	0	0	0	0	0	0	0	0	37	130	0.52	0.00	
	Sparus aurata	327	380	470	660	1200	1460	2528	3154	5491	7921	10913	43.30	12.82	
Weight		2551	2820	3933	3848	4750	5936	9218	10832	16003	21294	25206	100.00	100.00	
No of Taxa		4	8	8	8	8	8	9	8	8	9	10			
Molluscs															85.37
Mussels		65956	72668	75167	74936	94950	100137	103897	98031	104966	112242	105639	71.53	87.59	
Oysters		9347	9737	11073	10307	8259	11274	10743	13194	13152	15020	15034	10.18	12.41	
Oysters	Crassostrea angulata	4166	4659	5919	5243	5211	9589	10664	13149	13134	15000	15000	10.16	5.53	
	Ostrea edulis	5181	5078	5154	5064	3048	1685	79	45	18	20	34	0.02	6.88	
Clams		0	1	30	287	1936	7118	16102	20003	26403	26402	27003	18.29	0.00	
	Veneridae	0	0	0	2	2	2	2	3	3	2	3	0.00	0.00	
	Tapes spp	0	1	30	285	1934	7116	16100	20000	26400	26400	27000	18.28	0.00	
Weight		75303	82406	86270	85530	105145	118529	130742	131228	144521	153664	147676	100.00	100.00	
No of Taxa		1	2	4	4	4	2	2	2	1	2				
Total		77854	85242	90219	89398	109950	124549	140035	142126	160600	175038	172980			100.00

Table 4 Weight and species composition of aquaculture production from the Atlantic, North Sea and Baltic

		1984	1985	1986	1987	1988	1989	1990	1991	1992	1993	1994	%1984	%1994	% Category
Crustacea															
Weight		0	3	51	146	93	48	166	82	33	30	63			0.01
No of Taxa		0	1	2	4	4	4	2	2	2	1	2			
Fish															42.26
Eel	European eel	15	20	25	52	47	70	143	123	585	98	98	0.04	0.03	
Non classified		0	0	0	0	0	0	971	201	254	707	685	0.00	0.20	
	freshwater fishes	0	0	0	0	0	0	0	0	0	0	1	0.00	0.00	
	marine fishes	0	0	0	0	0	0	971	201	254	707	684	0.00	0.20	
Seabass		35	53	83	90	81	29	33	95	151	453	471	0.08	0.14	
Mullets		280	396	585	510	497	74	121	65	60	63	69	0.67	0.02	
Other		40	88	87	110	166	303	1639	1153	2012	2481	3176	0.10	0.92	
	cod	0	0	0	5	15	10	10	20	25	23	80	0.00	0.02	
	turbot	5	53	50	67	112	289	1627	1130	1982	2454	3094	0.01	0.90	
	sole	35	35	37	38	39	4	2	3	5	4	2	0.08	0.00	
Salmonids		41668	56254	79794	93493	143964	184599	231605	258000	223685	280300	338548	99.04	97.98	
	Arctic char	0	2	2	1	7	40	161	150	150	200	200	0.00	0.06	
	Atlantic salmon	27182	38448	59123	66527	107681	154644	201292	226403	191884	245323	301320	64.61	87.20	
	American salmon	0	70	80	80	85	85	80	17	0	0	0	0.00	0.00	
	rainbow trout	14486	17734	20589	26885	36182	29830	30072	31430	31651	34752	37028	34.43	10.72	
	sea trout	0	0	0	0	9	0	0	0	0	25	0	0.00	0.00	
Seabreams		35	172	191	176	222	364	670	1368	2044	2303	2498	0.08	0.72	
	gilthead bream	35	172	191	176	222	363	670	1368	2044	2303	2498	0.08	0.72	
	non classified	0	0	0	0	0	1	0	0	0	0	0	0.00	0.00	
Weight		42073	56983	80765	94431	144977	185439	235182	261005	228791	286405	345545	100.00	100.00	
No of Taxa		8	10	10	11	12	12	12	12	11	12	12			
Molluscs															56.98
Clams		0	0	1000	1300	900	325	390	600	700	800	400	0.00	0.09	
	non classified	0	0	1000	1300	900	325	390	600	700	800	400	0.00	0.09	
	hard clam	0	0	0	0	0	0	0	0	0	0	0	0.00	0.00	
Decapods	common cuttlefish	0	0	0	0	0	0	0	0	0	0	0	0.00	0.00	
Mussels	blue mussel	408436	443775	428729	441520	411810	368984	357587	346030	307048	258887	317589	78.45	68.17	

Non classified	marine molluscs	0	0	0	0	0	0	100	50	76	0	0	0.00	0.00	
Other		0	0	0	0	0	2102	2343	2691	2494	2655	3472	0.00	0.75	
	European abalone	0	0	0	0	0	0	0	0	0	0	6	0.00	0.00	
	common cockle	0	0	0	0	0	1302	1497	1638	1639	1704	2333	0.00	0.50	
	periwinkles	0	0	0	0	0	800	800	1003	807	900	1110	0.00	0.24	
	Japanese carpet shell	0	0	0	0	0	0	46	50	48	51	21	0.00	0.00	
	carpet shell	0	0	0	0	0	0	0	0	0	0	2	0.00	0.00	
Oysters		112170	140766	144322	134631	134805	126284	138974	123799	129103	138834	143744	21.55	30.85	
	Portuguese cupped oyster	0	0	0	0	100	100	70	155	380	585	1063	0.00	0.23	
	Pacific cupped oyster	106501	135819	138639	128331	128204	119996	132170	118009	121694	131021	134708	20.46	28.91	
	cupped oysters	0	0	15	525	933	1033	1182	1384	1561	3003	2225	0.00	0.48	
	European flat oyster	5669	4947	5668	5775	5568	5155	5552	4251	5468	4225	5748	1.09	1.23	
Scallops		0	150	179	277	300	230	187	219	454	638	686	0.00	0.15	
	queen scallop	0	0	4	128	172	100	54	61	62	62	38	0.00	0.01	
	common scallop	0	150	175	149	128	130	133	158	392	576	648	0.00	0.14	
Weight		520606	584691	574230	577728	547815	497925	499581	473389	439875	401814	465891	100	100	
No of Taxa		3	4	5	7	8	10	12	12	12	11	13			
TOTAL		566974	646783	663139	680181	703900	694576	740538	741444	675449	694218	817673			100.00

CHAIRMAN: LAURENT VAN DEPOELE, *EUROPEAN COMMISSION, DGVI, BRUSSELS, BELGIUM*

VICE CHAIRMAN: DANIEL TERRASSON, *CEMAGREF, ANTONY, FRANCE*

THEME 2

AGRICULTURE, ENVIRONMENT AND RURAL AREAS

Workshop 3
Agriculture as part of rural and regional development

THE DIVERSITY OF RURAL SPACE AND ITS TYPOLOGIES

ELENA SARACENO
CRES Centro Ricerche Economico Sociali
33100 Udine, Italy

ABSTRACT

Why do we need a typology of rural areas' diversity today, more than ever before? In the first place because the acknowledgement of the multiplicity of development processes requires some systematisation of the observed differences, even *ex post*, in order to understand better how and why territorial diversification is taking place, and in the second place because an adequate typology would considerably increase the efficacy of development programmes. Another by product would be that the discussion about the relevant criteria to be used would make evident the difficulties with long used dichotomies, such as rural-urban, centre-periphery, developed-underdeveloped, conceived as homogeneous types and considered comparable among each other. On the other hand functional criteria of classification such as local economy, labour market, and region appear increasingly as good first steps in order to understand how space is differentiating lately. To solve the problem of rural/territorial diversity we may have to proceed pragmatically because scientists are not ready, and everyday policymaking puts pressure for answers. In this case retracing the sequence of typologies used by the European Union, the difficulties that have been always present when trying to arrive at satisfactory definitions of rurality, and the work done by OECD on indicators are all good example of how to proceed. But in order to understand and explain we have to move farther than pragmatism allows: we need to understand better the new spatial patterns of economic and social development which are changing the traditional ways in which spatial differentiations were explained. The last section of the paper explores this difficult field. Two major and interconnected trends characterise spatial differentiation today: the shifting location of economic activities in response to the challenges of a global economy and the increasing diversification of some rural areas. These trends open new opportunities for rural development and at the same time explain why old typologies no longer work as they did in the past. It is a major and challenging task for research.

WHY WE NEED A TYPOLOGY OF TERRITORIAL DIVERSITY

The original assumption of convergence/irrelevance of territorial differences

Up to the 1960s economic development was understood as a unilineal process: territorial differences were conceived as disparities in development and each region or territory could be placed somewhere along an ideal developed/non developed line. Disparities in development were considered negatively: either spontaneously or with appropriate policies disparities had to be reduced or eliminated. Ideally, when this situation was reached, spatial differences would no longer be a relevant issue. The direction of change was conceived one way from the less developed to the more developed status and not vice versa. Within this framework all regions followed similar patterns or 'stages' of development. Differences among the less developed areas were considered irrelevant for development purposes, and policies to reduce disparities were substantially the same for any less developed area. This in turn justified top-down approaches to policy implementation. Since territorial differences were perceived as relevant only in terms of the development process this became the criteria for a typology which distinguished less developed areas (one undifferentiated aggregate) from developed ones (another undifferentiated aggregate); policy measures had to reduce disparities between the two, acting specifically on the first group. In my opinion there is no real explanation of territorial differences within this perspective but rather a simplification of its meaning, equated with the stage of development: once development was achieved spatial differences were no longer relevant.

The recent acknowledgement of the diversity of development processes

From the 1970s onwards the assumption of convergence and unilinearity has been questioned, and has had important consequences for the understanding of territorial differentiation and appropriate policies. It has been recognised that territorial differences are the result of a specific interaction between social, economic and institutional factors in an area and its links with the global market. This unique combination of factors influences its competitiveness not only during the initial stages of economic development but also at successive stages. This implies that there are no common or unilineal stages of development and that diversities (or territorial differentiations):

- reproduce over time both before and after the development process, thus remaining always relevant,
- should be understood as a positive factor defining the specific characteristics of an area rather than reducing any significant difference to negatively perceived disparities;
- become a crucial factor for the identification of the competitiveness of an area, which may increase but may also decline over time.

From this perspective, policies from any institutional origin should recognise and use territorial diversities as strategic factors to achieve competitiveness; this means that common recipes for less developed areas as well as top down approaches are inadequate, because they simplify and ignore contextual differences. Within this framework the bottom up approach, integrated actions and institutional partnerships become the new ways in which diversity is acknowledged for policy implementation.

The European Commission between disparities and diversities

With the reform of the structural funds, the principle that each Region or area, if it complies with certain requirements (one of the objective areas), is entitled to propose and negotiate its own approach to development with the Commission has been accepted; in this way the existing diversity among areas is automatically accounted for. However, when we want to evaluate or understand those diversities or the actions taken from a European perspective, we are confronted with a multiplicity of unique situations without any explicit link with each other except for the fact that they are either objective 1, 2, 5b or 6 and are located in some Member Country. This is still too broad a categorisation and leaves us with highly heterogeneous situations which need to be further differentiated according to significant homogeneous criteria.

Why we need territorial typologies

Creating a territorial typology, even *ex post* (after the realisation programmes), would serve to:

- help understand the internal articulation/differences of the various objective areas (refining eligibility criteria);
- identify the real processes under way in generalisable categories (dynamic aspects);
- find the similarities and differences among them (comparing homogeneous processes);
- identify patterns of structural situations and development (static aspects, structures);
- generalise lessons from single local experiences within the European context (policy purposes);
- establish criteria for the transferability of experiences and policy actions (what can be transferred where, how) (links among types of areas);
- establish associations between types of areas and best policies/practices (policy management).

Typologies to increase the efficiency of policy actions

Appropriate and relevant territorial typologies become a necessity if we want to monitor diversity and its evolution at any institutional level. Without such a categorisation we are unable to evaluate processes and policy actions at an aggregate level except on a case to case basis, taking policy decisions on the basis of individual local demands. The better we are able to categorise existing diversity into relevant typologies, the better we will be able to produce development policies of good quality, adapted to the real situations on the ground.

Clarifying and updating territorial differentiation criteria

The problem today is that the European Commission, as will be shown further on, started with a classical conceptualisation of territorial differences based on disparities and disequilibria identified with the criteria of development stage (less favoured areas, objective 1) and in the 1980s gradually shifted towards the recognition of diversities (bottom up approach, partnership in programming), without using consistent criteria of classification: the rural differentiation was introduced (objective 5b), industrial decline (objective 2), low density (objective 6); in official documents and current language we often find the concepts of disparity and diversity mixed up and heterogeneous criteria for categorising territorial diversity.

THE ISSUE ABOUT RURAL/URBAN OR CENTRE/PERIPHERY TERRITORIAL DIFFERENTIATION

Origins of the rural/urban significance

Since the industrial revolution the rural/urban differentiation has been a significant and relevant typology used to understand territorial diversity. The process of industrialisation, concentrated manufacturing, and workers in urban areas thus giving rise to intense movements of population and resources from rural to urban centres. The recognition that with economic development rural areas generally lost competitiveness in relation to urban areas and "specialised" in agricultural activities gave extreme relevance to this territorial typology and created an almost immediate association with the need for policy action for this type of areas.

Alternative criteria for spatial differentiation

Other typologies have been used quite frequently: the centre/periphery differentiation based on distance, the mountain/hill/plain based on physical resources, labour market areas based on commuting workers, local economies based on the peculiar characteristics of economic activities in a specific place. All these (and others that we can think of) are alternative ways of understanding spatial differentiation and are good examples of typologies. The issue is to choose the one that is most relevant/pertinent given the purpose or perspective of the analysis or the policy actions.

Present relevance of the rural/urban typology

The rural/urban categorisation is only one possible typology. Its significance is related to the observed persistence of ITS homogeneous character in relation to urban areas or to its continued relevance for policy purposes. In both aspects there have been important recent changes: some rural areas are no longer declining (losing competitiveness and resources) and some urban areas are no longer attracting resources on the one hand, thus differentiating former homogeneous behaviour and indicating that new processes are taking place (deurbanisation, rural 'renaissance'); on the other hand rural areas have greatly increased their policy relevance by becoming the beneficiaries at EU level of part of the financial resources formerly attributed to the agricultural sector.

This gives a contradictory outcome: while the traditional rural/urban dichotomy is losing significance in some areas of Europe, rural areas have increased their significance in terms of policy eligibility. This fact increases the relevance of the need to:

a) explain the new forms/typologies of spatial differentiation overlapping with traditional rural/urban ones;

b) identify the criteria which measure the current differention within rural areas in order to concentrate policy actions on those areas which are still losing competitiveness and resources.

Questioning the centre/periphery spatial differentiation

A similar analysis could be applied to the centre/periphery territorial differentiation: recently new centres, extremely dynamic but not contiguous to the old ones have appeared on the world exchanges thus questioning the significance of distance as a relevant criteria for understanding economic dynamism. The new industrialising countries (NICs), mostly in the Far East, have undergone very rapid growth on the global markets, competing successfully with older industrial centres (Europe, United States). The emergence of both rural areas no longer on the decline and NICS cannot be fully explained as the result of a decentralisation process from former centres of development towards the periphery but on the contrary as new forms of economic and social organisation emerging quite independently and in open competition with the old ones. Therefore centre/periphery dynamics might explain a part but not everything that is happening. This implies that new processes are emerging and should be accounted for. No relevant changes in the policies for peripheric regions have taken place recently, therefore it is only for conceptual and classification purposes that this typology is being questioned.

Diversification of the social and economic composition of rural areas

The emerging economic and demographic trends are indicating that the neat division of labour between the city and the countryside is progressively blurring. The former concentration of agricultural activities in the countryside and of manufacturing and service activities in the cities is evolving towards a less marked differentiation of the social and economic composition between rural and urban areas. Former urban residents more often choose to live in rural areas and commute to their places of work for the most different reasons (cost of living, lifestyles, security), and industrial and service activities show a similar propensity to more freely choose a location outside the urban markets. The result of these trends is that the distribution of employment (active population) is getting less polarised and often remarkably similar in rural and urban areas. This implies that a process of social diversification is taking place in rural areas, diluting the original relevance of farm families. This spontaneous process has been recently reinforced by the support given to the diversification of activities in rural areas by EU policies as a form of compensation for the decline in the support of agricultural activities.

National definitions of rural areas, the European Commission's eligibility choice and the recent OECD definition

The narrow scope of former national definitions of rurality

Recent changes in rural development and policies have revitalised the debate on the definition of rural areas. Countries had given national definitions of rurality well before the Commission put it in a strategic position within its policies at the end of the 1980s. It should also be underlined that not all countries, even if they somehow defined rural areas, had policies specifically directed at them. For example Italy, and Germany gave definitions of rurality for administrative or statistical purposes, but had chosen regional policies as a form of territorial intervention rather than rural criteria.

The criteria used by different countries for defining rurality are (sometimes more than one):
- size of resident population in communes (Spain, Italy);
- size of resident population in one agglomerated unit (France, Ireland, Greece, Netherlands);
- density of population (Germany);
- share of population active in agriculture (Belgium, Netherlands);
- administratively designated status (United Kingdom, Ireland).

This list, as was already remarked in the OECD analysis, considers the population (demographic size or density) as the most relevant criterion for discriminating rurality and not, as could be thought, the share of agricultural activities. The reason for this is probably that most countries preferred sectorial policies to support farming activities and attributed to territorial policies a function in the field of services, communications and infrastructures, thereby trying to make sure that small communes or low density areas were not too much discriminated against. Again it is the recent shift in EU policy in favour of rural areas in terms of the much wider scope of social and economic development which has given strategic relevance to the share of agricultural activities in rural areas.

The EU eligibility choice

The Commission at the time of the reform of the structural funds, established a policy for rural areas, avoiding however to give a definition of them. At the EU level, there are only eligibility criteria for objective 5b areas which supposedly represents (we cannot be sure until a definition is given) the part of rural areas which has undergone a decline in agricultural activities. In the procedure of finding a homogeneous and neutral criterion for defining rurality which does not prejudge the outcome or performance of rural areas, with its corresponding threshold for distinguishing what is rural from what is not, the Commission has skipped these preliminary steps in arriving at indicators of eligibility for support among rural areas.

These eligibility criteria are based, for the period 1994-1999 and for areas at NUTS III level or smaller, on the low level of economic development (measured in terms of GDP per inhabitant below the EEC average) and also on two of the following three criteria:

a) high share of agricultural employment in total employment;

b) low level of agricultural income, in particular as expressed in terms of agricultural value added per agricultural work unit (AWU);

c) low population density and/or a significant depopulation trend.

There are other eligibility criteria based on very heterogeneous criteria such as: peripherality, islands, impact of the Common Agricultural Policy (CAP), structure of agricultural holdings, pressures on the environment, less favoured mountain areas, fisheries restructuring impact.

These criteria, ascribed to declining rural areas, coexist, as was mentioned above with other territorial eligibility criteria. In fact objective 1 areas, even if based on different indicators (per capita GDP below 75% of Community average) do not refer specifically to rural areas but to regions lagging behind on the basis of economic development (disparities), and includes both rural and urban areas: in fact if we look at a map of objective 1 areas we see that whole countries are included; this is the case of Spain, Portugal, Greece, and the new Landers of Eastern Germany. The most recent objective no.6 has been constructed specifically for Scandinavian countries with problems of extreme low density population and desertification not necessarily linked with farming activities.

Therefore, a territorial typology based on rural/urban differentiation is just one, among others, with different criteria used at EU level for dealing with typologies for policy purposes. In identifying the criteria for rural policy eligibility, the EU has not used those previously defined by Member States within the framework of their territorial policies.

The OECD rural indicators

The OECD has been working for several years on a general definition of rurality with simple, neutral and homogeneous criteria; after a universe of rural areas has been established, their performance is studied through indicators which tell us which ones are, for example, losing population, declining in farming activities, providing poor services and communications, etc.

The homogeneous criterion chosen is population density calculated at two different levels, regional and communal level. This double step serves the purpose of establishing a link between the rural area and its regional context. The threshold for defining rurality is 150 inhabitants per km^2. The result of this exercise identifies three typologies of rural areas at regional level (usually corresponding to NUTS III areas in Eurostat's designation):

a) predominantly rural areas: > 50% of the population of the region lives in rural communes;
b) significantly rural areas: 15 to 50% of the population of the region lives in rural communes;
c) predominantly urban areas: < 15% of the population of the region lives in rural communes.

The problems with this typology is that in some cases high density does not correspond to a lack of significance of farming (associated with rurality): for example the Mezzogiorno of Italy where farm day labourers live in cities and commute to the estates where they work, or The Netherlands and Japan, also with high population densities and, were they to use such a threshold, they would find themselves with no rural areas. The heart of the matter is the inability to distinguish clearly the notion of rurality (territorial) from that of agriculture (sectorial) since common sense and tradition, as well as the expectations of support from policies, push towards an overlap of the two concepts.

Territorial typologies for policy purposes in the new context of diversity: a pragmatic approach

Having said that new forms of spatial differentiation are overlapping with traditional rural/urban or centre/periphery typologies on the one hand and that the definition of rurality is still an open issue which has not yet been satisfactorily resolved on the other, leaves us with a big unresolved question of how to move ahead in reading and interpreting territorial diversity. Finding new generalisable typologies such as the rural/urban one is not an easy task and cannot be done in the short term, precisely because it is not just a question of finding good indicators but more substantially finding the new logic of the spatial differentiation which is occurring in the real world. Without dismissing the rurality concept for the timebeing, but keeping in mind its difficulties, we should look for the explanation of the new trends which seem to be operating, what are the emerging aspects that seem most significant and easiest to measure. In this transitory phase it is extremely illuminating to try to read the new trends using the old territorial categories. For example, map the rural areas which instead of losing population as expected are attracting new population, using the more appropriate indicator of migration balances rather than the less precise overall demographic trend.

An example: small and medium industries (SMEs) in different types of policy areas

An extremely relevant criterion to look at is what is happening with small and medium industries (SMEs). One of the explanations that has been given for the current economic restructuring and relocation of resources is that the demand for goods and services is moving towards a growing segmentation with the maturing of industrial society. Suppliers have had difficulties, because of their large and rigid production structures to adapt to such segmentation of demand. This gap has created an opportunity for small and medium enterprises and for regions/countries to gain competitiveness. Location advantages thanks to the advancement in communication technologies are no longer limited to urban environments or old industrial centres.

If this is true then the growth of SMEs and the markets for their products become good indicators of the new trends. It is then relevant to monitor this trend: why, where, in which sectors, which entrepreneurs establish or expand the activities of SMEs? Findings could then be read against currently used territorial typologies: rural/urban, centre/periphery differentiations to start with, as well as the objective areas of the EU Commission or other territorial differentiations used for example in documents such as Europe 2000 (fragile, intermediate, urbanised rural areas)

In this last case the immediate use of the resulting typology would allow a better articulation of the objective areas in relation to relevant criteria of territorial diversity. A very rough idea of the result in terms of policy actions adapted to specific types of areas would be:

- in objective 1 areas, supporting SMEs would aim at the diversification from agriculture of local small size markets into any sector of activity (manufacture or service);
- in objective 5b areas (or intermediate areas), SMEs could aim at an integrated diversification, supporting links and cooperation among SMEs either following the logic of sectorial specialisation of industrial districts, or the links between enterprises along one production chain, not necessarily located in the sam area, following the logic of the filière;
- in fragile areas with very low density, the aim would be to stimulate entrepreneurship particularly for environmental protection and leisure activities;
- in urbanised rural areas, the aim could be to promote decentralisation and subcontracting of activities.

This example indicates how, without (very pragmatically) going into a redefinition of territorial typologies in general, we can take into account new dimensions, such as the support of SMEs, which appears as a new priority coherent with the emerging trends in the economy and also provides feedback on the association between types of areas and appropriate policy actions. This feedback contributes to an understanding of patterns of bottom up approaches.

A NON-PRAGMATIC APPROACH FOR UNDERSTANDING NEW SPATIAL DIFFERENTIATIONS

A critical analysis of any specific context may help us put the right questions and then explore what appears to produce major changes in the traditional forms of spatial differentiation, so difficult to understand conceptually: regional shifts in the location of economic activities both within a country, between countries, and between Europe and the rest of the world on the one hand, and, on the other, the processes of territorial diversification in general, including rural areas, are, in my opinion, the most important processes currently under way and are responsible for the changing competitiveness of rural areas.

If it is true, as it seems, that there are indeed alternative production systems exist, the spatial dimension of the "productive" environment becomes a crucial concept, since it becomes the unit of reference where a specific combination of factors (social, economic, technological, institutional) are organised in a particular way and determine its competitiveness. We should keep in mind however, that political economists and neoclassical theory have constructed a conceptual framework, the perfect market competition, which abstracts the enterprise from any spatial or historical anchorage, making it extremely difficult to go back to the real world and conceptualise reasons for alternative production systems. For the time being, the new conceptualisations of recent trends have in a way suspended or left aside the theoretical problem and are proceeding with single concepts that appear to better understand ongoing changes in spatial differentiation.

Industrial economists who have described the recent development of small and medium size enterprises, have generically indicated their territorial diffusion, opposing it to the urban concentrated pattern of traditional large scale industry. The association of this type of development with the rural-urban differentiation has never been considered an important dimension in the literature. It is the regional or local economy in which the small and medium enterprises operate, which has to be described as whole, *i.e.* its multisectoral aspects and interrelationships, its links with society, institutional organisation and capacity to reproduce over time, and its integration with other local, national or international economies. Among these aspects, the presence of one large urban centre, several small cities or the absence of them, and the nature of the exchanges between rural and urban areas is relevant and necessary information, but in itself is not sufficient for discriminating one regional or local economy from another or identifying its competitive advantage.

This relative indifference should not blur the fact that within the recent process of internationalisation of enterprises, the rural location may have *greater or fewer chances* to acquire competitiveness, and this alternative seems much better than the fatalistic decline to which rural areas were bound in the previously perceived concentrated urban development; however, these better chances are not predetermined or generalizable to all rural areas as such, as is often thought; on the contrary, it is the local context of each specific rural area which determines its capacity to play an active role in the local economy. The presence of an urban centre is a possibility, but some industrial districts are located in rural areas which have developed quite autonomously from major urban centres.

In the most recent situation, the maintenance of diversified activities in rural areas, the presence of small farm systems, simple forms of cooperation among enterprises, independent and artisan work, are no longer to be considered as obstacles to the modernisation of rural areas, but rather as a precious human capital that should be sustained in its evolution and permanence. The presence of a non agricultural rural population plays a central role in the process of diversification of activities and in the generation and circulation of entrepreneurship. This implies that in today's less developed areas, instead of expecting the decline of peasant farming and the abandonment of rural areas, we might aim successfully at a change in the sector of activity without a decline in the rural population, since the location of non farm activities need not necessarily be urban.

The concept of competitive advantage, applied to a productive environment, which could describe a specialised area in a specific sector (industrial or agricultural districts) or a mixed area with various sectors of activity with different degrees of interrelationships and integration, allows for a less predetermined evaluation of different spaces, whether defined as rural or as local economies. Comparative advantages on the other hand, do not appear to explain the changing nature of advantages over time but rather a static and predetermined reproduction of the same advantages, typical of the classical theorisation of development.

Shifting regions and the diversification of rural areas to explain changes in competitiveness

Observers and statistics have pointed out the presence of a heavy, long term trend, consisting of a movement of industrial activities out from older industrial areas towards either different regions in the same country, or to other less developed countries. This geographical redistribution has not been interpreted in general as an economic decline, partly because the service sector and different types of industries have been gradually substituting the activities that moved out and partly because the welfare system and policies have acted as shock absorbers for the unemployed. I have found evidence that France, England, Denmark, Germany and Austria have measured this kind of mobility within their national boundaries during the 1980s and early 1990s. It appears as part of the same process at another spatial level the emergence of new industrial countries (NICs) in the Far East area. In the U.S. former leading industries, such as steel, moved first to other States and then to Third World countries.

The most frequent explanation for these shifts has been that mature industries find lower labour costs in less developed countries. We could also add that emerging countries are the only markets for certain goods with low quality and prices, that some technologies are very easily sold to aspiring local entrepreneurs by developed countries, that it is certainly preferable, from this point of view, to export capital than to sustain the social and economic costs of importing labour. But a "picture" of emerging regions based on their poor, second best nature would be a highly distorted one. Higher segments of certain markets are solidly held by some of them (the example of the NICs is illuminating), and they have achieved competitive advantages in many goods destined to the advanced countries. The multiplication of producers in a growing number of countries reduces the size of the potential market that any one enterprise might have, and this cuts the ground off large multinational firms. The list of factors affecting the regional shift of industrial activities is certainly far from complete, but there is no doubt that this type of spatial mobility has not met with any of the barriers that were imagined in the 1960s and 1970s by theorists of development. As a result the achievement of competitive advantages on the global market is no longer a privilege of the older industrialised countries, and the nature of the links between areas is an extremely complex matter that should be studied without hierarchical prejudice.

In developed societies the spatial impact of regional shifts in the location of economic activities has affected urban centres more severely than rural areas. This is due to the simple fact that population and non agricultural jobs are concentrated in large urban centres and metropolitan areas of early industrialised countries. The regional shifts might have a temporary or less noticeable impact on rural areas: economic activities might move first to rural areas and then to other countries, in a rural or urban location.

The shifting location of economic activities is tightly connected with the changing nature of rural areas. Before exploring this matter, a point of clarification is useful. Most of what has been said in the past about the relationships between developed and less developed countries, is often found mechanically repeated and applied to urban and rural areas. The difference between the two pair of concepts is however quite substantial. The rural-urban polarisation distinguishes homogeneous categories. All rural areas are assumed to have similar ideal characteristics: low density of population, relevance of agricultural activities, traditional value systems and a little differentiated social organisation. At the other end, urban areas base their homogeneity on a complementary set of characteristics: high concentration of population, prevalence of industry and services, presence of modern value systems and highly articulated social structure, place of institutional power and bureaucratic organisation, setting for cultural contacts and exchange. The developed-underdeveloped dichotomy does not refer to a homogeneous set of criteria but to a functional area (or region) as a whole, comprehensive of its rural and urban parts. The criteria of spatial differentiation is in one case the degree of economic development while in the other it is the physical concentration of people. Movement between the two extremes of the polarisation responds in each dichotomy to a specific dynamic. If this is true, why are the two confused? Because preindustrial societies are agrarian societies, and because rural areas in developed contexts are left, after a process of urban industrialisation with a declining population and "modern" farms which need protection in order to go on producing. Both underdeveloped areas and rural areas were always thought to need policies in order to achieve general economic development in the first case and to maintain a "comparable income" in the second. These are probable reasons for the confusion between this two quite different concepts and territorial differentiations.

It is worthwhile clarifying the meaning of this concept because the shifting location of economic activities and the process of diversification of rural areas redefine the advantages and disadvantages of old and emerging regions as well as the division of labour between rural and urban areas. Emerging areas do not seem to reproduce the classical pattern of urban industrialisation but, as was seen in the Italian experience, more freely use all the space available, more closely following the spatial origins of the entrepreneurs than any other more sophisticated location criterion. This seems "irrational" if looked at through the eyes of location theory, however, it becomes quite reasonable in the light of what has been said about the local economy's competitive advantage, acquired through an original integration of local capacities and resources (the contextual knowledge) integrated with information about and relationships with the global markets and technologies (the formal codified knowledge) (Becattini and Rullani, 1993). The process of diversification of activities in rural areas is therefore the result of a quite flexible and not necessarily urban interpretation of the best location for non agricultural activities (industry and services).

We can identify two distinct periods in the diversification process. In the post-war period, rural areas in advanced economies lost significant numbers of farmers. However the decline in rural population has almost everywhere been less severe. This implies that part of the farmers left their activity but were able to find a job in other sectors or to commute, thus maintaining their residence in rural areas. This gradual and modest diversification process can be measured by the decreasing percentages of agricultural employment in relation to total employment in rural areas. While in the 1950s a rural area might have had over half of the active population on the farm, today we are impressed by the degree of rurality if this indicator is just a quarter. Since this is a relative measure, it also indicates the growing incidence of the non farming activities in rural areas. This long and gradual process of spontaneous diversification of the activities of the local resident population was largely ignored both as a real process and as an interesting policy orientation, against all evidence, until the European Commission recognised the trend and its usefulness for rural development in the second half of the 1980s (EC, The Future of Rural Society, 1988). This long blind period corresponded to the political relevance of the rural-agriculture myth and to the strong assumption that farmers had no income alternatives in rural areas, since the recognition of diversification would have implied a radical change in the compensatory approach.

This first type of "endogenous", modest, long term and invisible diversification, which often took the form of pluriactivity, was followed, from the 1970s onwards by an acceleration of the process. Of course this general statement simplifies the significant territorial differences in the intensity of the acceleration. The shift in the regional location of activities, in the same period, brought about new economic opportunities in rural areas. The diseconomies of agglomerations in urban areas increased the demand for rural housing and for transportation facilities to urban labour markets. The growing sensibility towards environmental protection and sustainable forms of development added a new dimension to more traditional rural-urban exchanges. The segmentation of demand included tourist and leisure activities in the countryside. These changes created a new demand for rural space from the non rural population which created new opportunities. In some cases these benefited the original local population, in others only the newcomers and non local enterprises. In any case this external and visible demand for space, living, working and transportation facilities brought about a social restructuring of rural communities, sometimes perceived as a welcome integration, sometimes bringing quite opposing "views of the world" into contact and giving rise to social conflict.

The acceleration of the social and economic diversification of rural areas blurs their homogeneous rural character in relation to its opposite, the urban one, and questions the relevance and significance of the rural-urban dichotomy. The more the rural is diversified and integrated not only with the urban, but with the global economy, the "less" rural in ideal terms it becomes. A new label would satisfy conceptual clarity but runs against the need to retain the rural ideology, which often is at the origin of the new demand for rural space. It should be made clear in any case that for the purpose of analysis and policy formulation the incidence, dynamics and type of diversification are crucial variables to take into consideration. It is possible that the process of diversification will turn out to be the real modernisation of rural areas, interacting in still unexplored ways with the "old" modernisation of just the agricultural sector.

Past processes of development will inevitably influence the response of rural areas to regional shifts and diversification. Little research has been done with this perspective in mind. The next section elaborates on the identification of possible alternative situations. For the future it should be clear that the two processes just described are currently operating to significantly change the logic of spatial differentiation until now considered "normal".

The resulting changes of competitiveness and the increasing heterogeneity of rural areas

The combined effect of the two processes described above has produced a relative improvement in the competitive advantage of rural areas. A point of clarification is useful about the conceptual framework for understanding the real processes that have actually taken place. What has been difficult to explain using as reference the logic of economies of scale, the assumption of unchanging comparative advantages and a uniform, mass demand for goods, becomes possible - at least in principle - when we consider a different conceptual framework, based on what has been called “economies of scope” or diversification (Chandler 1994), changing comparative advantages over time, and the growing segmentation of demand. With this new conceptual tool the growing competitiveness of rural space can get a first tentative general explanation. In order of time, the presence of alternative logics of production appeared first with the empirical acknowledgement of the success and modernity of small and medium industries in a global market, which was supposed to belong mostly to big enterprises; only afterwards - in front of an open disobedience of facts (Becattini and Bianchi 1987) - dominant theories were questioned and reformulated. The new conceptual tools provide the framework for understanding those cases which, far from being exceptional, are concrete examples of the existence of new forms of competitiveness. Having established the new processes under way on the one hand and a first conceptual framework on the other, we may proceed to assess the more favourable - or neutral - impact that these processes may have on rural areas in relation to previous ones, which privileged urban the urban location. In other words rural areas now have a chance of development without necessarily following the pattern established by urban areas in the early industrialisation period.

We should not make the mistake of thinking that because we have found cases of spontaneous rural development and new tools to explain them we may expect rural development everywhere. Specific historically accumulated conditions and human capacities, as well as the impact of the first logic mentioned, will continue as always to oppose formidable constraints or opportunities to rural change. But what is new is that the evolution of the industrialisation process has determined a reorganisation of production, still under way, which does not discriminate against the location of economic activities in rural areas as previously. Rural development becomes a real option in this new scenario, and it may also be promoted using different policy instruments from those of “classical” development. Hence rural areas can compete autonomously, if they still have the resources or can attract them - with urban areas on a more equal footing just as industrial districts can compete with big firms. There is no justification for hypothesising *a priori* a hierarchical subordination of one to the other. By the same logic that they can compete, they can also cooperate as a local integrated rural-urban economy in the first case and as networks of enterprises in the second. A symmetrical analysis may be applied to emerging regions which are attracting economic activities and may successfully compete with advanced regions.

Within the logic of economies of scale, rural areas were conceived as born losers, while now they can become winners (as they have in fact done in some cases) adopting the logic of economies of diversification. These economies are particularly suited for rural areas. The competitiveness in this case is sought with the production and distribution of different products by one or more enterprises instead of seeking it on the basis of one product by a large enterprise.

In the past years an intense debate about the recognition and evaluation of the increasing competitive advantage of rural areas has taken place, often touching highly emotional chords. Some have started talking about a renaissance, “les nouveaux ruraux”, diffused industrialisation, counterurbanization while others have recognized the change but have seen it limited to exceptional cases (the Italian experience), as a temporary passage (the United States), or as the result of decentralisation processes looking for cheap labour, involving the black economy, bad working conditions and on the whole not able to modify the overall declining trend of the rural world. The first group has been labelled as optimistic, the other as fatalistic or conservative. The possibility that rural areas might have a chance of development, not so much from traditional rural policies but from quite spontaneous processes under way has been quite difficult to accept and has caught many observers and policymakers unprepared. Worst of all, the increase in competitiveness was based on the diversification of rural activities rather than on agricultural specialisation, and this has been quite painful to acknowledge for those who spent the post-war period trying to create a rural world based on economies of scale - with obvious European limitations - in the agricultural sector. Furthermore this process proceeded with the support of increasingly costly policies.

The debate should obviously proceed on less emotional arguments. After all if we maintain a global perspective, what we consider an optimistic position turns out to be much less so. Behind the relative increase in the competitive chances of rural areas, not automatic and difficult to achieve in any case, the process of overall economic restructuring might end up with the old advanced societies having a smaller share of the global market than they had in the past, with all the negative consequences that this possibility will have on employment, average income, and capacity to generate surplus to redistribute through policies. This may happen if the regional shifts in the location of economic activities end up outside national or European boundaries and create new and more successful competing subjects. In the second place the growing competitiveness of rural areas is balanced by the difficulties of cities in sizing down. Third, the social restructuring taking place in rural areas as a consequence of diversification may be expected to produce more social conflict and a significant transformation of its Gemeinshaft character. In the fourth place, rural diversified areas still have to start thinking about the difference between the preservation of the environment from agricultural activities and from non agricultural activities, which are no longer just an urban problem. As anyone can see, there is little room for optimism. It is necessary to recognise and clearly define the real problems of the new situations rather than persisting in the old way of looking at them.

CONCLUSION

In conclusion, we need a territorial typology more adapted to new emerging territorial differentiations which question the relevance of older and established typologies such as rural/urban dynamics or the fragility of areas based on remoteness. In following this direction we may act pragmatically, constantly checking new trends against the established typologies.

While waiting for the elaboration of more adequate typologies in the medium term, we can think in terms of diversities which reproduce over time rather than disparities that should be eliminated, reconsider territorial analysis using the concept of changing competitiveness in the global market rather than guaranteed economic development forever, and collect information on emerging aspects which will help us link policy actions with types of areas.

A minimum requirement is that:

a) the region be considered the maximum level at which territorial typologies acquire some significance and the local economy as a new level of territorial differentiation;

b) an unprejudiced comparative analysis be made of the causes of success and failure of policy actions, and patterns and criteria of transferability between similar and different types of areas be identified;

c) an evaluation be made of best practices within their territorial context (partnership styles, quality of programming, etc.)

The themes and paths to follow in order to explore and know more about the emerging trends in territorial differentiation are:

- monitoring processes of internal diversification both in rural and urban areas (in less diversified and diversified areas);
- measuring the exchanges between the local economy and the global market (type of external integration);
- identifying patterns in the forms of partnership and programming (the institutional representation of collective interests at local level).

The implications of what has been said up to here have relevant consequences for rural thinking and policy making.

First, we should definitely abandon the usual crying game about the terrible fate of rural areas: some of them are well and alive. A lot more can be learned from an unprejudiced analysis of how these areas have arrived at such a level of competitiveness and how they have managed to maintain it. It is argued that very often new forms of external integration between the local/rural areas and the world markets, and internal diversification of economic activities are an important part of the explanation of their success.

Second, once it is acknowledged that rural areas may be competitive within the present trends in the economy, a major task lies before us: critically reconsidering traditional spatial hierarchies and finding rural indicators which measure current trends.

A few words about this task. Present spatial differentiation is embedded in functional or structural categories of thought, characterised by hierarchical relationships such as core-periphery, developed-underdeveloped, marginality-centrality, territorial disequilibria, dependency, disadvantages; these are all concepts which transform the process of change into a mechanical exercise where time and space have disappeared as relevant dimensions. The important thing is not to predetermine the path and the outcome of the socio-economic transformation of rural areas and to admit the reversibility of acquired advantages as a normal part of what has to be explained.

Current indicators of rural development should abandon the assumption of measuring rurality through agricultural activities, poverty and distance. As has been proposed in a recent work, rural-urban differentiation should always be analysed within the type of regional context in which it is found (OECD 1994). In fact rural indicators should not presume *a priori* decline, but rather - and more positively - what is considered distinctively rural, which might refer to more than one type of area. From what has been said above, information about the degree and type of diversification, both of activities and population should be central. Types and multiplicity of links with the outside world should be more important than distance. After all, physical mobility is not the only type of exchange with the non local markets: new forms of communication and exchange should be included. Entrepreneurship, turnover, type of sectors present, forms of organisation and cooperation among enterprises, participation in networks, market access also. Population density is obviously important, but should be kept as a descriptive rather than a classificatory variable. The threshold between rurality and non rurality becomes totally arbitrary if measured with density, since it is a function of the type of development (more or less concentrated) and of demographic evolution which have been influencing variations in density much more than rural-urban constitutive characters. We know very little about the non farming population in rural areas. The percentage of residents active in agriculture could have a different meaning if considered in the light of the possible increase of industrial and service employment associated with absolute variations in the number of farmers. In sum, indicators in general should measure the whole rural world instead of a declining aspect of it.

Third, the implications for policy of the different view of rural changes that has been presented are quite far reaching. If the indicated trends are accepted the whole construction upon which the compensatory philosophy for rural areas has been built should be revisited. The main strategies for rural areas should emphasise economies of diversification in relation to agricultural activities, local entrepreneurship, bottom-up approaches, small and medium enterprises, integration with the global economy, coherence with a segmented demand, maintaining flexibility and competitiveness. This policy profile closely recalls the LEADER programme of the European Union. Maybe it is this coincidence with current trends which lies at the basis of the success of the initiative.

Rural policy should not be agricultural policy in disguise. It could be asked why at the end of the 1980s the Common Agricultural Policy shifted towards a rural perspective. A plausible explanation is that the rural policy has been "invented" in order to compensate the farmers for the reduction of agricultural prices. This view would be supported by the constant confusion of agriculture and rurality that is found in many documents. But it appears restrictive: documents such as "The Future of Rural Society" point towards a mixture of reasons, some more instrumental, others with a higher inspiration, but still in conflict.

Finally, it seems like an ironic paradox that those rural areas which benefited most from the past agricultural policies, became most productive and efficient, non pluriactive and professional, are now the more fragile rural areas both in relation to the new situation created by the reduction in price support and to current trends in favour of economies of diversification. And vice versa for more diversified "disadvantages".

The very synthetic and broad reconstruction that has been attempted should be considered as part of work in progress which needs more thinking and empirical/comparative attention. At the same time that it proceeds in various disciplines, a parallel work on critical evaluation of traditional conceptualisations must take place. The original work that might be present in this paper deals much more with a time consuming effort of interdisciplinary bricolage than with a single theoretical perspective. Since agrarian economists have understood their work in sectorial terms, in order to adopt a territorial perspective they are obliged to go through this exercise.

The hypothesis I would like to sustain is that the rural-urban reading of spatial differentiation was meaningful until the processes of urbanisation and industrialisation functioned in the classical concentrated way that was typical of the first generation of developed countries. However with the decline of the spatial coincidence of both processes (diffused industrialisation, new leisure functions of rural areas, decentralisation of public services) the original homogeneity of these analytical categories has been progressively blurred, cancelling the neat division of labour between rural and urban areas, and thereby reducing its explanatory capacity. As long as rural areas remained in the domain of modern agricultural activities, their homogeneous nature was evident both within a single country and between different countries; they did not need to be contiguous to each other since the logic of their aggregation was found in the similarity of their characteristics - the predominance of agriculture. This opposed them to urban areas, which were also non contiguous to each other and whose similarity was based on the concentration (high density) of population and the presence of industrial and service activities. We have demonstrated that both from an economic and a social perspective the redistribution of population and activities in areas that were formerly considered as rural obliges us to reconsider the relevant spatial unit of analysis.

AGRICULTURE AT THE LOCAL AND REGIONAL LEVEL

ANDRÉ VALADIER
Cooperative Jeune Montagne
Route de Saint Flour
12210 Laguiole, France

I am a cattle breeder in the mountainous plateau of Aubrac. I am also the national leader of the INAO Dairy Products Committee and, at the regional level, I head the Agriculture and Rural Economics Commission for the Midi-Pyrénées region.

In 1960, when we were wondering about the future of our mountainous terrain, a CNRS team came to have a look at the way we did things. Our production methods and practices had not changed one iota since time immemorial. In our isolated, landlocked region, our ancestors were able to combine local know-how and practices with human and natural factors to survive, and then later on to cater for economic demand.

Grasses were the only useable vegetation. We produced a cheese that would keep, that came from cows often on the move in search of more palatable grasses. (transhumance). The cheese, as it was ripening, grew more and more tasty. After the animal had given its last "grass-driven power", it was culled. Every part of the animal was used, even the horns; they were made into knife handles.

Producing for the pleasure of producing worked for a while, but... Aubrac chose to enter the post-war period with determination. When the CAP moved in, our way of thinking took us sure-footedly into the market. But after 20 years we started wondering about the value of inflicting industrial organisation dominated by strictly produced-oriented criteria on our local scene. At that same time, we realised the dangers of the European compensation policy for local handicaps lasting forever. We felt that it was not advisable for a zone in difficult conditions to found its future totally on exploiting this handicap to the hilt, by perpetuating errors, inadequacies and oblivion — by planting bananas on the mountains where the yield would be zero so that we could capitalise on a handicap 'til kingdom come'!

Since genetic research opened the way, in 1976 we decided to revive the "genes" of our local cattle breed, a symbol of our identity. We decided on an approach. It wasn't a step backwards, but rather a look at everything that had been done in the past. The Aubrac breed, Laguiole and Aligot cheeses, the knife industry, these were the things that had kept our region going. Why not try them again?

At that time, the work by CNRS which I mentioned earlier, allowed us to examine our past without, however, becoming its prisoners. Now the results speak for themselves. In 1978 we had 5,000 calves from the Aubrac breed, by 1997 we'll have about 25,000. And for the cheese production, in 1997 we should reach the 1883 level of 700 tons again. As for the knife industry, ten years ago it only employed three people, now that figure is up to 190.

Tourism is another field that is developing rapidly, along with local restaurants and food products.

Further, some elements of the past, that seemed ready to enter the Musée des Arts et Traditions Populaires, turned out to be major assets for our economy.

We knew that we could not win in the competitive game of making products at the lowest cost, so we turned elsewhere, beyond nutrition and product safety, and tried to highlight other values by bringing out qualities that appealed to the senses, culture and emotions. This is how we made 20 to 30% added value using an economic approach that might be called "serenity". What I call a "serene" market is in opposition to a "tense" market, one which requires special sales events or demonstrations in big supermarkets to push prices up.

In the main, I could say that we somehow followed instructions from our ancestors on how to survive: *"When a blizzard starts, stop early before the snow covers everything and makes it impossible to turn back, then walk in the footprints you left behind you until you reach the road".*

We did all this without needing wooden clogs. Everything we did helped us build a "human dike" to stop the demographic, economic and cultural haemorrhage. Aubrac is on a dynamic limb, and its population is growing younger. This is because we did not give up. We played on everything that was positive. We also managed because our approach was coherent with our identity and culture, and because we sought to respond with foresight to a demand from consumers, urban-dwellers and a society trying to find its bearings.

AMENITIES IN RURAL AREAS: RESPONDING TO A NEW SOCIAL DEMAND

MARKUS F. HOFREITHER
Universitaet für Bodenkultur
Gregoe Wendel Str. 33
A-1180 Vienna, Austria

ABSTRACT

Improved private welfare as well as information about the state of the environment has increased the demand for rural amenities in the form of landscape features as well as intact habitats and ecosystems. Modern farming systems, however, are sometimes at odds with this demand. The property rights structure is seen as an important element for an efficient provision of socially demanded rural amenities. A couple of important open questions can only be answered by intensifying research activities in related fields.

RURAL AMENITIES - WHO DEMANDS WHAT?

According to standard economic measures of well-being our life has improved a lot during the last decades: we do earn more money in real terms although our working hours have decreased. This development has important consequences: among them, higher personal incomes make private goods less scarce, thus shifting our focus of interest more and more towards an adequate provision of public goods. In addition, the increase in leisure time also results in an increase in the actual use of public goods as an input for consumption activities. An important form of public goods in this respect is the environment in its broadest meaning.

Amenities in rural areas are in most cases linked to a stock of natural or semi-natural resources providing a (potential) flow of services generating utility for consumers. An illustrative activity in this respect is outdoor recreation. This not necessarily implies a physical use of these resources, but also simply enjoying the attractiveness of a particular area. In general, deriving utility from the rural landscape and environment falls into the category of *use value* of natural assets. However, defining rural amenities in such a narrow sense would not be sufficient to cover the whole range of this subject, as one can also enjoy rural amenities without directly making use of a natural resource. In economics these forms are summarised within the *option value* and *existence value* concept. Here utility is derived from the fact that some natural endowment is maintained in its present state, either to protect it for potential later use or to ensure its existence per se. Summing up, we can assume that utility associated with rural areas mainly is derived from the following elements:

- aesthetic characteristics of the countryside (demand for *landscape protection*);
- wildlife and plant diversity and maintenance of genetic resources (demand for *habitat protection*);
- sustainable water and soil management (demand for *ecosystem protection*).

Many elements of farming systems influence one or more of these environmental assets. In many cases the elements of rural amenities are interlinked, making proper action already from a purely natural science point of view quite difficult. Moreover, in economics many rural amenities have to be treated as externalities and public goods, which poses serious problems when it comes to determining the value of such goods. Strategic behaviour of potential consumers due to the well known characteristics of non-rivalry and non-excludability make a direct assessment of such values impracticable. Hence, the market mechanism is not a feasible instrument to induce optimal provision of such services. Only in a few cases a COASE an outcome may assure the desired level of provision. Sometimes such services are provided on a voluntary basis due to altruism (Colman, 1996; Weaver, 1996). The majority of activities in this respect, however, is brought about by legal regulations and monetary incentives provided by public authorities. Hence, attempts to equate demand and supply of rural amenities are in most cases part of the political decision process. However, as in many cases the demand side is covered by a veil of ignorance; this enhances the risk of political ad hoc decisions.

PRACTICAL DOMAINS OF RURAL AMENITY PROVISION

As stated above, rural amenities can be derived from three sets of environmental assets: the rural landscape, habitats of animals and plants, and ecosystems. All three domains are more or less influenced by agricultural activities. Many semi-natural habitats of Europe embodying amenity as well as habitat and environmental services are the result of sometimes millennia of agricultural production in these regions. However, empirical evidence suggests that the rapid changes of agricultural production techniques in the last decades, on the one hand, but also the potential cessation of agricultural activities, on the other do imply the most severe threats for the future provision of that service.

By and large, the **rural landscape** is created and managed by agriculture. The consumer of this public good appreciates primarily the visual attributes of the rural countryside in certain areas, hence the maintenance of particular landscape features adds to making an area attractive for residents and tourists. However, it is reasonable to maintain that there are important exogenous prerequisites for such an appreciation in terms of use values. The most fundamental prerequisite certainly is that such rural amenities are a positively valued element of the preference structure of consumers. Another is the possibility to actually have access to rural amenities, *e.g.* in an area where farm land is spread across a flat scenery up to the horizon, attempts of farmers to structure this landscape are less visible than in a mountainous region like the Alps, where the spectator can enjoy a great variety of different views. Hence, the valuation of an identical landscape activity may differ widely according to the area where it is performed.

The implications of agriculture on **habitats** is a mixed story. On the one hand, from a historic perspective, agriculture has broken up the previous forest cover of Europe, thus creating a number of different ecosystems by combining different management practices with different soils, climates, and gene-pools. This increase in variety has favoured the development of different species of animals and plants. The tendency towards "homogenisation" of European agriculture due to technical and socio-economic changes in the very last decades poses serious threats to habitats. There is a strong empirical indication that plant diversity has diminished significantly since the beginning of the "industrialisation" of agriculture (Nowicki, 1996). However, under specific conditions agriculture still can contribute to biodiversity through changing production practices, *e.g.* meadow management.

The impact of agriculture on **ecosystems** is mainly caused by the use of soil, water and air. Well known examples on the negative side are the pollution of ground and surface water through nitrates and pesticides. Positive contributions come in the form of *e.g.* wetland management and soil conservation practices. Yet, the net effect of present agricultural production activities on the various ecosystems seems to be negative, at least in regions with intensive agricultural production.

In general, any incentives to induce the provision of rural amenities have to compete with strong technical, economic, and socio-economic driving forces that continuously pushing agricultural production to improve efficiency and cost effectiveness. Hence, the farmer weighs his/her willingness to comply against the opportunity cost of performing such measures. We frequently encounter situations, where detrimental effects of a basically commodity-oriented design of agricultural policy have to be offset by particular counter-measures. In quite a few cases this implies a significant loss of economic resources, as budgetary outlays of environmental programs partly have to be applied to offset the detrimental environmental impacts of budgetary outlays directed to commodity programs. Sometimes an erroneous distinction between positive and negative contributions of farming enhances the risk of an inefficient provision of the desired state of ecosystems through overcompensation of farmers (Nellinger 1996).

Obviously, a better integrated agricultural policy would improve efficiency in reaching stated objectives. The essential prerequisite to outline efficient programs for rural amenity provision is detailed information about the state of nature and landscape (OECD 1994). Similar to other countries Austria is quite active in this domain (Brandenburg *et al.* 1996, Österreichische Bundesregierung 1995, Tiefenbach 1993, Paar 1995). Table 1 provides selected information about particular environmental measures in Austria. For a better integrated agricultural policy however, a set of operational and agreed objectives also has to be given, as is shown in Article 39 of the Treaty of Rome.

Table 1- Selected Measures Provided by Agri-Environmental Programs in Austria (1995).
Source: BMLF (1996), Sinabell (1996), Wagner (1996).

Measures	Required Activities / Restrictions	Area (1000 ha)	Particip. farms	Budget outlays (mn ECU)	Costs per unit ECU/ha
Elementary support	reduced stocking rates, fertiliser management	2,302,968	169,955	118.62	51.5
Organic farming	Reg. 2092/91 (plant production), Codex Alimentarius Austriacus	197,952	15,844	50.74	256.3

	(livestock production)				
Reduction of agricultural chemicals (whole farm coverage)	mineral fertilizers, pesticides, growth stabilizers prohibited)	309,729	37,718	50.02	161.5
Extensive grassland (farm coverage)	pesticides prohibited, no silage fodder	111,647	10,848	20.33	182.1
Winter cover crops	minimum acreage covered by crops in winter, maximum share of cereals and maize	908,764	53,987	98.63	108.5
Extensive bread grains production	maximum amount of fertiliser, special grain varieties	249,215	30,982	46.01	184.6
Reduction of ag. chemicals (arable land)	maximum amount of agri-chemicals	330,202	78,675	24.75	75.0
Reduction of ag. chemicals (grassland)	maximum amount of agri-chemicals	246,571	45,841	33.81	137.1
Erosion control measures	vineyard management, seeding technology, etc	10,100	3,120	1,3	129,4
Control cost organic farms	contract with auditing firm	218,282	15,844	5.05	23.2
(measures excluded)		...	...	...	...
TOTAL (all measures)		2,633,081	179,478	562,2	216,2

PROPERTY RIGHTS AND RURAL AMENITY PROVISION

The stock of environmental assets in rural areas is primarily determined within two policy domains: agricultural policy and environmental policy. All other legal and economic influences - *e.g.* rural zoning restrictions, entitlements to make use of privately owned roads - are of minor importance. Agricultural policy is the dominant regulator of the supply side of rural amenities because rural areas in Europe, as well as the property rights concerning the use of these areas, are almost exclusively with farmers.

On the one hand, agricultural policy influences farm activities via commodity and input price administration, restrictions concerning the volumes of inputs and outputs, but also via information as well as support of production and infrastructure investments. The particular measures chosen in the past had mostly unintended, but nevertheless quite harmful effects on the rural environment. This comes as no surprise, as the main objective has been to support the farming community. As long as the state of the environment has not been a matter of great concern, this negative side-effects have been accepted. Quite amazingly, however, also with the increase in public awareness concerning the environment, things did not change fundamentally. This startling fact may be caused, on the one hand, by deficiencies concerning our knowledge about the environmental impacts of modern agriculture due to the complex and dynamic characteristics of the problem, but, on the other, also by an asymmetric distribution of information between farm interests and the public.

This seems to be at odds with the observation that European agricultural policies features an increasing number of measures related to the environment at all levels. The advent of these measures partly was connected to the formal commitment of agricultural ministers to a substantial and progressive reduction of support and protection at the outset of the Uruguay Round. The intensified pressure in the direction of decoupling agricultural support facilitated the creation and implementation of environmental measures, yet, without automatically optimising their environmental impact.

At present, environmental services sometimes are treated by farmers as minor modifications of agricultural production methods which have to be remunerated by the public, without providing a guarantee that the alleged objectives are positively attained. So for the 'median' farmer, the options provided by the national programs based on Rg. 2078/92 can be simply a menu of choices for enhancing his/her level of income. Admittedly, no explicit responsibility exists for the individual farmer beyond carrying out the operations specified by the program. This responsibility lies fully with the politicians, particularly with agricultural policy makers. Due to the lack of adequate formal links to environmental policy the incentives for agricultural policy makers to establish a set of strictly "goal oriented" environmental measures is fairly limited. Hence, the presently given "measure oriented" menu of choices is - at least sometimes - the result of an optimisation process to serve agricultural interests according to political-economic considerations.

In this process the status quo of property rights related to land and natural resources plays an important role. The current situation is still dominated by the notion that the only "serious" task of farmers is the production of food, feed, and fibre. The fundamental changes concerning economic conditions and relative scarcities so far have not induced adequate modifications in the domain of property rights. Contemporary agricultural policy still builds upon this obsolete structure of property rights. This means, among other things, that farmers have the power to enforce compensations for - sometimes quite insignificant or even not quantifiable - improvements of the environmental situation. This success of farmers in simultaneously capitalising not only the value of agricultural programs into land value, but also the property rights in land into favourable agricultural as well as "environmental" programs is a serious threat for an efficient provision of rural amenities (Bromley and Hodge 1990).

CONCLUSION AND OUTLOOK

Agriculture is the sector with the most comprehensive use of land in Europe. Historically, this utilisation of land has caused a net increase in biodiversity. However, during the last decades this land use has become more and more intense, thereby increasing the potential of harming environmental assets linked to it. At the same time, however, the demand for these environmental assets undoubtedly has increased, although the magnitude of this increase is quite difficult to estimate. In order to equate supply and demand of rural amenities, comprehensive and detailed answers to the following questions are urgently needed:

- What kinds of rural amenities are positively demanded by society?
- How has this demand to be ranked in view of severe budget constraints?
- Who can contribute to the provision of these services in an efficient way?
- Are property rights adequately defined to enhance rural amenity provision?
- How can these services be financed technically?

We have to admit that a substantial part of the information required to respond to these five questions is not available partly because of lack of research, but partly, also because these requirements hit fundamental problems of welfare economics and perhaps can never be resolved (Colman 1996, 303). On the other hand, however, practical decisions are taken regularly by politicians, be they in the area of agriculture or environment or some other field of politics. The fundamental question is, how can sound political decisions be taken without a proper information base? A conceivable explanation is that there will be a tendency to substitute missing information by power of interest groups. In order to improve the efficiency of the process of rural amenity provision, research activities in this area have to be further enhanced. Besides the broad attempts within particular fields of the natural sciences, economics also has a comprehensive research agenda to cope with. Distinctive efforts should be directed to:

-engaging in *interdisciplinary research* to improve our knowledge about the interrelationship between the physical consequences of agricultural activities and the environment as well as the economic costs of these activities and potentially useful alternatives;

-improving the methods for estimating *economic values* of agricultural externalities as well as provided public goods for (potential) consumers to enable society to move in the direction of an optimal provision and allocation of such services;

-gaining deeper knowledge concerning the *political-economic aspects* underlying the sometimes second-best like reactions of agricultural policy in the face of new social demands.

In the last few years these questions have gained in importance within agricultural research strategies at the community level. However, to be successful in the sense of coping with regional and local peculiarities, these topics also have to set foot in national and subnational research interests. By tenaciously attacking these questions it should become possible in the near future to satisfy new social demands for rural amenities provided by agriculture in an efficient and cost effective way. Last, but certainly not least, such a strategy would significantly add to the acceptance of agricultural interests by society as a whole.

SELECTED REFERENCES

BMLF (BUNDESMINISTERIUM FÜR LAND- UND FORSTWIRTSCHAFT), 1996. Österreichisches Umweltprogramm 1995. *Umsetzung der Verordnung (EWG)* Nr. 2078/92. Bericht an die Europäische Kommission, Vienna.

BRANDENBURG CH., GRIMM K., PROKSCH TH., SCHRAMAYR G., AND ZWICKER E., 1996. *Der Landschaftsplan. Stand und Empfehlungen. Monographien Bd.* 69, Umweltbundesamt, Bundesministerium für Umwelt, Jugend und Familie, Vienna.

BROMLEY D. W., AND HODGE I., 1990. Private property rights and presumptive policy entitlements: reconsidering the premises of rural policy. *European Review of Agricultural Economics*, 17, 197-214.

COLMAN D., 1996. Ethics and Externalities: Agricultural Stewardship and other Behaviour. *Journal of Agricultural Economics*, 45, 299-311.

NELLINGER L., 1996. Zur Weiterentwicklung der EU-Agrarreform unter einkommens-, markt- und umweltpolitischen Gesichtspunkten (Teil 1), *Agrarwirtschaft*, 45, 295-302.

NOWICKI, P. L., 1996. *An Overview of the Environmental Benefits from Agriculture in Europe.* Report for the OECD Seminar on Environmental Benefits from a Sustainable Agriculture: Issues and Policies, Helsinki, 10. - 13. Sept. 1996.

OECD, 1994. *Environmental Indicators*, Paris.

PAAR M., 1995. *Important bird areas' in Österreich. Monographien Bd.* 71, Umweltbundesamt, Bundesministerium für Umwelt, Jugend und Familie, Vienna.

ÖSTERREICHISCHE BUNDESREGIERUNG, 1995. *Nationaler Umweltplan.* Bundesministerium für Umwelt. Vienna.

SINABELL F., 1996. Austrian Country Report for the Inventory on Landscape and Nature Conservation. *In*: Umstätter J., Dabbert St. (ed.*): Policies for Landscape and Nature Conservation in Europe*, Proceedings of the EU workshop on Landscape and Nature Conservation, Stuttgart, 26.-29. September, 5-20.

TIEFENBACH M., 1993. *Naturschutzgebiete Österreichs. Band 5: Zusammenfassende Darstellung. Monographien Bd.* 38 E. Umweltbundesamt, Bundesministerium für Umwelt, Jugend und Familie, Vienna.

WAGNER K., 1996. Regional differenzierte Wirkunges des ÖPUL. *Der Förderungsdienst*, 7/1996, 207-214.

WEAVER R. D., 1996. Prosocial Behavior: Private Contributions to Agriculture's Impact on the Environment. *Land Economics*, 72, 231-247.

CHAIRMAN: JACQUES MAIRESSE, *INSEE, PARIS, FRANCE*

VICE CHAIRMAN: PIERRE-BENOÎT JOLY, *INRA, GRENOBLE, FRANCE*

THEME 3

AGRICULTURAL RESEARCH, INNOVATION AND SOCIETY

Workshop 1
Research, technological change and economic development

LESSONS FROM A MACRO-HISTORICAL ANALYSIS: THE FUTURE OF EUROPEAN AGRICULTURAL RESEARCH

AD M. VAN DER WOUDE
Rural History Dept.
Agricultural University
Wageningen, The Netherlands

INTRODUCTION

The development of agricultural research in the 21st century will be closely interwoven with the development of agriculture in general during that period. Therefore, let us start with a prognostication of the development of this branch of economic activity, fully recognising, however, the uncertainties we have to cope with. In the field of European agricultural history, a degree of consensus has been reached about the existence of long secular trends. It might be useful to project the development of agriculture in the European Union during the next half-century in light of the findings of this historical research.

When the historian sets out to investigate how long-term developments in the future can be illuminated by those of the past, he should start with conceptualising the different levels of causation that might influence the object under discussion. To this end, he will find it helpful to utilise the Braudelian trinity of historical processes.

In the *first* group, he put those developments which are situated outside the scope of human decision-making or which, in any case, can only be influenced by human action in the long-term: geological and climatic changes, or the appearance of new illnesses. Most of the time, the influences of these changes will affect social life so slowly that they will pass unnoticed. Only the historian may observe them, afterwards, just as he may observe the gradual adaptation of cultures to the new circumstances.

A *second* group of phenomena comprises developments that take place more quickly, though the word quickly has to be understood in a relative sense: measured on a human scale, these phenomena seem to move rather slowly. Examples include most of the developments pertaining to the economic, social, and cultural life of societies. Mental attitudes change but not so quickly. Economic life exhibits developments and has its cycles, but it is exceptional in that it can change overnight. The same is true of demographic and social phenomena. Population growth and population structure are not static phenomena. It takes decades, however, if not generations, for fundamental changes to take place. Occupational structures, levels of income, social classes and the relations between them are all subject to processes of constant but relatively slow and gradual change.

The *third* group of the phenomena are the most volatile. They are the outcome of the struggle for power over people, *i.e.* developments caused by political decisions, diplomatic actions and military events. At this level the course of history may be changed suddenly by new treaties, military defeats and victories, and political coalitions. But most of the time there is uncertainty about how long the victor will enjoy the fruits of his victory. Coalitions may break down, parties split, elections be lost, new powerful adversaries unexpectedly arise, or agreed-upon measures be revoked. And each of these developments can exert far-reaching influences within a relatively short term.

The course of history may be analysed with the help of this three-level scheme, in which the developments of the first and second level tend to act as limitations on the developments in the third. This, however, does not exclude the possibility of influences in the opposite direction. In the long run, even climate can be influenced by human action, and political decisions can have a deep impact on social structures. But the historian who wants to look into the future with the help of this analytical tool will be in a stronger position if he first pays attention to developments on the first and second level, rather than at those of the much more uncertain third one. This is most probably true for a forecast about agricultural developments in Europe.

ANALYSIS

One important element in the Braudelian scheme of historical forces, *change of climate*, is immediately relevant. There is much discussion about the so-called greenhouse effect. Higher mean temperatures may deeply influence long-term developments in agriculture. If it proves to be true that we may expect a rise of 4-5°C, the effects will be so far-reaching that it will be very difficult to assess correctly its impact on agriculture and vegetation in general. But the question is not only whether such a fundamental change will take place, but also what its speed will be and how far ahead we want to forecast the future. Climatologists have detected a mean rise of about 0.5°C over the last hundred years and are inclined to forecast a further rise of 1-2 degrees by the middle of the next century. There is more uncertainty about what will happen after that date, especially as far as the speed of change is concerned. For us, therefore, it seems wise not to extend a forecast of agricultural developments beyond the middle of the 21st century.

If the change in climate between, let us say, 1900 and 2050 is no more than a mean rise of temperature of 1.5-2°C, the consequences for European agriculture will not be disastrous. Although the factor of humidity, which is less easy to evaluate, may spoil our optimistic expectations; on the whole a small rise in the temperature will probably have more advantageous than disadvantageous effects on agricultural productivity in Europe. Vegetation with extended leaf-growth will especially profit, and a double harvest in some crops will become possible. Also the zone of possible arable farming will extend further north.

We have indeed to reckon with the possibility that in north-western Europe the consequences of the change of climate between now and the middle of the next century will be felt more in the costs of structures to prevent flooding because of the *rising sea level* than in any fundamental change of agriculture.

On the whole, agricultural products are bulky commodities. Consequently, *transportation costs* are an important factor in consumer prices. The change in ocean shipping from the wooden sailing vessel to the iron steamer, with its much lower transportation costs, was one of the decisive causes of the agrarian depression in the last quarter of the 19th century. But the transportation revolution that began around 1840 with the railway and steam vessel seems to have come to a standstill after the introduction of the mammoth tanker and container traffic. The next century will most probably bring revolutionary developments in electronic communication but not in traffic. On the contrary, we may presume that, among overall price components, transportation costs will exhibit an upward tendency. The same is true for the costs connected with the greater care we will be demanding in the near future for the *environment*. Under these circumstances, with rising price tendencies caused by water management, and by transport and by environmental policies, we do not expect the share of the producer in the consumer price to rise, and certainly not as long as European production prices experience difficulties in competition with world market prices. On the whole, European farmers would do better not to count on improvement in income. The influence of long-term changes in climatic, ecological, and geographic factors on farmers' incomes is slight and tends to be negative.

The developments at the second level of the Braudelian scheme are decidedly negative. A central factor here is *demographic expectations*, *i.e.*, the demand for agricultural products. Here we touch on old, long-debated historical questions about the relation between population and agricultural production. We need only to mention the names of Malthus for the 19th century discussion, and of Ester Boserup for modern development economics.

The disparity between the number of mouths to be fed and the quantity of food available has also been a much discussed issue among economic and demographic historians over the last 40 years. They have discovered that for the last 900 years of Western history, alternating periods, with a length of 100-300 years, can be observed in which population growth has outstripped or fallen behind food production. During the periods 1100-1350 and 1475-1650, population growth was a rather general phenomenon in Europe. Sooner or later this provoked a Malthusian tension with regard to agricultural production, and prices of agricultural products showed a clear tendency to rise. During the periods 1350-1475 and 1650-1750 this tension was much smaller because population was static or even in decline, while at the same time agricultural production was expanding. During these periods of contraction, the prices of the most important agricultural products showed an unmistakable tendency to fall.

From 1750 until recently, population growth became a general Western phenomenon. But there also was an unprecedented growth in production and in the availability of goods in the Western world, thanks to technological innovations and/or expanding import opportunities. This is true of the agricultural market too. Therefore, during this time the standard of living has risen formidably, despite the rapid population growth. Hunger has been banished from the Western world, and food has absorbed a much smaller part of the budget than it did 200 years ago. Indeed, notwithstanding 200 years of demographic expansion, the surpluses of agricultural products in the Western world have never been so large as they are at the present time. But at this very moment of abundance in agricultural home production, Europe seems to be at the brink of a new period of demographic contraction. According to demographers, we may expect a decline of the European population by about 10% between now and the middle of the next century. Consequently, the agricultural home market for food products will shrink; in fact, all the more so because sometimes hidden cases of undernourishment have ceased to exist as they did in the past.

The population decline will be felt in three cumulative ways. It will cause a decline of about 10% in the demand for food, all other things being equal. But other things will not remain equal. It is well known that, as a result of the declining birth-rate, the population structure has been deeply affected, and the European population is ageing. At the beginning of this century about 6% of the population was 65 years and older. This is now about 13%, and it will have risen to about 26% by the year 2040. By then, as much as 15% of the population will be 75 years or older. Older people eat less than younger ones. Although no exact information is available, a guess of a loss in demand for food of 5%, caused by this process of ageing, seems to be on the safe side. A third cause for declining demand can be presumed because better eating habits will continue to develop over the next generations. A decline in the daily ration of calories from, say, 2,600 to 2,500 calories, may seem like a trifling amount, but it means a further decline in food consumption of 4%. These three developments taken together suggest a decline in the demand for food in Western Europe of about 20% between now and the middle of the next century.

But for about fifteen years, the European farmers have been producing surpluses! The fact that Europe could develop from net-importer into net-exporter of agricultural products has to be considered as a very important historical fact. Productivity will exhibit a further substantial rise during the next decades, while population developments in Europe will most probably confront us with decreasing numbers and further ageing of the population. This means a decreasing number of consumers with, on average, a lower level of food consumption, just when agricultural productivity is growing. If European farmers do not find a solution for this overproduction, either by developing new export markets or by changing the crop mix to include a larger share of industrial crops in their production, they will be in serious difficulty. However, both these solutions seem to be rather improbable.

Here we arrive at the third level of the Braudelian scheme: that of *political measures*. Will economic policy during the next half century keep intact the system of protection? That cannot realistically be expected. The negotiations within the framework of GATT involved an effort to demolish tariff barriers around the world as much as possible. The world community is becoming too small for lasting protectionist systems. Here too, European farmers have more to fear than to hope as they belong to the most expensive producers in the world.

All the information and insights into the most reasonable expectations for the next fifty years put together *we have to conclude that a long protracted period of depression for European agriculture seems to be at hand.* In fact we are already engaged in it at this moment. It seems to be a period showing remarkable resemblances to farmer long periods of depression in agriculture, *i.e.* the secular agricultural contraction periods between 1350-1475 and 1650-1750. Declining demand by smaller populations combined with expanding supply because of increasing agricultural productivity is the common mark of these contraction periods.

Historians have done a lot of research during the last forty years on the characteristics of these periods. Let me present a number of the most relevant findings for our meeting:

1. Agricultural overproduction not only brings unfavourable terms of trade for farmers but is accompanied by relatively sharp short-term price fluctuations for the most depressed agricultural products;

2. Sharp drops in land and farm rent prices are also characteristic of protracted periods of agricultural depression;

3. Tenant-farmers have been better off than farmer-owners during periods of secular agricultural contraction;

4. Secular agricultural depression has brought about enlargement of the remaining farms and was accompanied by the creation or expansion of a class of expropriated farm labourers and cottagers (nowadays: part-time farmers);

5. During former periods of agricultural depression there were attempts to enlarge the use of agricultural raw materials for industrial purposes (nowadays: especially wood?);

6. The decline of agricultural acreage must be considered one of the most characteristic developments in periods of protracted agricultural depression (nowadays: set-aside regulations);

7. A decline in agricultural science and technology can also be observed during periods of secular depression.

This last topic is at the centre of this conference. Although I tried to show that this is only a part or an aspect of a much broader process, I nevertheless shall conclude with some special observations on this subject concerning the future of agricultural research in a world of agriculture in depression.

Agricultural research, in the sense we use this expression today, did not exist during former secular phases of contraction. Even agricultural science, conceived as interest in agricultural topics unaccompanied by systematic empirical observations but being expressed in writings about those issues, is seldom encountered before the middle of the 18th century. What we do see, however, is a diminution of innovations in practical farming and a standstill in technological progress concerning land reclamation, manuring and water management during periods of secular depression. As those periods were first of all depression periods in arable farming it was especially in this branch of agricultural activity that the introduction of new tools or techniques reached only a low level. The situation in the field of dairy farming was only a little bit better. Inventions in this branch of agricultural production were mainly directed to issues such as hay storage and labour saving arrangements in milk churning. Reprints of old treatises on agricultural topics were more common than the publication of new ones.

CONCLUSION

A new secular period of agricultural expansion started by the middle of the 18th century. The same is true with the interest in agriculture and with the development of the agricultural sciences. I need only to recall names and concepts like the Physiocrats; authors on agriculture and forestry like Duhamel du Monceau, Von Thünen, Arthur Young and many, many others; the foundation of agricultural societies with their regularly and often even yearly publication series in nearly every important country; the birth of modern agricultural sciences with the special impact of German chemists like Justus von Liebig leading to the introduction of artificial fertilisers, and so on. After well over two hundred years we are so accustomed to the existence of this historically new situation that we automatically live with the idea that it will continue forever. Indeed, none of us can seriously believe that agricultural science will not continue to exist as long as human civilisations do. And that idea is correct.

Nevertheless, nearly ten years ago already, I defended the statement that we cannot expect the governments to continue to spend huge amounts of public money on an economic activity suffering from structural overproduction but offering employment to only a very small percentage of the total labour force. Can we expect private companies, in the long term, to continue to invest in research and development of technologies in a sector characterised by a downward pressure on the relative price level and with small margins of profit? Will we witness the fact that the study of agricultural sciences remains attractive for young people, especially for those with above average aptitude for scientific research? If agriculture in Europe has entered a protracted, I mean a secular period of structural difficulties, we may have reason for doubt.

Although it remains very difficult to extrapolate into the more distant future by way of historical comparison, some developments that can inspire thought are already underway. Using data from my own university as an indication of more general trends, we see a decrease in State paid subsidies of as much as about 60% in real value per student between the beginning of the 1970s and the middle of the 1990s. For every one hundred guilders in real value spent per student around 1973, there were only about forty in 1995. It may be that the trend in other Dutch universities has not been very different. But for the year 1997 the agricultural university is the only one among the Dutch universities to be confronted with new cuts in the budget.

During the last ten years the number of students has gone down in nearly every Dutch university, but nowhere as much as in the Agricultural University (>25%). A most remarkable fact may be discovered in the almost total disappearance of students in the plant sciences, traditionally one of the corner-stones of the agricultural sciences. Relatively new branches of agricultural research, like nutrition and food preservation have behaved much better. Environmental studies also gained in importance. It is not too dangerous to presume that this development foreshadows the development of research and scientific publications during the next half century: most probably the total number of publications in agricultural sciences will go down or at best will remain stagnant. Between the different branches, however, there will be a shift in relative numbers: the traditional branches will show important losses, while the branches directly connected with the well-being of the ageing population (*i.e.* with ecological and nutritional issues) will gain in numbers.

AGRICULTURAL RESEARCH AND TECHNICAL CHANGE IN AGRICULTURE: LESSONS FROM MICRO-ECONOMIC STUDIES

ROBERT E. EVENSON
Economic Growth Center
27 Hillhouse Avenue
P.O. Box 208269
New Haven CT, USA

INTRODUCTION

The agricultural sciences have faced major challenges and experienced considerable institutional development over the past century. In the early part of the century important "inter-science" institutional change occurred. In the latter part of the century when global food challenges became more important, "inter-sector" and "international" institutional changes occurred. These changes affected the performance of the agricultural sciences in terms of producing farm productivity improvement and hence increases in the supply of food and fiber. Current systems face continuing food supply challenges (especially in Africa) as well as new challenges associated with the environment and with potential global climate change. These new challenges must also be faced in the context of international market changes and rapidly changing income levels and industrialization in much of Asia.

INRA has been a major part of the international agricultural research system for the past half century. It has participated in the inter-sectoral and international institutional changes of the period. Its performance has been vital to the French economy and to the international economy. It will have a major role in meeting future challenges.

In this paper I have three objectives. First, I will review the major institutional changes in agricultural research systems in recent decades. My second and major objective is to review a large number of micro economic studies of the economic consequences of agricultural research programs over recent decades. I will also review a recent projection of economic consequences over the next 25 years for agricultural research.[1]

1. See Seckler and Rock (1995), Brown and Kane (1994), Mitchell and Ingca (1993), and Crosson and Anderson (1992), for recent examples of such projections. I will rely on projections detailed in Evenson and Rosegrant (1996).

INSTITUTIONAL CHANGES IN AGRICULTURAL RESEARCH

There are three major institutional changes that I will focus on in this section. The first is the "inter-science" changes which were initiated in the first half of the current century. The second and third are the "inter-sectoral" and "international" changes that have dominated the second half of the century.

Inter-Science Institutional Change

The emergence of the "Agricultural Experiment Station" model in Europe in the middle of the 19th century with its formal field experiments and with the development of statistical methods provided a design for much agricultural research that remains intact today. Colleges of agriculture were created as farm groups expressed the need for more formal and more scientific agriculture. These colleges in turn enhanced the demand for research, and with the experiment station model at hand, experiment stations were built in many regions throughout the U.S. and most other countries in the latter part of the 19th century and the early part of the 20th century.[2]

During the latter part of the 19th and the early part of the 20th centuries in the U.S. and elsewhere, considerable tension existed between practical farmer interests and scientists' interests. This tension between farmer interests and the interests of scientists has continued over the years and is present today. It has generally been a healthy tension. Healthy because it reflects clientele (farmer) demand mechanisms which in turn have been transformed into demand for "pre-invention" science. This latter demand was firmly established only after a period during which invention opportunities were perceived to be relatively exhausted. Research leaders and administrators were able to sell farmers on the necessity to develop research programs in the related sciences to "recharge" the invention possibility pools.

The resultant institutional development produced the system depicted in Figure 1. It depicts 6 layers/activities of specialized activities in a hierarchical fashion. Layer III can be seen to be the central focus of most agricultural research centers. Technological invention is their job. But invention does not lead to productivity change if researchers are isolated from "downstream" contacts with extension workers and farmers. Prospective technologies must be tested and monitored. Good extension systems maintain contacts with researchers upstream from them and with farmers downstream from them. Effective systems are thus responsive to the demands of their clientele, - producers, government, and consumers - the final users of their products.

However, applied technological researchers (Layer III) cannot be effective without "invention potential". Invention potential, *i.e.*, the "germplasm" from which new inventions are made, includes literal (*i.e.*, biological) germplasm which serves as a "parent" to further crop and animal improvements. It also includes other inventions that inspire and "parent" further inventions. It includes the "tools" and methods (and instruments) of the field. And most importantly it includes intellectual germplasm as communicated in graduate schools and scientific journals. This is why figure 1 upstream links to the pre-technology or pre-invention sciences (Layer II) are important. Applied research programs without such links can find themselves in an "exhausted" state with little invention potential. In agricultural research

---

2. Huffman and Evenson (1993) discuss these developments in more detail. This section is based on this work.

Figure 1: Hierarchical Specialization in Research and Development Systems for Agriculture.

Layer/Activity	THE R&D SYSTEM				
	Mathematical Sciences	Physical Sciences	Biological Sciences		Social Sciences
I. GENERAL SCIENCES (University and public agency research primarily)	Mathematics Probability & Statistics	Atmospherical & Meterological Sciences Chemistry Geological Sciences Physics	Bacteriology Biochemistry Botany Ecology	Genetics Microbiology Molecular Biology Zoology	Economics Psychology
II. PRE-TECHNOLOGY SCIENCES (University and public agency research primarily)	Applied Math Applied Physics Engineering Computer Science	Climatology Soil Physics & Chemistry Hydrology & Water Resources	Plant Physiology Plant Genetics Phytopathology	Animal & Human Physiology Animal & Human Genetics Animal Pathology Nutrition	Applied Economics Statistics & Econometrics Political Science Sociology
		Environmental Sciences			
III. TECHNOLOGY INVENTION (Public and private research)	Agricultural Engineering & Design Mechanics Computer Design	Agricultural Chemistry Soils & Soil Sciences Irrigation & Water Methods	Agronomy Horticulture Plant Breeding Applied Plant Pathology	Animal & Poultry Science Animal Breeding Animal & Human Nutrition Veterinary Medicine	Farm Management & Marketing Resource Economics Rural Sociology Public Policy Studies Human Ecology
		Integrated Pest Management			
IV. PRODUCTS FROM INNOVATION (Agri-industrial development)	Farm Machinery & Equipment Farm Buildings Computer Equipment/ Software	Commercial Fertilizers Agricultural Chemicals Irrigation Systems Pest Control Systems	Crop/Plant Varieties Horticultural/Nursery Species Livestock Feed	Animal Breeds Animal Health Products Food Products	Management Systems Marketing Systems Institutional Innovations Health Care Child Care
V. EXTENSION (Public and private)	Resources & Environment ⇄	Commodity Oriented	Management & Marketing	Public Policy ⇄	Family & Human Resources
VI. FINAL USERS/ Sources of Clientele Problems	Producers		Governments	Consumers	

systems, these pre-invention sciences were created in response to perceived exhaustion of invention potential. The pre-technological sciences are differentiated from the general sciences by their responsiveness to demands from below. They do "science for inventors". General scientists do "science for scientists" (see Huffman and Evenson, 1993 for further discussion of these interrelationships).

Few developing countries have complete systems as depicted in Figure 1. Some have systems stressing extension and sub-invention.. Few have institutionalized pre-technology sciences. And today many of the National Agricultural Research Systems (NARS) in developing countries rely on the International Agricultural Research Centers (IARCs) as germplasm suppliers. And, as will be noted in a later part of the paper, this leaves them vulnerable to exhaustion because the IARCs were not designed to undertake pre-inventive science.

Inter-Sector Institutional Change

Industrial firms have become increasingly important as producers of technology for agriculture. The R&D conducted in the agricultural implements, agricultural chemical, animal feed, seeds and other service industries produces inventions that are "used" in the farm sector. Evenson (1993) and colleagues at Yale University have developed the Yale Technology Concordance (YTC) based on Industry of Manufacture (IOM) and Sector of Use (SOU) patent assignments in the Canadian patent office. The YTC links IOM and SOU assignments on a probabilistic basis to International Patent Classes (IPC). The YTC can thus be applied to patent data bases from any country with IPC assignments reports YTC based inventions by agricultural SOU for the U.S., U.K., France and Germany for the 1969 to 1987 period. For each agricultural SOU a matrix of inventions by source country is reported.

For each industry two further indexes are reported (in parentheses). The first is the total number of domestically produced inventions of use in the four granting countries. The second is the ratio of total patents granted in the four countries to domestically produced inventions in the four countries. This ratio can be regarded to be an index of "transferability" for the technology in question. When this ratio is high it indicates that inventions originating in one country have a high degree of usefulness in another country.

The crop production farm sector received considerably more invention attention (2.4 times as much) than the livestock production sector. The services to farmer industries also received a high level of inventions, more than received by the implements and chemical inventions combined.

The transferability indexes indicate that inventions used in the agricultural chemicals industries are most transferable. The lowest levels of transferability (*i.e.*, the most location-specific) are for inventions used in the livestock production industries and in services to farmers.

Table 1 also reports import and export ratios for the four major invention countries. Since the denominator of these ratios is the same, the differences in the ratio is a kind of net trade ratio (note imports from all countries are counted in the import ratio but only exports to the other three countries are counted). The US., it may be noted, has a favorable ratio of exports to imports relative to other countries except in agricultural implements and to a lesser extent in agricultural chemicals where Germany has a more favorable ratio.

The country of origin data show that the US. is the leading originator (*i.e.*, domestic producer) of inventions in most industries and that Germany is the second leading originator of inventions. The US. is also the leading "consumer" or user of inventions in all industries.

International and Inter-Sectoral Institutions

The agricultural sector is increasingly integrated with other sectors as economic development proceeds and economies become industrialized. The economic growth and industrializing processes themselves are dependent on the technological infrastructure levels of different countries. Technological infrastructure encompasses legal systems and other inventive systems for private sector R&D conductextension and education as well as public sector investment in research. The system of IARCs was developed to compensate for perceived inadequate technology infrastructure capacity in developing countries.[3]

To clarify what is meant by technology infrastructure and to organize evidence on investment and economic consequences I will resort to a classification of technology infrastructure levels modified from Weiss (1950). These are summarized in Figure 2 in terms of 3 classes of level 1 developing economies and 3 classes of level 2 developing economies.

The Type 1 countries include approximately 75 developing countries with a total population of roughly 1 billion people. The Type 1a (Traditional Technology) countries (16 countries - 100 million people, all small except for Zaire, have little or no technological infrastructure). The 19 countries (220 million people) in Type 2a (First Emergence) have some capacity for undertaking research in agriculture but little in other sectors. The 38 countries in Type 1c (more than 700 million people including Pakistan and Bangladesh) have developed quite significant capacity for agricultural research and engage in some industrial R&D in public sector units. None of these countries, however, has significant R&D capacity in industrial producing firms.

The 20 Type 2 countries, on the other hand, have made sufficient investments in technology infrastructure, but not all have managed to put the requisite policy environments in place to achieve rapid growth. The four East Asian NICs (Hong Kong, Singapore, Taiwan and South Korea) constituted Type 2c in the 1980s. Thailand, Malaysia and China are now realizing NIC type follower growth, and Indonesia and possibly Chile are as well. India is positioned to realize such growth but remains committed to "inward-oriented" regulatory and industrial policies that have blocked its progress. Several Latin American members of Types 2a and 2b have pursued disabling macro-economic policies.

Table 2 reports average "investment intensities" (R&D/GDP) for research, extension and science for the six types of developing countries and for recently industrialized countries (Greece, Portugal, Spain) and for other developed OECD countries. The patterns here are quite clear. For agriculture there is a public sector research capacity in most Type 1 countries and all Type 2 and more developed countries. This capacity increases as the general level of technological infrastructure rises. Scientist/GDP ratios (Panel IV) are actually higher in the Type 2 developing countries than in the more developed countries because real costs per scientist are much lower in developing countries.

3. See Evenson and Westphal (1995) for a full discussion.

Figure 2: Synthesized levels of technological development for LDCs.

Level 1 – Developing Countries

Countries ranging from those having little capacity to improve technology to those with partial capacity to design and develop new technology.

Level 1a: Traditional Technology (Yamen, Laos, Surinam, Bhutan)
Level 1a economies lack basic infrastructure. Government influence on the country is limited. The manufacturing sector is of limited size. Infrastructure is poor. Literacy levels are low. No effective technology policy is formulated.

Level 1b: First emergence (Nepal, Papua New Guinea, Haiti, Ethiopia, Burkina Faso)
These countries are building infrastructure. Many have parastatal organizations. Some direct foreign investment has taken place. Many small enterprises have little access to technology. Modern technology is usually the result of foreign management. Macro-economic and trade policies have been developed. Educational institutions have begun to establish scientific capabilities and graduates dominate government bureaucracy.

Level 1c: Partial Modernization (Sri Lanka, Tunisia, Jordan, Indonesia, Jamaica, Peru, Kenya, Ivory Coast)
In these countries modern agricultural practices have been established. Modern small scale industry is developing. Most modern equipment is imported. Local construction capability is well established. Domestic negotiating skills for larger firms have been developed. First IPRs and technology transfer laws are developed. Most younger workers are literate. Most countries have well established universities with scientific disciplines. Graduates beign to work outside government.

Level 2 – Developing Countries

Countries that have established basic competence to screen and implement technology, but differ in their capacity to imitate, invent, and design.

Level 2a: Mastery of Conventional Technology (Malaysia, Turkey, Colombia, Argentina)
These countries have engineering capabilities in most industries. Manufactured exports are important. Larger enterprises have skilled staff. Market skills are well developed. Technology policy is given serious attention. Demand for skilled manpower is increasing. Education in disciplines is well established. Consulting and engineering organizations are developed.

Level 2b: Transition to NIC-hood (India, Thailand, Hongkong, Mexico)
These countries produce most goods and have imitative reverse engineering capability. Capital goods sectors are well developed as are consulting and engineering organizations. Local brand names are important. Advanced technology is applied. Sub-contracting becomes important. Technology policy includes some strategic investment. Science teaching is expanded. Expatriates are replaced by local staff. Technology policy becomes important.

Level 2c: NIC-hood (Singapore, Korea, Taiwan)
These countries have achieved export competitiveness. Standards are well developed. Large scale firms reap scale economies. Widespread reverse engineering and imitative capacity exist. Adaptive invention is broadly achieved. Technology exporting begins. R&D becomes a major activity. IPRs are used to stimulate R&D. Strategic technology policy is implemented. Literacy of work force is universal. Scientific and engineering competence is seen in all sectors.

Source: Adapted from Charles Weiss, "Scientific and technological constraints to economic growth and equity" Tables 2.1 and 2.2 In Robert E. Evenson and Gustav Ranis, ed.s., *Technology: Lessons for Development Policy*, Boulder: Westview Press, 1990.

Roughly the same pattern holds for agricultural extension. Most developing countries have extension programs. And the comparisons of field extension workers per unit of GDP show that staffing intensities in extensions are considerably higher in developing countries than in developed countries, again because of large differences in real costs per extension staff members.[4]

The IARCs are generally servicing most developed countries except for the Type 2c countries. Private sector R&D directed toward agriculture is practically nil for Type 1a and 1b countries and very low for Type 1c countries. For level 2 developing countries, it is roughly one-third of public sector investment and for developed countries, private sector investment is of roughly equal magnitude to public sector investment.

For industries (*i.e.*, where the technology is used in manufacturing sectors) investment intensities in the public sector are relatively low at all levels of development reflecting the perception that industrial experiment stations have not been as successful as agricultural experiment stations. Perhaps of more relevance is the fact that Type 1 developing economies do not have private sector industrial R&D capacity. This capacity is more developed in the Type 2 developing economies. Even though investment intensities in the NICs are lower than for developed economies, R&D staffing intensities are approximately as high for countries in the NIC-Hood phase as for the most advanced countries.

Table 3 provides international qualitative ratings of intellectual property rights. Type 1 developing countries have by and large not developed functioning IPR systems. Type 2 developing countries have until recently maintained weak, underfunded and underadministrated IPR systems from an international perspective. Only the more successful NICs have used IPRs as an incentive system for encouraging more R&D.

EMPIRICAL STUDIES OF ECONOMIC CONSEQUENCES

Returns to Investment Studies[5]

Two methodological approaches to research and extension program evaluation have been pursued in empirical studies. Both are based on Total Factor Productivity (TFP) growth accounting principles. The first is based on direct imputation and is an application of project evaluation methods. The second approach is statistical and entails construction of variables derived from investments in research, extension, schooling, infrastructure, and other TFP enhancing activities. These variables are typically expressed in stock (service-flow) terms, with appropriate temporal and spatial weights to

---

4. It should be noted that it is not possible to fully adjust for skill differences between types of countries in staffing ratios. We know from survey data that the proportions of research staff with Ph.D. degrees rises from roughly .2 for Type 1 developing countries to the .5 and higher range for Type 2 developing countries. It is probably .9 or higher for developed countries. Timilarly, most extension staff in Type 1 countries do not have college level technical training in agriculture, while most extension staff in Type 2 countries do, and virtually all extension staff in developed countries have relatively high levels of technical expertise.

5. This summary is based on Evenson (1993).

reflect time lags, depreciation, and spillovers. These variables are sometimes termed meta variables to distinguish them from conventional input variables. Statistical frameworks used have included:

TFP decomposition using hedonic regression specifications, where TFP measures are regressed on meta variables of the kind just discussed; production function specifications where meta variables are included together with conventional inputs in a production function framework that is usually Cobb-Douglas in form; and profit functions or output supply-input demand systems which include meta variables and rely on the duality theory and the assumption of competitive markets to obtain estimates of production function parameters.

The key issue in the direct imputation studies is typically the identification of an appropriately matched sample of before-and-after or within-and-without observations relating to technology or program use. The classic study of Griliches (1957) demonstrated the basic methodology. Griliches utilized data on the first generation of hybrid corn varieties developed both by private firms and public experiment station systems. Experiment station and farm level data enabled Griliches to estimate the yield advantage of hybrid corn varieties over the older varieties in each state. These data were used along with adoption data to compute year-by year benefit values, given by the change in producer plus consumer surplus. The resulting cost and benefit time series were used to compute benefit-cost and rate of return measures.

The statistical studies employing meta variables have in some cases estimated both the temporal and spatial spillover weights utilized in constructing these variables. Temporal weights estimated for agricultural research programs indicate that the TFP responses generally begin one or two years following expenditures and rise, reaching a peak after 7 to 10 years. Agricultural extension programs have faster and shorter-lived impacts.

Spillover weights are designed to capture the value contributed by research programs outside of the region. Early studies used climatic indicators as simply proxies for technological distance. More recent studies (Evenson 1994, Da Cruz *et. al*, 1994) have specified technology distance spillover measures.

The estimated coefficients on the meta variables in these studies are used to compute the marginal benefits from investment. The coefficients have temporal and spatial dimensions which are taken into account in deriving benefit-cost and rate of return measures. Some studies provide parameter estimates which can be used in computable general equilibrium (CGE) models to examine the distributional consequences of technological investments.

Table 4 summarizes results of 156 studies estimating returns to agricultural research programs. Most of the agricultural studies surveyed utilized secondary data (district-level data by year in India, for instance) and were to some degree based on cross-section variation in the meta variables. Cross-section variability in research and extension inputs has been quite important in permitting the identification of their impacts; very few studies based on simple time series have been able to identify these impacts. These TFP determining variables include measures of research, extension, schooling, roads, markets, prices, and related variables.

The summary by region shows first that the median or average rates of return reported are very high. It should be noted that these are "social" rates of return as opposed to private rates of return (see below). The EMIRRs are highest for the IARCs, and next highest for Asia. They are lowest for Africa. Returns to research are as high in developing countries as in developed countries.

The summary by period of study shows that the earliest period had lower return than later periods. There is no indication of a decline in returns to research in recent periods.

The summary by method shows slightly higher EMIRRs for statistical methods - especially the TFP decomposition methods - but there appear to be no strong method biases in these estimates.

Table 5 reports EMIRRs for agricultural research by type of technological infrastructure. Here we note a tendency for returns to be lower in Type 1 countries than in type 2 countries. No studies are available for the poorest economies.

Table 6 reports EMIRRs by commodity for studies that were commodity specific. Here we find evidence for commodity differences. EMIRRs are highest for cereals and oilseeds, lowest for tree crops. EMIRRs are higher for crop research than for livestock research and are also higher than for forestry research.

Several of the studies estimated the separate contributions of pre-technology science research and of downstream applied research. Several also estimated the contributions to agricultural TFP growth of private sector R&D by firms supplying inputs to the agricultural sector. This contribution constitutes a pecuniary spillover from industry to agriculture, one which occurs because supplying firms capture only part of the return to their R&D through higher prices for improved inputs. Table 7 reports EMIRRs by type of program and for private sector R&D.

Fifty-three of the 156 studies reported estimates pertaining to entire agricultural research systems rather than to individual commodity research programs. The distribution of estimated rates of return in these studies did not differ from that for studies focused on specific commodities. The similarity between the distributions of system-wide and commodity-specific programs suggests that the latter studies do not suffer from a serious selectivity bias, that is, that they have not focused only on the best programs. Nonetheless, it remains possible that there has been some failure to report estimates for all types of studies that are not deemed "high enough" to report.

Surveys of returns to private R&D in developed countries show that investments in R&D, when evaluated *ex post*, yield private returns that are at least as high as returns to other investments. Mansfield *et al.* (1977a) report on 17 case studies of innovation for which the median private rate of return was 25%. Griliches (1980) reports rates of returns for large U.S. industrial firms ranging from 30% to 50%. Mairesse and Sassenou (1991), reviewed a number of studies giving statistical estimates of the impact of research expenditure on firm-level productivity covering several advanced countries (France, Japan, and the U.S.), and found that all implied positive and highly significant elasticities, with approximate rates of return ranging from 14% to 24%. They found corroborating evidence in another set of firm-level studies that gave direct estimates of rates of return, leading them to conclude that, for the countries covered, private rates of return to R&D were at least as high as those for other forms of investment.

Social rates of return should exceed the private rates because of the individual firm's inability to appropriate, or capture, the full benefits from conducting R&D. Even in the presence of strong IPR protection, a private firm's rents from licensing or product sales generally represents only a fraction of the real value of the invention to the economy, that is, of the invention's social return. According to the previously cited study by Mansfield *et al.* (1977) social rates of return (median, 56%) were in most cases more than double the private rates. Griliches (1991) has reviewed a number of empirical studies to estimate spillovers from R&D and concludes that spillovers are of considerable importance, which is consistent with the evidence that social returns are considerably in excess of private returns.

Very few studies have estimated returns to industrial R&D in LDCs. Deolalikar and Evenson (1989) reported effects of R&D on factor demand but stopped short of computing returns to investment. Two studies of industrial R&D in industries supplying agriculture have reported high rates of return as measured by the impact on agricultural productivity (see Rosegrant *et al.*, 1993).

Returns to Agricultural Extension

Experience in Asia and Latin America in the 1950s and 1960s indicated that large investments in extension and rural development programs had relatively small impacts in many countries. T.W. Schultz (1964), in his classic monograph on traditional agriculture, argued from this and other micro evidence that traditional farmers were "poor but efficient", having exhausted the potential of the best suited technology. Thus it was generally accepted in the 1970s that the gap between the average and the best productivity levels was relatively small so that extension could be productive only after local research programs generated new technology.

Perceptions have changed somewhat in recent years. A new approach to extension, the Training and Visit (T&V) system, was developed in World Bank projects in the late 1970s (see Benor and Baxter, 1984). This system imposes a formal structure linking extension workers to technical specialists and entailed a fixed schedule of extension worker visits to farmers and farm groups.

Some of the early studies to investigate the return to extension relied on variables measuring extension worker contact with farmers as indicators of extension provision. Since extension contact is at least partly determined by farmers' behavior, such variables are endogenous, and positive correlations between them and farm productivity can not be used to claim the existence of a causal link between extension and productivity. Later studies have overcome the problem of endogeneity by using extension supply variables. Technological and price information is diffused to farmers through a broad range of channels, with farmer-to-farmer communication being especially important.

Birkhauser, *et al.* (1991) reviewed 40 studies of returns to agricultural extension programs. Few of the early studies subject to endogeneity bias showed significant returns. But, of the more recent studies, 15 of the 26 that provide estimates of rates of return report values in excess of 50% (see Table 8). Two recent studies (Bindlish and Evenson, 1993; Bindlish, *et al.*, 1993) of T&V extension in Kenya and Burkino Faso also report very high rates of return and suggest that countries in Africa still have considerable scope for reducing inefficiency even when new technology is not being made available to farmers.

Estimate of Distributional Impacts

A principal result from nearly all Computable General Equilibrium (CGE) and micro studies is that major gainers from new agricultural technology are the consumers of agricultural products. For urban consumers, improved agricultural technology leading to lower prices is an unmitigated blessing no matter where or how the gains are realized. Farmers and rural workers also gain as consumers, but may lose as workers and owners of rural assets. (Among the relevant models are those constructed for the Philippines by Habito (1990) and Quisumbing, *et al.* (1993) and for India by Quizon, *et al.*, (1991); table 8)

Subsistence farmers tend to be insulated from such changes because they consume most of what they produce (Barker and Herdt, 1985). In an open economy facing the highly elastic world demand for the product, total farm revenues increase and farmers as well as workers gain.

Many of the micro studies were motivated by a concern that advances in agricultural technology harmed the poorest rural families, small farmers and landless peasants. There is a consensus that this is generally not true and that losses, where there they have occurred, have accrued to landowners in areas that were circumstantially unsuited to adopt the new varieties. Barker and Herdt (1985) review studies for rice showing that small rice farmers adopt new technology about as rapidly as do larger farmers and thus share in the gains to early adopters. In turn, a recent study for rice at IRRI examined wage differentials within seven countries cross regions which are differently endowed with respect to the ability to adopt the new varieties and found that they have been largely eroded by labor mobility. Instead of wage differentials, land rent differentials have emerged (David and Otsuka, 1990).

Contributions to Economic Growth

Returns to investment in research and extension imply economic growth. Relatively few of the studies that are reviewed above have actually undertaken the implied growth accounting associated with the estimates obtained. Table 9 reports a recent effort (Rosegrant *et al.*, 1993) for India. The study first computed Total Factor Productivity indexes for most Indian Districts. As the table indicates, TFP growth accounted for more than half of Indian agricultural growth over the periods. It is quite clear that without TFP growth India would have experienced declining food production per capita during the periods studied. India's population more than doubled over the 1955 to 1988 period. Cultivated land expanded by less than 15%.

The accounting for sources of TFP growth shows that private sector R&D, particularly foreign private sector R&D was an important source of TFP growth. This reflects the modernization process. As modern inputs are used in a rational manner they convey technology to developing countries. India's public sector agricultural research and extension system contributed more than half of Indian TFP growth in all periods.

RESEARCH AND EXTENSION PROSPECTS

We have now experienced 50 years of modern (post World War II) agricultural research and extension. Agricultural production has changed markedly. Industrialization and the income and price changes associated with economic growth have caused much of this change. The technology produced by research programs and diffused in part by agricultural extension programs has also caused changes in productivity and in other features of production. There is concern in both developed and developing countries that research programs may be experiencing diminishing returns and that potential invention "pools" may be becoming "fished out" or "exhausted". This concern is particularly relevant to plant (and animal) breeding programs where it is noted that most potentially valuable genetic combinations may have already been discovered.

Actually, there is a great deal of evidence for "exhaustion" and research program failures that is not reflected in the economic studies reviewed in the previous section. There are two reasons for this. The first is simply that all of the research (and extension) programs studied are aggregates of numerous projects and combination of projects using different techniques and approaches. Many of the projects and sub-programs actually fail and some of these failures are due to exhaustion. We don't observe these failures at the aggregate level because they are mixed with successes. This type of failure and exhaustion is simply part of doing research.

The second reason that the studies reviewed above provide a misleading picture is that when research programs are dominated by project and sub-program failure we find few studies reporting this. There are, in fact, many NARs, especially in Africa, where we cannot point to successes. There are also commodity programs that have low success rates.

The rice research study (Evenson, *et al.*, 1995) included three components:

1)Crop loss estimates from plant diseases, insects and abiotic stresses were obtained.

2)A scientist survey research potential using particular research methods (management or agronomic improvements, conventional plant breeding, wide-crossing and hybridization and transgenic breeding). Scientists were asked to rate remaining research potential for reducing specific crop loss by each research method and to estimate "time to achievement" of research potential. They also rated increased yield potential from new plant design and other factors affecting biological efficiency.

3)The crop loss and biological efficiency potential data were combined with research potential and timing probability estimates to produce optimal portfolios for particular regions over future years.

This optimal portfolio was then converted to productivity projections for the 1995 to 2020 period. Table 10 shows these productivity projections for South Asia. These projections show that the contribution of conventional plant breeding is expected to decline (from 76% per year to 22% per year) over the next 25 years, but that these declines will be offset by increases in the wide-crossing-hybrid methods and in transgenic methods.

Table 10 also reports other productivity sources for South Asia rice. These are based on the studies reviewed above and detailed in Rosegrant and Evenson (1996). Similar projections for other commodities were made and used as the basis for projections of prices, production and trade by IFPRI.

Table 11 shows the IFPRI model projections for area, production and world price levels for the major crops for the year 2020. The 1990 levels, and two 2020 scenarios are presented. The first is the optional public and private research investment policy. The second is a "neglect" scenario in which the International Agricultural Research Center support declines and the IARCs are phased out over the next 15 years. Biotechnology programs are not pursued and international aid support for NARs is also phased out.

Two points are obvious. The first is that under the "optimal" scenario a favorable result will obtain. Cropped area will expand by about 9%, most of this in Africa. For the rest of the world multiple cropping expansion will dominate. Production will increase by almost 60% and real cereal grain prices will decline to about 82% of their 1990 levels. The second point is that the "neglect" scenario is much less favorable but does not produce a world food crisis. Cropped area will expand by 13% instead of 9%. Production will expand by 50% instead of 60%. Real world food prices will rise by 7% instead of falling by 18%.

Table 12 provides further projections by regions. They show that with the "optimal" scenario world trade in grains will double with almost all developed countries expanding exports (Japan will expand imports.). Most developing regions will expand imports, (China will be the major importer but will not have crisis imports). The "neglect" scenario predicts even greater world trade because this neglect affects the poor regions and has little effect on developed countries. This point is made clearer in the welfare simulations. The percent of children (0-6) considered malnourished is projected to fall from 34% to 25% for all developing countries. For parts of Africa, however, even under the optimal scenario this measure falls very little. For the neglect scenario, this index falls from 34% to 30% for all developing countries, but actually rises for Bangladesh and East Africa and is roughly constant for the rest of Africa. There is little doubt that the burden of public policy neglect for agricultural research (and extension) investment will fall on the poor.

As noted above, this review has not fully addressed the problem of NAR and commodity programs that have achieved little. We have much better evidence for successful programs than for failed programs. As a rough approximation, I would offer the following judgments as to failed NARs programs:

- There is little evidence that any of the level la (see Figure 1) NARs have been productive.
- Very few of the level lb NARs appear to have been productive.
- ANARs in countries with significant military conflict (many of which are 1a and 1b) generally do not produce during periods of conflict (*e.g.*, Vietnam, Iran, Iraq).

There are some broader non-achieving commodity programs that can be attributed to real scientific difficulties that appear to challenge even the best managed and skilled programs (*e.g.*, in spite of best effects we have not found an effective cure for AIDS, and this is true for some agricultural problems).

Studies of research management (*e.g.*, by ISNAR) probably give us some sense for the reasons for non-achievement, but I have not seen a comprehensive treatment of the problem. The fact that non-achieving programs are generally located in the lowest technology infrastructure settings (1a and 1b) suggests that policies to improve agricultural research may have to deal with a broader set of institutions and organisations than those strictly devoted to agriculture. One suspects, however, that these general principles of management and design could markedly improve the performance of many non-performing programs.

These principles or guidelines would be of two types or classes. The first includes those principles that apply to any program. These include the setting of objectives and priorities, of planning programs, of monitoring performance and of continuing evaluation of program performance. They include financial responsibility and accounting. Many NARs failures can be attributed to failures to meet these guidelines.

CONCLUSION

But, in an important sense, research is different from most other activities, and thus there are some guidelines that are specific to research. The first sense in which research differs is that it is not a repetitive or duplicative process. A factory or farm can "learn by experience" (or doing) over repeated production runs (or seasons). Researchers "learn by experience" also but it is more demanding than the learning by doing of producers. Many projects fail and will not be repeated. Even success are not repeated in research. The researcher must carefully assess both failures and success and design the next step. This process requires rigor and method and communication with others engaged in similar programs. A well functioning research system imposes standards for analytic rigor and for scientific reporting and communication.

The fact that many weak NARs do not produce scientific reports, even simple standard experiment evaluations, and often can't even come up with financial reports would appear to be a serious flaw. There simply are minimum standards of performance to be successful.

SELECTED REFERENCES

AGCAOILI, M.C. AND M.W. ROSEGRANT, 1992. *Regional Trends and Projections for Food Supply, Demand and Trade in Asia.* Washington, D.C., IFPRI.

BASANT, R. AND B. FIKKERT, 1993. *Impact of R&D, foreign technology purchase and technology spillovers on Indian industrial productivity: Some estimates, mimeo.* Maastricht, The Netherlands: United National University Institute for New Technologies.

BENOR, D. AND M. BAXTER, 1984. *Training and visit extension.* Washington, D.C., The World Bank.

BINDLISH, V. AND R.E. EVENSON, 1993. Evaluation of the performance of T&V extension in Kenya, *Agriculture and Rural Development* Series No. 7. Washington, D.C., The World Bank.

BINDLISH, V., R.E. EVENSON AND M. GBETIBOUO,1993. *Evaluating the performance of T&V based extension in Burkina Faso, mimeo.* Washington, D.C., The World Bank, African Region, Technical Department.

BIRKHAUSER, D., R.E. EVENSON AND G. FEDER, 1991. The economic impact of agricultural extension: A review, *Economic Development and Cultural Change*, 39:607-650.

CROSSON, P. AND J.R. ANDERSON, 1992. Resources and Global Food Prospects: Supply and Demand for Cereals to 2030. *World Bank Technical Paper* No. 184. Washington, D.C., The World Bank.

EVENSON, R.E., 1993. *Estimated returns to research and extension: A review*, unpublished manuscript. New Haven, Connecticut: Yale University, Economic Growth Center.

EVENSON, R.E. AND D. GOLLIN, 1991. *Priority setting for genetic improvement research.* Unpublished manuscript prepared for workshop on Rice Researh Prioritization. Los Banos, Philippines: International Rice Research Institute.

EVENSON, R.E., R.W. HERDT, AND M. HOSSAIN, EDS., 1995. *Setting priorities for rice research.* Los Banos, Philippines: International Rice Research Institute.

EVENSON, R AND C. PRAY, 1991. *Research and productivity in Asian Agriculture.* Ithaca: Cornell University Press.

FOOD AND AGRICULTURE ORGANIZATION, 1986. *The FAO World Food Model: Model Specification.* Supplement to FAO Agricultural Commodity Projection to 1990. Rome: FAO. GOLLIN, D. AND R.E. EVENSON, 1991. *The economic impact of the International Rice Germplasm Center (IRGC) and the International Network for the Genetic Evaluations of Rice (INGER), mimeo.* Los Banos, Laguna, Philippines: International Rice Research Institute.

HUFFMAN, W.E. AND R.E. EVENSON, 1993. *Science for agriculture.* Ames, Iowa: Iowa State *University Press.*

MAIRESSE, J. AND M. SASSENOU, 1991. R&D and productivity: A survey of econometric studies at the firm level, *STI Review*, 8:9-43.

MANSFIELD, E., 1961. Technical change and the rate of imitation, *Econometrica*, 29:741-766.

NORTON, GEORGE W. AND JEFFREY S. DAVIS, 1981. Evaluating returns to agricultural research: A review. *American Journal of Agricultural Economics* 63(4):685-699.

ROSEGRANT, M., R.E. EVENSON AND C. PRAY, 1993. *Sources of agricultural productivity growth in India, research report.* Washington, D.C.: International Food Policy Research Institute.

SECKLER, DAVID AND MICHAEL ROCK, 1995. *World Population Growth and Food Demand to 2020.* Winrock International Institute for Agricultural Development.

SIEBECK, W., R.E. EVENSON, W. LESSER AND C.P. BRAGA, 1990. *Strengthening protection of intellectual property in developing countries,* World Bank Discussion Paper no. 112. Washington, D.C.: The World Bank

Table 1: International Comparisons by Industrial Use Field

Granting Country:	Trade Ratios: Import	Export	Inventions by Country of Origin: U.S.	UK	France	Germany	Japan	Canada	Others
A. Crop Production (21,251, 1.94)									
U.S.	.35	.33	8544	421	270	470	446	29	1056
U.K.	1.68	.57	945	2256	317	892	196	31	1422
France	1.24	.26	913	352	3754	1384	258	17	1701
Germany	.64	.46	959	523	386	6697	231	13	2143
B. Livestock Production (8,753, 1.57)									
U.S.	.30	.30	3541	152	99	234	146	25	405
U.K.	1.09	.44	373	1099	102	227	93	6	401
France	.81	.21	306	158	1644	361	83	6	423
Germany	.56	.33	367	172	142	2499	92	4	631
C. Services to Farmers (14,918, 1.58)									
U.S.	.26	.33	6717	270	195	384	292	54	581
U.K.	1.26	.43	988	1863	150	416	164	16	605
France	.99	.24	570	257	2441	649	146	13	788
Germany	.55	.37	668	274	236	3897	193	13	765
D. Agricultural Implements (6,219, 1.69)									
U.S.	.41	.47	2518	140	105	304	240	22	223
U.K.	1.43	.65	430	663	103	357	110	14	280
France	1.58	.41	332	129	852	478	81	3	330
Germany	.28	.66	432	159	145	2187	141	4	476
E. Agricultural Chemicals (7,462, 2.39)									
U.S.	.66	.51	4011	501	212	789	508	5	636
U.K.	1.98	1.23	866	801	184	540	323	9	476
France	3.13	.71	733	325	471	478	279	8	520
Germany	1.25	.86	930	410	215	2116	382	7	695

Table 2

International Technology Investment Indicators: Investment Intensities
(R&D/GDP, Extension/GDP, Science/GDP, 1990)

Indicators	Technological Infrastructure Type							
	Type 1 Developing Countries			Type 2 Developing Countries			Industrialized	
	1a Traditional Technology	1b First Emergence	1c Islands of Modernization	2a Mastery of Conventional Technology	2b Transition to NIC-Hood	2c NIC-Hood	Recently	OECD
I. R&D/GDP								
Agriculture								
Public-NARs	.002	.004	.005	.006	.007	.010	.010	.015
Public-IRAs	.0005	.0005	.0005	.0005	.0005	0	0	0
Private	0	0	.0002	0	.003	.005	.005	.015
Industry								
Public	0	.0001	.0001	.0002	.0005	.001	.003	.003
Private	0	0	.0002	.005	.007	.010	.015	.023
Services								
Public	0	0	0	.0001	.0005	.0005	.0001	.003
Private	0	0	0	0	.0005	.0001	.0002	.0005
II. Extension/GDP								
Agriculture								
Public	.005	.005	.010	.010	.010	.010	.010	.010
Private	0	0	0	0	.001	.002	.005	.010
III. Science/GDP								
Public	.0002	.0002	.0005	.0005	.001	.002	.0025	.003
Private	0	0	0	0	0	.0002	.0004	.0005

Table 3

International Institutional Ratings

Institution	Technological Infrastructure Type							
	Type 1 Developing Countries			Type 2 Developing Countries			Industrialized	
	1a Traditional Technology	1b First Emergence	1c Islands of Modernization	2a Mastery of Conventional Technology	2b Transition to NIC-Hood	2c NIC-Hood	Recently	OECD
Intellectual Property Rights								
International	0	0	1	2–3	2–3	2–3	4–5	5
Domestic	0	0	0	1	2	4	4	5
Regulations	1	1	1	2	2	4	4	5
Foreign Direct Investment	2	2	2	2	3	4	4	5
Trade Policy	1	1	1	2	2	4	4	5

Table 4

Estimated Internal Rates of Return to Agricultural Research, EMIRRs by Region, Period and Method

	Number of Studies	EMIRR Range					
		Non-significant	1-24	25-49	50-75	75+	Median
By Region							
Africa	10	1	2	3	3	1	(41)
Latin America	36	2	14	22	13	13	(46)
Asia	35	2	7	20	23	25	(57)
IARCs	4	0	0	0	1	5	(81)
U.S.	44	3	13	41	14	22	(50)
Other Developed Countries	24	2	8	13	12	7	(40)
By Period							
Pre 1968	12	1	4	7	5	1	(46)
Inc 1960s	12	0	1	11	3	5	(52)
Inc 1970s	43	2	10	24	14	20	(52)
Inc 1980s	57	3	20	34	31	30	(52)
By Method							
ES-Imputation	77	5	27	44	34	29	(43)
Meta-PF	46	1	7	34	15	17	(51)
Meta-TFP	22	2	5	16	18	31	(61)
Meta-Production	5	0	0	4	4	0	(50)

Table 5

International Comparisons, Rates of Return to Agricultural Research

	Technological Infrastructure Type							
	Type 1 Developing Countries			Type 2 Developing Countries			Industrialized	
	1a Traditional Technology	1b First Emergence	1c Islands of Modernization	2a Mastery of Conventional Technology	2b Transition to NIC-Hood	2c NIC-Hood	Recently	OECD
Number of Studies	0	8	8	21	28	0	0	71
EMIRRs Range								
Not significant	na	1	0	3	0	na	na	5
1-24	na	2	7	5	5	na	na	21
25-49	na	3	9	9	16	na	na	54
50-75	na	2	2	11	13	na	na	26
75+	na	1	6	15	8	na	na	29
Median	na	(39)	(45)	(55)	(51)	na	na	(48)

Table 6

Summary of Estimated Rates of Return EMIRRs by Commodities

	Number of Studies	EMIRR Range					
		Non-significant	1-24	25-49	50-75	75+	Median
Rice	23	0	3	7	12	13	(62)
Wheat	22	1	4	7	5	7	(51)
Maize	15	0	3	7	1	6	(52)
All Cereals	69	1	13	23	21	27	(55)
Soybeans	12	6	1	5	5	6	(60)
All Oilseeds	16	0	1	5	6	9	(64)
Cotton	5	1	0	2	1	3	(56)
Sugarcane	6	1	0	4	0	1	(39)
Legumes	2	0	1	2	1	2	(46)
Tree Crops	5	0	4	1	0	0	(26)
All Agricultural Crops	117	45	22	43	30	46	(53)
All Livestock	20	2	7	12	5	6	(43)
All Forestry	15	1	5	5	3	4	(44)

Table 7

Summary of Estimated Ratio of Return EMIRRs for Public and Private Research

	Number of Studies	EMIRR Range					
		Non-significant	1-24	25-49	50-75	75+	Median
All Applied Agriculture (Aggregate)	53	2	6	33	18	12	(49)
All Pre-Technology Agricultural Science	5	0	0	3	4	1	(56)
All Private R&D (agricultural)	5	0	0	3	3	2	(58)
All Private R&D (non-agricultural)	35	0	0	3	0	5	(44)

Table 8

Summary of Estimated Return to Agricultural Extension

	Number of Studies	EMIRR Range					
		Non-significant	1-24	25-49	50-75	75+	Median
Africa	3	1	0	0	0	2	(56)
Asia	5	0	3	1	1	0	(30)
Latin America	9	0	1	1	3	4	(64)
Developed Countries	6	0	1	0	3	2	(63)
All Agricultural Extension	23	1	5	2	7	8	(55)

Table 9

Growth Accounting Indian Agriculture

Growth Source	1956-67	1967-76	1976-87	1856-87
Conventional Inputs	1.10	1.31	1.02	1.14
Markets	.013	.023	.016	.019
Irrigation	.026	.131	.086	.081
Extension	.510	.134	.401	.348
Public Research	.213	.231	.663	.369
High Yielding Varieties	.010	.116	.015	.048
Private Sector R&D				
Domestic	.050	.112	.049	.071
Foreign	.708	.280	.232	.408
Total Non-conventional	1.530	1.027	1.462	1.344
TFP	1.270	1.490	1.140	1.310

Table 10: The South Asia rice non-price yield (base) projections (expressed in percents).

	95-00	00-05	05-10	10-15	15-20
1. Public Research					
A. Management	.216	.216	.216	.216	.216
B. Conventional breeding	.763	.654	.436	.327	.218
C. Wide-crossing, hybrids	.100	.200	.300	.250	.150
D. Biotechnology	.158	.316	.474	.682	.790
Total Public Research	**1.237**	**1.386**	**1.426**	**1.425**	**1.374**
2. Extension - Schooling	.470	.570	.597	.593	.569
3. Private research	.100	.150	.200	.200	.200
4. Markets - Infrastructure	.150	.150	.200	.200	.200
Total Base Case	**1.957**	**2.256**	**2.423**	**2.418**	**2.343**

Table 11: Area, Production and Price Projections 1990-2010 by Commodity

	Area 5			Production			Prices		
Commodity	**1990**	**BASE**	**Weakening**	**1990**	**BASE**	**Weakening**	**1990**	**BASE**	**Weakening**
Wheat	221	235	244	530	840	826	156	133	180
Rice	146	155	162	349	566	529	210	164	245
Maize	126	142	145	476	743	698	109	84	112
Other Grains	203	221	230	358	529	532	89	67	93
Soybeans	57	65	66	106	188	181	247	222	242
Roots - Tubers	47	54	58	582	879	836	148	121	161

Table 12: Grain Trade and Welfare Implications of Alternative Productivity Scenario

	Grain Trade			Welfare	% Malnourished	
Region	**1990**	**BASE**	**Weakening**	**1990**	**BASE**	**Weakening**
USA	88.8	122.2	114.5			
Econ on Europe	23.9	30.4	4.8			
Japan	-27.8	-35.6	-37.1			
Australia	13.0	21.6	24.9			
Eastern Europe	-1.9	15.3	18.9			
CIS	-28.7	-.0	5.0			
Developed except Japan	91.2	188.2	221.2			
Latin America	-13.3	-15.0	-18.6	20.4	14.0	15.3
Nigeria	-.4	-.8	-1.2	35.4	29.5	31.5
N Africa	-2.9	-8.7	-8.6	31.4	27.9	31.2
C & W Africa	-3.0	-5.5	-5.4	22.7	21.1	22.3
S Africa	-2.8	-7.0	-7.2	24.8	21.2	22.6
E Africa	-.7	-4.1	-5.0	25.5	24.8	26.4
India	2.0	2.3	-9.6	63.0	45.5	49.5
Pakistan	-.7	-15.8	-23.2	41.6	32.4	35.1
Bangladesh	-1.2	-4.7	-3.5	65.8	52.8	67.6
Indonesia	-.2	-3.7	-7.5	14.0	7.8	8.8
Thailand	5.1	4.3	2.5	13.0	5.3	5.8
Malaysia	-2.5	-5.3	-5.0	17.6	9.9	10.6
Philippines	-2.1	-4.1	-3.1	40.0	32.9	39.4
China	-13.2	-22.3	-20.0	21.8	13.8	15.7
Developing	-91.2	-188.7	-221.2	34.3	25.4	29.7

THE AGRO-FOOD INDUSTRY AND INNOVATION

CHRISTIAN GODDE
Astra Calvé
23, rue François Jacob
92842 Reuil Malmaison, France

BRAND NAMES THAT SATISFY

The agro-food industry cannot develop without famous brand names that satisfy consumer expectations in all fields and that are based on a high degree of innovation.

This presupposes thorough knowledge of what the consumer needs and expects in a variety of fields:

- concepts of intrinsic quality (quality of taste, appearance, presentation) or perceived quality (return to regional products, old family recipes, strong tasting products, etc.) indicate that France still has a strong Epicurean tradition, despite the growing popularity of the "fast foods";
- concerns for nutrition and health are becoming increasingly visible, especially with an eye on prevention — as long as it does not mean sacrificing taste;
- awareness of environmental protection is growing and will become part of a non negligible emotional choice — once the other criteria have been met;
- concern for the quality/price ratio varies enormously between the so-called "basic products" or "staples" and the "pleasure, impulsive, festive" foods. This idea is gaining ground because the present bullish economic context offers a wide range of very inexpensive products of acceptable quality;
- interest in practicality and facilitating home chores is growing, since time is becoming more and more scarce for everyone. This is a field in which innovation has made the greatest efforts and had the most success.

None of these fields can be considered alone. The success of a product is always the result of a judicious combination of all or some of these factors at a given moment in time. This combination will change with time just like the consumer's concerns change along with changes in the environment.

THE NEED FOR RESEARCH

Against this backdrop, the agro-food industry cannot live without research in various domains:

- raw materials: research is expected to provide better quality in order to facilitate improved preservation and even enhance nutritional and gustatory characteristics to make up for certain current dietary imbalances and revive the pure tastes of traditional agriculture without giving in to the demands of intensive agriculture;

- processing and ancillary production: research scientists and everyone else involved are trying to make these processes less aggressive, more "natural", respectful of the initial raw materials, and, at the same time, ensure food safety and shelf life adapted to our present-day modes of distribution and consumption;

- packaging systems: this field is especially important to guarantee practicality, originality, product preservation and easy distribution.

Since the agro-food industry cannot work in all these fields at the same time, it has to make choices. As a rule, agricultural aspects are handled at the initiative of the production side (seed suppliers) or together with them and public sector research. Because of this situation, industry has to define its medium- and long-term needs clearly, bearing in mind that, by nature, research has a time horizon that is infinitely greater than that of the competition-pressured agro-food market.

IMPORTANCE OF PSYCHOLOGICAL ASPECTS

A clear definition is not a passport to guaranteed success. As mentioned early, the consumer's mindset is extremely important. And everything linked to this psychological domain is particularly difficult to control and predict.

Let's look at a few examples.

Nutrition is a field that arouses great consumer feeling.

In the early 1970s, when nutritionists severely condemned the erucic acid level in rapeseed, agricultural research managed to develop cultivars with low levels. Since then, nutrition studies have played up the role of rapeseed in preventing cardiovascular diseases. Efforts to improve rapeseed have focused on its nutritional features but have ignored its tendency to oxidate, which makes it smell too foul to be of any use in frying foods. The law actually recommends it only be used in seasonings. Furthermore, neither agriculture nor industry has improved the consumer's overall - adverse - impression of rapeseed; it is very difficult to change this perception.

The job could become even more difficult now that rapeseed grown on set-aside lands is increasingly used for fuel, as some towns advertise on their buses. Would you be thrilled at the prospect of making salad dressing with motor oil?!

During that same period of time, there was another, opposite case. Sunflower cultivars were adapted to French climatic condition. In just 20 years, sunflower seed oil has upstaged groundnut oil as the reference in France. This was the result of research by the Unilever Group in the 1960s that showed that the high linoleic acid content in sunflower seed oil was effective in preventing cardiovascular diseases.

What made this possible was the excellent reaction of sunflower seed oil when used as a substitute for groundnut oil, and the consumer's awareness of the link between cardiovascular diseases and the characteristics of certain fats. The European economic policy also contributed by subsidising the sunflower crop.

This example clearly shows the close interdependence among agricultural, scientific, economic and psychological factors. All of them contributed to the success of sunflower seed, in contrast to rapeseed, which never managed to overcome its original handicap.

Health safety is another important sector.

Close cooperation between agriculture, industry and the public authorities has led to considerable progress in the quality of dairy products.

Things started to change once it had been recognised that strict hygiene on dairy farms was essential in preserving the quality of milk during processing. Processing conditions were improved and extended product life was guaranteed when milk was paid for according to quality: refrigerated vans were used for transport, and fast, reliable microbiological tests were developed. The organoleptic qualities of the dairy products also improved. A huge number of fresh dairy products and cheeses were put on the market, without having to cope with the health risks caused by the use of fragile raw materials.

This evolution created very strong consumer confidence which industry banked on to introduce innovations in many fields. One example is the development of yoghurt ferments used in natural, less acidic, smoother or more nutritive products, such as the whole range of yoghurts - Bio, BA, LCI, etc.

In another vein, unpasteurised cheeses are very popular among the French. Our foreign friends often make alarming, critical comments about the related health risks although they recognise the exceptionally good taste.

The listeria epidemic made things worse. As a response, researchers in the Astra Calvé Group developed a certain protection through the discovery of a strain of pediocines that stopped the growth of, and eventually killed listeria. Our product was certified by the French authorities. It is prepared through lactic fermentation and is used to protect unpasteurised cheeses against accidental contamination by listeria. This means that thanks to a better understanding of food safety conditions we can continue enjoying our good cheeses, made with unpasteurised milk, and cast a conniving wink at our foreign friends.

CONSUMER CONFIDENCE: MOST IMPORTANT

These examples bring out the importance for agricultural research and the food industry to build up consumer confidence in product quality and safety. This is a *sine qua non* precondition to introducing innovative improvements to products being offered, and to developing outlets.

Everyone recognises the importance of biotechnology in facing the enormous increase in food required to keep up with the growing world population. These technologies, if well mastered and used, could accelerate final research in developing plant varieties that have decisive economic or qualitative advantages because of their agricultural, organoleptic or nutritional characteristics and because they can be easily processed.

It is broadly held that the most touchy operation is winning consumer confidence through flawless scientific ethics and benefits, even if indirect, that the end user can perceive. The wager facing industrial and agricultural research on agro-foods is not essentially technological. It first and foremost involves the capacity to give the consumer a satisfactory explanation about the consequences of using the technologies that make him worry.

Attention cannot be limited to solving scientific problems. In today's world, much more attention has to be given to communications, choice of words, listening to "anxieties", rationalisation of stakes, and scientific popularisation in the noble and demystifying sense of the term.

Agriculture needs products that are less vulnerable to all sorts of pests, that are more respectful of the environment, and whose higher yields generate higher profits.

The agro-food industry would be very interested in obtaining raw materials, ingredients and processing aids for production as soon as possible and in the most favourable economic conditions so that it can expand its product line. This said, it has to be very clear that industry, fortunately, does not have the power to impose products that the consumer doesn't want, even if they are more "advantageous" for him. The future of biotechnology will be built together with the consumer, not against him.

The common responsibility of agricultural research and industry is to be professional and transparent in approaching problems, without releasing uncontrollable forces. Public authorities should join the movement by serving as strict guardians to ensure the safety of techniques and the resulting products and as impartial educators to fight false ideas and immoderation, and to explain opportunities.

Only then will the creative, innovative talents of science and industry emerge as a positive force at the service of the consumer and his ever-changing needs.

AGRICULTURAL RESEARCH IN FRANCE: A USEFUL TOOL IN EFFECTIVE, PERVASIVE AGRICULTURE

MAURICE RIGAUD
Secrétaire général
Assemblée Permanente des Chambres d'Agriculture (APCA)
9, avenue George-V
75008 Paris, France

ABSTRACT

French agriculture has to face new challenges that include sharper competition for the major markets, greater consumer demand for quality and transparency, and society's expectations for environmental protection.

Research, in France, should respond to these challenges by optimising the special capacity of French agriculture to benefit from regional diversity and at the same time fulfil the requirements of the consumer and the society.

Strong national and regional consultation facilities and a new agriculture-society interface provide an appropriate setting for approaches based on partnerships geared to accommodate economic and professional expectations.

To meet this challenge will also require a coherent European commitment to theme-specific research poles.

INTRODUCTION

During the last fifty years, a gradual increase in resources has equipped research to contribute more to progress in agriculture.

In the coming years, research should continue serving as a tool to cope with the economic and social challenges facing agriculture.

WINNING THE AGRICULTURAL WAGER

There are many elements that constitute a break from what agriculture was in the period from 1960 to 1980. Let me mention three of the main ones:

- economic considerations: both foreign sales and market interdependency are on the increase. The present tendency for the major agricultural and agrofood markets to become interdependent will be further accentuated by Europe's financial future, European expansion to include Central and Eastern Europe, and negotiations within the framework of the International Trade Organisation.
- during this same period of time consumers in France, and in all the developed countries, will be paying closer attention to the quality of agricultural and food products and demanding not only transparent production conditions, but also unambiguous information on the connection with local lands and resources, and on animal welfare.
- more generally, society is stipulating conditions for protecting the quality of water and keeping up agricultural lands and forests.

The prime function of agriculture will continue to be production, first of all food production, and second to serve industry and to produce energy for markets that are growing as a result of special interests in renewable biomass. But performance targets must be kept compatible with market constraints, consumers' demand for quality, and the obligation to protect the environment.

Going beyond this wager, agriculture will have to meet society's expectations better bringing out the full potential of regional diversity, on adequate numbers of farm holdings, and by contributing more to rural land management and maintenance, bearing in mind the need for more jobs and for territorial balance.

RESEARCH ADAPTED TO THE STAKES

Individually, agricultural units and small/medium agro-industrial enterprises do not have the resources to conduct the research they need to foresee or cope with present stakes.

That is why France has public research organisations (INRA, CEMAGREF, universities) and professional applied research organisations (technical agricultural institutes, technical agro-industrial centres, chambers of agriculture, certain cooperative organisations).

All these organisations should be able to help in meeting socio-economic requirements through:

- research that allows scientific innovations of various origins (biology, physics, etc.) to be adapted to agriculture. We know that innovations in the field of human health and that biotechnologies can substantially impact animal health and plant production techniques. These innovations can often improve the performance of production by reducing the production costs. They can also help upgrading industry, especially in the field of energy production.
- research that heeds the expectations of society in order to weigh the consequences of technological choices, without compromising the performance of agricultural and agro-food enterprises, all the while recognising that the French society is not necessarily ready to accept every sort of innovation and that French and Anglo-Saxon expectations may differ. Genetically Modified Organisms provide a good example.

- research, as effected by agricultural policy on farm structures, with priority being given to human-size agricultural enterprises where responsibility is individual, to the assets of the French *terroirs* (village lands), and to soil fertility maintenance; innovations must respond to the expectations of the farming systems and protect the specificity of French products.

Agricultural research must take all these French problems and approaches into account.

NEW APPROACHES BASED ON PARTNERSHIPS

Research can only be innovative if it has broad contacts with national and international scientific circles.

But to meet economic requirements and society's expectations it has to:

- make political and professional decision-makers well aware of the major wagers, (coherence and contradictions) and the most appropriate solutions so that aggregate research resources can be directed to priorities of French agricultural policy. Research should play an important role in providing the foresight and perspective needed by the policy-makers and professionals;
- strengthen consultation facilities at the national and regional level by including economic and professional considerations when ranking priorities and selecting research programmes;
- mediate more effectively between agriculture and society by providing the scientific information needed for the consumer and society to understand the requirements of agricultural activity as connected to environmental constraints, and, at the same time, by finding ways in which agriculture can better fulfil societal expectations.

As French society becomes increasingly urban, its direct contacts with the rural world, and thus, mutual understanding, are fading. New demands for quality and food safety require a capacity for transparency founded on an uncontested scientific basis.

On the other hand, the farmers now have to achieve an economic performance level that no longer allows them to tend the rural land areas without charge. The cost of this function, *i.e.* maintaining rural lands that can welcome the urban world, has to be calculated objectively. The same applies if environmental protection measures require costly changes in cropping and stocking practices that risk jeopardising the zone's productive function in agriculture.

IN THE EUROPEAN FRAMEWORK

To justify collective funding, it is essential to anchor agricultural research in the political, economic and societal spheres.

This is increasingly true as public resources become increasingly scarce.

Budgetary rigour and the globalization of agricultural markets make trans-European cooperation among research organisations more important than ever before. This cooperation could be obtained by building up European poles on the basis of (major) theme-cum-expertise. France should be able to play a central role here because of the resources it allocates to agricultural research and the international clout of its agricultural research organisations.

These poles will have to correspond to the ultimate aims of agriculture; agriculture will have to be competitive on both the production and the export front, and not merely serve to keep up the rural lands.

French agricultural organisations are well aware of how much research contributes to agricultural advancement.

Strengthening dialogue and partnerships between the world of science and agriculture is essential if research is to continue impelling the development of high performance French agriculture that also contributes to rural enhancement.

CHAIRMAN: JEAN **GABOLDE**, *EUROPEAN COMMISSION, DG XII, BRUSSELS, BELGIUM*

VICE CHAIRMAN: MICHEL **SEBILLOTTE**, *INRA, PARIS, FRANCE*

THEME 3

AGRICULTURAL RESEARCH, INNOVATION AND SOCIETY

Workshop 2
The evolution of partnerships

FIRST PANEL DISCUSSION:

SCIENTIFIC AND INDUSTRIAL PARTNERSHIPS

Michel Callon, on behalf of **Jean Gabolde,**
Director of Research Policy Coordination, European Commission

Interaction between research activities and various sectors of socio-economic activity is constantly growing. This is what the economists now call the research and innovation system. Schumpeter's linear model is being faulted because it cannot represent current activities in research, technical development and innovation. Yet many of the practices used in research, like programming, are still based on the linear model. Attempts to climb out of the rut dug by a half a century of R&D policies, measured by categories described in the Frascati manual, have met with especially resistant administrative and financial red tape. The task forces that have been set to work in Brussels during the last year actually constitute an effort to introduce interactive planning, in other words, to define projects that serve several EU countries and interact in their operations with other programmes, rather than being carried out alone, concomitantly. These task forces have been established to hold consultations with specialised groups composed of research scientists, end users, services that participate in preparing regulations, and the citizens' representatives.

Nevertheless technical R&D activities still require proper sequencing. Fundamental research will always be upstream of industrial development, and application of innovations will always be downstream. We have seen that systemic mutation does not annul linear practices, nor does it replace them with panacea, such as interactions and networks. Even partnerships can be seen as a linear system, because they are based on a relationship between two partners. What we are interested in now is the variability of its configuration. The search for partners is always an interactive operation. Future partnerships will bring together different disciplines, and different socio-economic sectors. The linear chain, on the contrary, has encouraged actors in these disciplines and sectors to intervene in succession, separately, rather than together or simultaneously, as required in the interactive model.

The example of agricultural research is especially interesting. The research-development policy, particularly in an institute such as INRA, is in constant interaction with national policies regarding agriculture, health, environment, land planning, development, etc. This is not at all surprising since, at the scientific level alone, agriculture interacts with a wide range of disciplines, from biomedical sciences to the science of man and society.

The position of this particular workshop, in-between two panel discussions, one on technical questions and the other on society, clearly illustrates the systemic nature of research activities in organisations such as INRA. The importance of bioethics and food safety for the everyday citizen resounds as an echo to scientific and technical development.

The research that has the greatest impact on society often emerges at the interface between disciplines and between sectors of application. This is the case in the agro-food chain as concerns applications from genetic engineering.

The serious problems encountered by European countries and EU institutions with regard to contaminated growth hormones, adulterated products, polluted groundwater tables, and the BSE epidemic were deplorable, but they are instructive examples that can serve as a starting point for preparing the partnerships we want to establish in the future. These partnerships should not only involve innovators; the users should also take part. In general, they should have the European dimension needed to ensure their effectiveness. They should reflect Europe's openness to other regions of the world, whether these regions are competitors or partners. These partnerships will be essential if we are to keep up with developments in research to reliably inform the public, and at the same time, to predict future difficulties.

Mr Lambert, *Director of Research, Groupe Besnier, France*

The agro-food sector is composed almost entirely of SMEs, and where only 3% of the companies have more than 500 employees. The SMEs represent 63% of the sales in this sector. Private companies seldom spend more than an average of 0.3% of their turnover figure on research, although this figure varies widely, since 95% of the SMEs do not have their own R&D service.

What limits are put on research in this sector? The products of research are used in highly competitive markets, characterised by a decreasing demand for general consumer products and new consumption models in which the consumer is looking for more practical products. This is a sector where the supermarkets are enormously powerful, which makes competition between companies even more fierce, and forces them to optimise quality, product safety and cost price, and to innovate by using R&D, even if resources are scarce.

How can these companies carry out research and development? Remember that most of them are SMEs. They have three possibilities. The first is to work via the market. The second is to create an in-house capacity, which often helps the company to make better use of technological potentials in their immediate environment, in particular by detecting new production opportunities. The third is to establish partnerships with public research organisations such as INRA, CEMAGREF, advanced agricultural schools that are under the Ministry of Agriculture, and laboratories such as INSERM and CNRS. These three approaches give a clear idea of the complementarity between in-house R&D and R&D conducted elsewhere.

Mr Freyssinet, *Scientific Adviser at Rhône-Poulenc, France*

Rather than speaking about innovation-targeted partnerships, as the workshop theme calls for, I would prefer talking about partnership-targeted innovation. At Rhône-Poulenc, our customers, the farmers, constitute the starting point from which research programmes and market products are defined.

These products should be effective at small doses, be highly targeted in order to avoid risks to the environment and to health, and be sold at a reasonable price. Also, considering progress in genetics, we should not delay studying complementarity between the genetic and the chemical approach.

After defining the product's characteristics, we try to identify the technological challenges. The first challenge is to increase chemical product diversity and clearly identify targets (insects, fungi, weeds) so as not to destroy anything else. The second challenge is to analyse how the active molecule penetrates and moves through the plants, *i.e.* its bioavailability. The third challenge is to limit the effects on the environment and be capable of detecting residues.

Then, we try to build up the most effective partnerships, such as the partnerships within industry under Eureka, the European programme created at the initiative of the producers with input from university laboratories.

Some partnerships bring industry together with research organisations, such as in the BioAvenir programme, or are composed of a variety of laboratories. Timebound individual partnerships may be developed for just one project. What is important is flexibility, adaptability and reactiveness, qualities that are not always readily present.

Mr Magnien, *DG XII, Head of the Biotechnology Unit, European Commission*

Without going into detail on all the possible partnerships described in the EC documents, I would like to speak about four of their main attributes.

A CULTURAL APPRENTICESHIP

First of all, these partnerships are the result of cultural evolution, not the product of technocrats, or even politicians. The history of EC programmes shows that, before 1985 multi-partner research was unknown in the Community where subsidies were extended to laboratories individually. The European Laboratories Without Walls, with its inter-disciplinary collaboration, was created in 1986, and, in 1990 manufacturers were invited to participate in Community programmes, at least as far as the life sciences were concerned. Now their participation is one of the criteria in the project screening process. Industrial platforms, which are informal flexible conglomerates attended by all types of companies interested in a scientific problem, like the one on lactic acid bacteria, have only existed for the last five years or so.

EVOLUTIVE DYNAMICS

The coordinator's effort in taking these types of initiatives is decisive. INRA, which headed the research project on the bacteria Erwinia, a pathogen that causes fruit and vegetables to rot, taught the Commission a lesson in partnership by requesting that half of a subsidy it had been granted be allocated to a neighbouring country which would also be participating in the research.

What makes partnerships interesting is their ability to adapt. One should be able to move easily and quickly from one partnership to another, and also be able to organise a sort of reservoir. We have seen that before joining partnerships, many companies were part of an industrial platform that served as an especially good observation post.

EXTENDING A HAND

The third essential element in starting a viable partnership is the human factor where the role of the partnership instigators and leaders must be accompanied by internal and external communications. Between 1993 and 1995, the percentage of manufacturers in the fermentation sector who participated in EC research rose from 40% to 100%. This quantum leap was largely the outcome of effective communications. Using a different language, and introducting the term "cell factory" provided the necessary impetus.

MORE THAN A PROJECT, AN EVENT

The fourth element is that partnerships must be inherently structured for success. That the underlying line of logic is not always interpreted in the same way by all the partners does not matter. For scientists, success lies in the results of their research, in particular if they are applied. For the public authorities, success means using the taxpayer's money well. Partnerships offer an opportunity to optimise every Ecu spent on an EC programme, since besides the expected results, there is also a fair share of unexpected and unforeseeable results. The more flexible, open, and versatile a partnership, the greater the role of the scientists. For manufacturers, it is essential to have access to functional "turnkey" knowledge about biological systems, the kind that only multi-party partnerships can really provide. There are numerous examples, from yeast to plant transgenesis.

Mr Callon *Ecole des Mines, Paris, France*

Cooperation between industry and public research is not new but, during the last few years, its form has changed. What is new is the inclusion of partners and actors who used to be excluded.

Mr Legrand, *Secretary General of the INRA Environment Delegation and member of the Economic and Social Council, France*

Actors who previously had minor roles have become full partners. Relations between the man in the street and research have also changed.

The popularisation campaigns of the 19th century evolved as the relationship between the citizen and the scientist became interactive. The citizen asks questions, which are not always unjustified. One example is the question on the potential consequences of genetic engineering on the environment.

People want scientists to have a clear position, a status of their own; this status still needs to be defined. They also want to participate, indirectly, in defining standards, and energetically insist that precautions be taken. Besides wanting to know more about how research is applied, they most certainly are having their say about what direction research should take in the initial alchemy that makes certain futures possible, and others not.

Who are the new social actors? Some are environmentalists, some are farmers whose progress forces research to include so many more themes that within them unimaginable convergence of interests can be found between manufacturers and environmentalist on questions such as genetic engineering. Convergence of this sort causes traditional interest groups to fall apart, but also constitutes the basis for new alliances.

Another change concerns where decision-making and negotiations should take place. One of the vocations of the regional parks as a structure, for instance, is to co-ordinate functions related to land and space. It is easy to understand that this radically changes the problem-oriented approach used in research, as well as inter-partner relations. New institutions have their place in partnerships at both the local and the regional level, and are apt to become partners in their own right, since they sometimes occasion very specific work. Finally, we see a new type of multifunctional rural development structure appearing on the scene. Here, I am thinking in particular of the company created by Nestlé and Vittel with a group of farmers.

All this has its effects on the themes and approaches adopted by research, and the downstream application of results. These non-linear, multidisciplinary partners want the stage to be set anew. The time has come to turn the page on the more narrow, sector-specific, single-purpose partnerships and move on to a more systemic vantage point. If these partners are not among the chosen, they will elect themselves. We should prepare for the future.

SECOND PANEL DISCUSSION:

FUTURE RELATIONS BETWEEN AGRICULTURAL RESEARCH AND AGRICULTURAL EXTENSION SERVICES

MICHEL SEBILLOTE
INRA, Paris, France

This discussion on relations between agricultural research and agricultural development, or extension, calls for a few introductory remarks. Agriculture is going through a period of radical change. In the past it was the farmer who decided what was going to be sold, whereas now the customer dictates what agriculture will produce, thereby making product quality a major concern. At the same time, new actors have come on the scene. Citizens and consumers have requirements that apply to both the act of producing, the direct function of providing food, and the indirect function of providing a healthy, protected environment and landscape. This explains why the issue of sustainable development is at the forefront of debate. Last, there are new, as yet ill-defined, functions for the farmer that will give him a multi-action status or may become his main occupation. Whatever changes occur, by virtue of rules and regulations, the public authorities are becoming increasingly involved at both the national and the international level.

What are the stakes and challenges of agriculture in today's world? First we need to learn to reconcile economic imperatives with new demands, to combine the logic of the market-oriented sector with that of territorial development, to handle relations between agricultural situations that differ from region to region within a single country or between countries in an ensemble like the European Union,and to combine respect for regulations in the search for greater freedom in business with voluntary decisions to undertake environment protection activities.

These changes have a great impact on relations with research. Farmers and manufacturers are expecting more and more from research in the provision of decision-support tools and, more importantly, tools to use in negotiations and even in their everyday life. Last, there is a need for research on the forms of development, *per se*.

New fields and themes for research, even new forms of research with more seriously organised expertise, are emerging. The system of individual commitment is being turned into a system of collective decision, or even the decision that no decision can be taken!

Mr Rawlinson, *Head of Research and Development, Home-Grown Cereals Authority, London, United Kingdom*

The future of most industries in the U.K. depends on research and development. This also holds true for agriculture which is a key sector in the national economy. In a country driven by market forces, the production, trading and processing of commodities must constantly become more efficient to keep up with the competition.

I want to speak about the major funder of agricultural research in the U.K, and leave it up to you to decide whether you think that this model can be transposed to other countries in Europe. We already focused on sustainable agriculture. Speaking by analogy, my concern is for sustainable research. What conditions will be needed?

The U.K. Government has gradually withdrawn from its former role as a major funder of near market research. Industry responded positively and has created mechanisms to fund such work; at the same time the relevant agricultural research council has been re-organised to focus on world-class fundamental science. Government has promoted wide-ranging consultation to establish the future research needs of the nation's agriculture sector and so provide guidance for all involved. Within this process, Government is also examining what would be the most appropriate status for research institutes; options include privatisation to create organisations that are no longer dependent on central funding.

The U.K. now has a range of agricultural research funders, each with a clearly defined role. Research councils address fundamental science, and Government builds on this by funding strategic research. Industry funding of applied work uses these foundations to generate technical innovation, new products and technology for farmers, traders and processors. In this system fundamental research seemsto be the weak link.

I am convinced that European agricultural development will rely increasingly on industry funding for research and on farmers contributions through levies. Collaboration across Europe will be needed for maximum effectiveness.

Mr Nielsen, *Director of the Danish Institute of Animal Science, Fouloum, Denmark*

Let me begin with a very short explanation of the organisation of research and extension services in Denmark.

Basic research, or fundamental research, is supposed to be carried out in the university sector and therefore is managed by autonomous universities, *i.e.* by scientists, and financed mainly by public money, *i.e.* the government.

Strategic and applied research is carried out in research institutes, like INRA, managed by boards with representatives from the users, that is, the customers. It is financed partly by government and partly by industry. The scientists in this sector are not free to choose their own projects as they are in the university sector.

Extension services are managed by the main farmer organisations, and financed by the farmers themselves, with some government support. These services are composed of a central unit and a large number of local advisors.

There is a lot of overlapping, which I think is necessary. We all agree that efficiently organised research and well-disseminated results are becoming more and more important for both agricultural industry and society. I believe that in the future, an increasing share of agricultural research will be financed by industry. However, with the present structure of primary agriculture, a large share of public funding is needed, since the relatively small agricultural holdings are not able to carry out firm-based research. It is also true that what we call agricultural research today is of more general interest to society as it is, since it focuses on subjects like animal welfare, environmental research, etc. With the assumption that in the future we will have a mix of industry- and public funding for agricultural research, the problem is how to organise people and institutions in such a way that we get the funding parties to work together, so that we get high quality, relevant research and well-disseminated results.

All kinds of research have to be measured by quality. The procedure for measuring quality is peer review in one form or another. Applied research should also be measured by its relevance. The concept of relevance implies high probability of a useful result, high probability of results being applied, and rapid exploitation by the customer. I see the following points as important in creating an efficient research policy.

First, it is important for customers (agricultural industry and consumers)to be involved in the overall formulation of the national agricultural research policy and thus have real responsibility for the research agenda. It is our experience that negotiations and agreements about split financing of research programmes are very important in securing quality and relevance. At the institutional level, it is also important that the users of research be strongly represented in the governing bodies. With strong user representation, some would fear that the research policy would focus too strongly on solving day-to-day problems. This has not been our experience. If you have responsibility you act accordingly.

Second, scientists should be actively involved in disseminating research results. That scientists are evaluated mainly by their production of scientific papers in internationally recognised journals creates an obvious dilemma. Less credit is given to lectures to farm audiences or articles in farming journals. It is my opinion that the researchers cannot be released from the dual obligation to publish in both scientific and farming journals, however difficult it may be. Contact with the customer is also essential in the formulation and interpretation of relevant research projects. Furthermore, involvement in the dissemination process stimulates the necessary development towards multidisciplinary research programmes.

Third, a central body of extension people should be involved in the formulation of research programmes. The best way to ensure the efficient dissemination of research results is to have customers waiting for them. If extension people are involved in the formulation of the research agenda, they will also be waiting for project results.

In conclusion, my message is that all parties formulating research policies, conducting research, and disseminating the results of research should interact closely, and therefore be given responsibility for the outcome of the process.

Mr Bros, *President of the National Agency for Agricultural Development, France*

The French agricultural development system can be broken down into two sectors: applied research with each production sector having its specialised technical institutes, and farmer support and guidance, with backing from the Chambers of Agriculture, associations, cooperatives, etc. The former works on a per sector basis, the latter on a territorial basis. They are complementary at all levels, national, regional and local. Agriculture seems to get satisfactory results from the system, although the weaklink is applied research, because itlacks the necessary ressources.That's where more efforts should be made.

Relations between agriculture and fundamental research, which are channelled essentially through the scientific boards of the technical institutes, are less smooth. What does the agricultural profession expect from research? First of all, there are concerns for the future because agricultural research seems to be splintered just at the time when it should be adopting a more global approach. Farmers are also worrying about the orientations of research being defined without due regard to repercussions at the field level. They want research to be close to their preoccupations, which include markets, product quality and presentation of the environment. They want scientists to recognise that progress is not always a top-down process, that it can also start at the grassroots level. They also expect the scientists to serve as specialists, and want research to be one step ahead, to detect tomorrow's potential problems today. It is too late to start working on BSE when the consumers have already turned their backs on meat. They feel that agricultural research should also include the social reactions of people involved. This is important in order to predict how an innovation might be received and how progress can be shared, especially since progress is increasingly group-oriented, rather than being geared to the individual.

Agricultural research and agricultural development are brought together through technical institutes, a link worth developing. This link can also be strengthened by organising contacts between scientists and farmers, and between scientists and agricultural technicians, especially at the regional level. New methods must be found. Farmers have welcomed the introduction of the scientific and technical interest groups (GISTs). A very effective pork GIST was created for everyone, from scientists to producers.Efforts must be made to remove the barriers and facilitate the formation of multidisciplinary partnerships that rally to a given project. More consultation is needed, especially at the regional level.

People should be able to move more easily between research, industry and production, and people working in the same area should work together.

CHAIRMAN: NOËLLE LENOIR, *CONSEIL CONSTITUTIONNEL, PARIS, FRANCE*

VICE CHAIRMAN: JEAN DUNEZ, *INRA, BORDEAUX, FRANCE*

THEME 3

AGRICULTURAL RESEARCH, INNOVATION AND SOCIETY

Workshop 3
Research and society: scientific expertise and public decisions

WHY ARE PUBLIC ORGANISATIONS NEEDED FOR AGRONOMIC RESEARCH?

PATRICK COHENDET
Professeur, Louis Pasteur University
67000 Strasbourg, France

PIERRE-BENOIT JOLY,
Directeur de recherche, INRA/SERD
Grenoble, France

INTRODUCTION

Tension around public budgets, the recognition of the limits of intensive agricultural production (together with the saturation of markets, worsening environmental problems, etc.), and the growing importance of private R&D efforts (especially in biotechnology) have caused public agronomic research to be called into question, in many countries. In the US, debate on the excessive weight of USDA and USAID research in the federal budget is particularly lively. In the UK, the Agricultural and Food Research Council has been abolished, basic research in biology has been integrated into the BBSRC, and near-the-market research has been privatised.

Such a situation is indeed surprising in light of look-back studies which generally show that investments in PAR were economically highly effective. The most proven econometric results indicate returns for agricultural research which far exceed the internal rates of return considered by private firms to be attractive. As Joly (1997) indicates, three hypotheses can be put forward to explain this paradoxical situation responsible for budget cuts in agronomic research in numerous countries: either public decision-makers are unaware of econometric studies, or they know about them but do not believe them, or they consider that such analyses of past situations are of no help in predicting the future. We believe that the latter assumption is the most likely one. This standpoint requires us both to reconsider the viewpoint of economic theories on the place and role of public research, and to investigate the concrete forms of public action in agronomic research, notably in France:

- Recent theoretical work in the Economics and the Sociology of Innovation (Nelson, 1995; Callon, 1994; etc.) call for a profound reconsideration of the traditional concept justifying public research. The basic idea that public research must be devoted to the production of scientific knowledge - a public good *par excellence* -, while private research (after having translated and appropriated the results of public research) is devoted to technological production with a view to improving and developing goods sold on the market, is increasingly contested by theoretical reflection and empirical observation alike. This challenge is all the greater since classical criticism of public research (principal-agent effect, bureaucracy, research lobby, difficulty of operating external control, etc.) is particularly fierce today. If we stop postulating the existence of that sacrosanct dichotomy around which the respective roles of public and private research seem definitively set, new conceptions of the place and modalities of action of public research emerge naturally. A large portion of this paper will therefore be devoted to an exploration of these new conceptions.

- Empirical observation of the evolution of public research organisations, such as INRA in France, confirms that new forms of effective action by public research organisations are taking shape. Three trends warrant emphasis: (1) public organisations benefit from a stable structure which enables them both to develop original scientific and technological skills with significant long-term effects, and to maintain and manage a pool of competencies. From this point of view, strategic basic research is the favourite field of agronomic research. This is clearly illustrated by research on the biological fixation of nitrogen. We note that in the US a tendency for federal laboratories to be financed by private firms (which thus have access to this reservoir of competencies without having to maintain it) is currently emerging. (2) Agronomic research is characterised by the dispersion of its users (in France, 800,000 farms and 4,200 firms with over 10 employees in the agro-food sector). Moreover, innovations generally involve knowledge from a variety of disciplines. These two characteristics urgently raise the question of the co-ordination of R&D activities. In this context a centralised organisation can, theoretically, solve such co-ordination problems effectively provided it is able to limit risks related to a bureaucratic mode of functioning. (3) In these "regalian" activities, the State has certain responsibilities (watching over public health, preserving environmental public goods, etc.) and has to be able to rely on independent capacities for producing scientific expertise. Public organisations should, in this respect, be one of the main sources of independent expertise, able to act as a "watchdog" for society as a whole.

Thus, the specificity of the role of public organisations is not questioned. What is, however, is their capacity to fulfil these roles effectively, with modes of action suited to the new context that is progressively setting in. The problem is particularly acute since the systems to which these public bodies belong have become extremely complex. The Colbert model of innovation, which accompanied the silent revolution in agriculture in France, is being replaced by a multi-actor system in which the construction of innovation networks has to involve new methods for the management of research.

THE VIEWPOINT OF ECONOMIC THEORIES

As Estadès *et al.* (1996) emphasise, the entry of science into the economic world was based on two divisions: the division between fundamental and applied research, and the division between public and private research. The separation between fundamental and applied research was grounded in the hypotheses of a linear vision of the innovation process where science ("open science" governed by the norms of the academic community) produces freely accessible and usable public information, while applied research, situated upstream from basic research, strives to transform basic results into applications. The separation between public and private research dates back to the hypotheses of Nelson (1959) and Arrow (1962) for whom the uncertainty of basic research and the difficulty with which it can be appropriated was the cause of under-investment by the private sector. The social returns on innovation, partly captured by imitators and consumers, are greater than private returns which belong to the innovator. In this context, the role of the State is to increase private returns so as to encourage investments in research.

The renewed conception of the innovation process, inherited from the work of Nelson and Winter (1982) and Rosenberg (1982) among others, proposes a radical change of outlook concerning these two divisions. In a sense everything is started again from scratch, be it the very conception of science - where any distinction between the creation and diffusion stages is increasingly criticised -, the public status of basic research, the substitutable character of internal and external research for a given organisation, the need for strong ownership rights, or whatever. Innovation is henceforth considered as a fundamentally interactive learning process of a highly tacit, localised and cumulative nature. More specifically, the more tacit knowledge is, the more localised and cumulative the learning processes will be. Even when knowledge is codified, it requires the user to have specific absorptive capacities. From this point of view, research as an activity has two complementary sides: not only does it contribute to the creation of new information and knowledge, it also constitutes a learning process which helps to increase the absorptive capacity. The classical postulate of the availability of information contained in innovation is rejected. Innovation is not based only on the exploitation of easily transmittable information; it is also based on know-how which remains partly non-formalised and which is embedded in the functioning of collective routines.

The theoretical results obtained thus authorise us to question in depth the hypothesis of the two divisions in which the justification for public action was grounded. This theoretical questioning is, moreover, largely supported by empirical observation:

- The separation between "open science" and applied research does not hold in light of the interactive innovation schema. Basic research interacts constantly with applied research. In particular, the construction of scientific themes is not independent from the themes and from the ways of mobilising resources developed by researchers. Nor is it independent from the absorptive capacities of those who strive to apply knowledge. Furthermore, there is no "specialisation" between basic research and applied research respectively on the nature of the knowledge produced. Basic research does not only produce information, but it mainly produces knowledge of which some is codified and some tacit. The orientation of this tacit knowledge and the construction of new research themes are structured by ties of proximity and historic relationships between heterogeneous actors built up over time, particularly in the case of so-called "strategic basic" research. "One important function of academic research is the provision of trained research personnel, who go on to work in applied activities and take with them not only the knowledge resulting from their research, but also skills, methods, and a web of professional contacts that will help them tackle the technological problems that they face" (Pavitt, 1994). In fact basic research laboratories manage different levels of confidentiality on their work. What industrial firms try to do is not so much appropriate the results of basic research, as negotiate access rights to knowledge belonging to the world of "open science".

- The separation between public and private research needs to be rethought when we dispute the idea that basic knowledge is a public good while applied knowledge is appropriable by private entities. One is currently as likely to see profit-orientated firms investing in basic research activities, and even trying to publish, as one is to see public institutions turning towards efforts at technology transfer on a new market spawned by so-called basic research (Hicks, 1995). Thus, "is it not better rather to recognise that the main difference is based on the specific systems of incentives and social norms (related, for example, to disclosure)? There would no longer be intrinsically 'technological' as opposed to 'scientific' knowledge, but rather knowledge which possessed a sort of institutional marking, the only way of classifying it and understanding its mode of production and distribution" (David and Foray, 1994).

In this context the new role of public research must no longer be defined as "sharing of responsibility" for unconnected knowledge. Rather, it should be seen in terms of the need to strengthen and support the continuous process of creation and circulation of knowledge in specific places where the absence of a public presence would help to undermine, or even to endanger, the viability of the process. The main arguments in favour of a public research activity in the new context are as follows:

- Public research is the only research which can benefit from a stable and permanent framework. This framework enables it to build up new or original scientific and technical competencies, or to maintain and manage a reservoir of complementary competencies required for the development of technological trajectories, thus enhancing the viability of these trajectories in the long term. This direction implies profound interaction and hybridization between public and private research, particularly to allow for the selection of research themes. Public laboratories must, moreover, intensify their interaction with private laboratories, in respect of diverse resources, research material, possibilities for experiments, access to certain pilot facilities, training facilities, etc. The existence of public facilities such as technological platforms, test centres or pilot plants, appear in numerous studies to be one of the key elements in the success of a process of creation and circulation of new knowledge.

Public research can naturally be solicited to ensure the coherence of dispersed research efforts. This role of providing coherence in the innovation process may be fulfilled in two ways:

- First, through the co-ordination of dispersed research efforts which would not be able to converge "spontaneously". The dispersal of research efforts on a specific theme at a given moment, the increasing need to integrate varied disciplines in order to innovate, or the need for access to secondary competencies in order to develop new technologies, may lead to the need to find co-ordinating authorities. On condition the bureaucratic risks are limited, public organisations clearly have a major role to play in this direction. Providing the coherence of research efforts may mean providing a large number of activities related to the process of creation and circulation of new knowledge, which would not "spontaneously" be provided by private organisations. This is the case with activities involving the memorisation of knowledge, or activities aimed at the codification of new knowledge in order to better ensure its diffusion within innovation networks.

- The second way in which public research ensures coherence in the innovation process is by countering the risks of new types of relationships between public and private research undermining the system of open science as a whole, through their effects of fragmentation. These risks have been mentioned by Dasgupta and David (1994), who suggest that the intensification of collaboration between public research and industry around themes oriented towards industrial goals may be cause to fear the weakening of all academic networks. In particular, it may spell a marked restriction on the free circulation of knowledge due to the adoption of the norms of private science by public laboratories. From this viewpoint, the role of public co-ordination bodies may be precisely to ensure that the circulation of knowledge in the global research system is not hindered, so that coherence and the ability to distribute society's "stock" of knowledge may be maintained.

- Public research must fulfil certain functions concerning control or expertise, which are naturally the State's responsibility. The need to act as society's watchdog becomes all the more desirable as certain problems peculiar to the agro-food sector - *e.g.* the necessity for quality control throughout the agro-food production chain - and global problems such as environmental threats or risks of pollution, increase. In order to play this part effectively, the State has to be able to rely on research capacities which are independent from private research interests. But beyond this role of a mere watchdog, the State can also rely on public research capacities to open new research fields oriented towards the solution of such complex global problems. These new directions imply the integration of widely diverse knowledge and know-how which generally do not come together spontaneously. The recording and enhancing of basic competencies needed to find effective answers to global problems, and the availability of the access to complementary competencies essential for technological development (even if the construction of these competencies is temporarily provided by the State), constitute the main areas of State action in the framework of its usual role as a watchdog.

Apart from its role as a watchdog in the medium term, the function of the State is also to act as a watchdog for society in a long-term perspective, particularly by guaranteeing the diversity of technological options. For the State this means maintaining a level of technological diversity which keeps the field of possibilities open. The aftermath of the oil crisis at the end of the 1970s clearly demonstrated the advantages for society as a whole to be able to draw upon the diversity of technological options, some of which had been supported exclusively by the State.

As far as diversity is concerned, the role of the State is not limited to preserving that of technological trajectories. The State is also the guarantor of the heritage of regional diversity, so that public research may be called upon to steer research towards promoting the development of local micro-markets. This public research function appears particularly important in light of the risk at local level of the current economic globalization resulting in the uniformity of modes of production and consumption.

EMPIRICAL EVIDENCE: THE CASE OF AGRONOMIC RESEARCH

As indicated in the introduction, the calling into question of public agronomic research is somewhat paradoxical. All studies on the subject, especially in the US, tend to show that the impact of public research is very important in terms of social returns on investments. Among these studies, that of Huffman and Evenson (1993) enables us to understand the contribution of the different types of research better. Drawing upon an analysis of the evolution of productivity in the livestock and agricultural sectors between 1950 and 1982, the authors show that the impact of research is far greater in the case of cropping than in that of livestock farming. Similarly, it is far greater for public research than for applied research (see also Evenson, this volume). The authors highlight the problem of over-investment in applied research which leads to duplication of experimental work with only marginal gains of negligible knowledge. As Joly (1997) writes, such a result can probably be explained by one of the characteristics of the research institution in the United States: applied research is carried out in Experimental Stations under the authority of different States and probably lacks good co-ordination. It is important to be very careful in any attempt to extrapolate results in the French case; it is, however, probable that the overall result, that is to say the high social returns for public agricultural research, would remain valid.

In light of the theoretical reflections above, the paradox of the challenge to public research must be resolved by accepting the hypothesis that these results (measured *ex-post*) were obtained in a context which has undergone profound change. The sharing of responsibility between public and private research which was grounded in the dichotomy of scientific and technological knowledge, and which was perhaps in the first analysis acceptable in an emergent phase of agronomic research, is no longer justified today. This sharing has been called into question while the actors in research have at the same time multiplied their interaction, and new technologies such as genetics have helped to scramble roles by imposing new research practices, particularly applied research. As innovation networks become increasingly complex, the effectiveness of investments in R&D is more the result of the existence of linked and convergent networks that have to be constructed, supported or strengthened at their weakest points, than of the exploitation of an existing scientific potential that needs merely to be translated into technological applications. If we reconsider the role of public agronomic research from the viewpoint of the new directions proposed by the theoretical discussion above, we can illustrate the different possible modes of intervention by means of the following examples.

- The case of the biological fixation of nitrogen illustrates the essential role of public research in providing for the long-term development of scientific and technological competencies, and in exploring all possible tracks and hypotheses in the name of a mere precautionary principle for society - without which society is exposed to the risk of finding itself in a dead end in the event of the energy reserves becoming significantly rarer (Denarié and Joly 1994). At a stage where there are two competing approaches - one by obtaining fixation with cereals through the establishment of symbiosis, and the other by a direct transfer of nif genes - and where these two objectives are necessarily long-term, only a public research organisation can guarantee the permanence of research efforts, independently from temporary oil price fluctuations, where the stakes for society are critical.

- The cases of hybrid maize and milk production illustrate the necessity for the State to co-ordinate dispersed research efforts. The former clearly illustrates the joint creation, by INRA and industrial laboratories, of a new research theme and an entirely new research activity. In the 1950s a small group of INRA researchers created a project to develop varieties which combined the production of American maize with the earliness of certain well acclimatised species in France. Since the 1960s innovation networks have become denser and firms have progressively taken over from public research. They have, however, thereby revealed the importance of the role of the State as a co-ordinator and the leverage it has to promote the modernisation of agriculture. The case of milk production illustrates the implementation of a vast genetic improvement programme initiated by the State to improve the quality of French herds without exclusively importing sperm or heifers. This type of project necessitated the formulation of breeding programmes based on scientific knowledge, but also on an original system of collection of large amounts of data needed for the selection of progeny. In particular, the collection and interpretation of all the figures (on the two million cows subjected to dairy control) in a single centre at Jouy-en-Josas, under the supervision of INRA, played a major part. Progress was spectacular. In 30 years milk production per cow doubled owing to a genetic gain of 80 to 100 kg per year. For some breeds such as the Holstein breed, France is no longer an importer; on the contrary, it exports Holstein sperm.

- The role of expertise and of State control in the agro-food sector is considered to be extremely important in view of recent events. In order to fulfil this role, research has to be independent from economic interests, which may contradict the need to construct partnerships in a perspective of industrial innovation. For the development of transgenic plants, it is necessary for independent scientific experts to be able to promote public confidence. The role of public organisations is less to promote innovations than to create a favourable context for diffusion by:

• enhancing the basic knowledge which allows for more efficient work, *e.g.* the case of work on the homologue recombination, which is to lead to the introduction of a single copy of the gene inserted on a predetermined site;

• favouring modelling which makes it easier to evaluate, a priori, a potential risk; that is the case for example of inter-institute experimentation (ITCF, AGPM, CETIOM, INRA) aimed at increasing knowledge on the risks of transmission to weeds of genes resistant to herbicides. Such experiments will permit a better assessment of the risks of the spread of rapeseed resistant to herbicides in terms of the modification of farmers' practices.

The recent series of crises (contaminated blood, asbestos, mad cow disease) show that public bodies are not yet properly organised to play this role of expert and watchdog effectively. One of the current challenges for public research is probably to construct the rules of independent and responsible scientific expertise.

- An examination of the role of guarantor of the diversity of the system requires us to distinguish two complementary facets of this function. First (and this is the more traditional meaning), technological systems tend strongly towards standardisation. This tendency is the result of economic advantages related to the exploitation of increasing returns (static and dynamic economies of scale). But, as many studies suggest, such a tendency may cause early blockages on technologies which, in the absolute, are not the most effective. By maintaining a greater diversity of available technological options, public research can correct this bias towards hasty standardisation, inherent in a private system. The same applies to the stakes involved in the dynamic conservation of biodiversity (when this is done in such a way that genetic resources are rapidly usable in the productive system). A case in point is the pursuit of parallel strategies to reach the same objective, *e.g.* plants may be protected from predator insects by means of chemical treatment, biological control, or the use of transgenic plants.

The second facet of the maintenance of diversity is related to the systems of production. This objective is particularly important for agricultural research since that is often how economic activity is maintained in disadvantaged areas. Research must then allow for innovations in traditional production, often peculiar to a given area, while preserving its essential characteristics. The agricultural world in France has a wealth of this type of experience where the professional world and public research were closely linked, *e.g.* improvement of the system of production of Roquefort cheese (genetic improvement of the herds, introduction of milking rooms, etc.), the production of Beaufort cheese, and the production of dried sausage in Corsica (development of a high quality farm product involving a series of actors and activities: technical development in collaboration with the producers; creation of a brand, of a commercial organisation, of terms of reference, of rules for the approval of products, etc.) (Casabianca and Sainte-Marie 1996).

CONCLUSION

The above examples clearly show the potential growth of hybrid systems of research where links between public and private organisations are not governed by the rules of the great divide in the creation and distribution of knowledge. Theoretical reflections open up new and original possibilities of intervention for public research, and numerous empirical observations confirm the direction of developments underway. Yet many points still remain unclear in current upheavals. Thus, more than ever, it is necessary to distinguish public financing from the implementation of R&D in public organisations. The public funding of private research organisations helps to solve certain problems such as that of the existence of a competent public service in those areas where expertise which is independent of commercial interests is necessary (transgenic plants, mad cows, etc.). But the question of knowing what the limits of these possibilities of public funding are, remains open. Similarly, the question of knowing which types of organisation of public research are the most adequate must also be added to research agendas. For example, should a system of financing scientists be opposed, as Stephan suggests, by research institutes (European model) or by funding agencies (American model) (Stephan, 1996)?

SELECTED REFERENCES

ARROW K., 1962. Economic welfare and the allocation of resources for invention, *In: The rate and direction of inventive avtivity*, ed. R. Nelson, Princeton, Princeton University Press

CALLON M., 1994. Is Science a Public Good?, *Science, Technology and Human Value*, 19(4), pp.395-424.

CASSIER M., 1996. Les contrats de recherche entre l'Université et l'Industrie : les arrangements pour la production de biens privés, de biens collectifs et de biens publics, *Sociologie du Travail*, 3/96, pp.377-390.

COHEN W.M. and LEVINTHAL D.A., 1990. Absorptive Capacity, A New Perspective of Learning and Innovation, *Administrative Science Quarterly*, 35, pp.128-152.

COHENDET P., 1996. Recherche industrielle et processus d'innovation : le rôle structurant des pratiques de recherche, *Sociologie du Travail*, 3/96, pp.365-376.

DASGUPTA, P. and DAVID P., 1994 "Toward a new economics of science", *Research Policy* 23 (1994) 487-521.

DAVID, P. and FORAY D., 1994. Accessing and expanding the science and technology knowledge base, *STI Review*, 16 (Paris : OECD).

DENARIÉ J.and JOLY P.B., 1994. La fixation de l'azote, *Biofutur*, 30-34.

ESTADÈS J, JOLY P.B., MANGEMATIN V., 1996. Dynamique des relations industrielles d'un grand organisme public de recherche : coordination, apprentissage, réputation et confiance, *Sociologie du Travail*, 3/96, pp.391-408.

HICKS D., 1995. Published papers, tacit competencies and corporate management of public/privé character of knowledge, *Industrial and Corporate Change*, 4/2, 401-447.

HUFFMAN W.E. and EVENSON R.E., 1993. *Science for Agriculture: A Long-term Perspective*. Ames, Iowa State University.

JOLY P.B., 1997. La recherche agronomique en quête d'un nouveau modèle d'organisation. *INSEE Synthèses*, 10, pp. 49-57.

NELSON R., 1959. The simple economics of basic research, *Journal of Political Economy*, 91, 907-918.

NELSON R., 1995. Recent evolutionary theorising about economic change, *Journal of Economic Literature*, Vol. XXXIII (March 1995), pp. 48-90.

NELSON R. and WINTER S., 1982. *An Evolutionary Theory of Economic Change*, Cambridge, Harvard University press.

ROSENBERG N., 1982. *Inside the Black Box : Technology and Economics*, Cambridge, Cambridge University Press.

STEPHAN, P., 1996. The economics of science, *Journal of Economic Literature*, Vol. XXXIV (September 1996), pp.1199-1235.

THIRD PANEL DISCUSSION:

SCIENTIFIC EXPERTISE AND PUBLIC DECISION

NOËLLE LENOIR,
Conseil Constitutionnel, France

Research and society is a very current theme, as we see from the "mad cow" crisis and the disparity in opinions about the recent importation of U.S. genetically modified soybean onto the European market.

Mr Brodhag, *member of the French Commission for Sustainable Development, and Vice President of the Research Committee of the Rhône-Alpes Regional Council, France*

I will be speaking about the triple-track problem of the link between scientific activity and political decision. First of all, the expert as an individual, a citizen. There is a large gap between pure science, known as a stock of objective, validated knowledge, and the more uncertain science of today, where societal values have an important role. Everyone knows that the scientist is increasingly solicited and that society urges him to take a stand. It is perfectly legitimate for him to give his opinion, like any other citizen. But an expert who wants to continue being an expert has to be careful not to transcend his field of competence. His behaviour should be guided by his sense of ethics. If he wants to take a stand, he is perfectly entitled to do so, but in that case he should no longer claim the status of expert.

The second category of problems relates to the research organisations themselves. Public research serves the interests of society. Private research serves perfectly legitimate economic interests, since the economy contributes to our well-being, albeit in a different way. Doesn't the internal logic and system of organisation of large integrated research organisations like INRA, for instance, tend to make them monolithic as the result of a style that may ensure efficiency but also makes it difficult for them to respect diversity in their approaches and to remain objective in their evaluations. How can a government and a society orient research when faced with such a highly structured organisation, so capable of mobilising funds? One solution may be to develop transversal approaches that bring together several research institutes, and thereby break away from monolithism and verticality.

My last point deals with the principle of precaution, a subject the French Commission for Sustainable Development has considered carefully. Scientists are often called upon in times of crisis, but their assignment should be to avoid crisis. Precautions should be mandatory for problems that may have very serious, irreversible consequences. The situation is different if the problem is less serious and the consequences are not irreversible. The real problem is to know exactly what position to take. The starting point is the hypothesis and then, at the other end, we have the established fact. Between the two, when science is busy at work, and hypothesis becomes probability, that is when the politician wants the scientist to give him his opinion because at that juncture, if science wavers, the decision must be very clear, and the scientists should not be afraid of the media or of public opinion. They should be ready to make their doubts known, so that at a later time, perhaps, they can speak with certainty.

Mr Dormont, *Physician at the C.E.A (Commissariat à l'Energie Atomique), EU adviser on BSE*

In 1996 certain number of scientists were expelled from the laboratory because of the mad cow crisis. The crisis was triggered by sudden public perception of the problem, after a British minister stated that certain information suggested that the disease could be transmitted to man.

This crisis is unprecedented, as far as I know. In the case of AIDS, we knew the causal agent. Here, we did not have a method for making a diagnosis, we had no way of knowing how many animals had been or were infected, and we had no definite proof that this cattle disease was transmissible to man, even though scientific research since March 1996 drew that conclusion. Even now there is still no irrefutable proof.

The problem took on special dimensions because it touched on an "orphan" sector of medical research, little researched by laboratories. Before the mad cow crisis, only a few scientists worked on bovine spongiform encephalopathy,(BSE) and their position had been marginal. All of a sudden they were thrown into the limelight and strongly solicited because they possessed the little information available on the subject at that time. That was the first problem.

The second problem was the role of the scientific expert vis-a-vis the public authorities. Informing the political and administrative authorities, and bringing them up to an acceptable scientific level of understanding was a tremendous job. Before offering any solutions, the scientists had to explain the present state of knowledge on the mad cow disease and the many gaps in that knowledge. The public authorities decided to set up an expert committee that would use collective wisdom to avoid individual misinterpretations. The expert committee operates democratically.

It answers precise questions from political and administrative authorities, and, on its own authority, can decide to study a problem which it feels has an effect on public health. It can also put questions to the responsible authorities. In the case of BSE, the committee recommended applying the principle of precaution, even before any information was received proving that the disease could be transmitted to man.

The expert should stay within his specific discipline. Which means that I, as a doctor, have nothing to do with veterinary techniques or agricultural research. This explains the importance of collective expertise, with specialists from various horizons, each of whom, as an individual, does not work beyond the limits of his particular discipline. The expert also has to recognise that via his opinions, he will, sooner or later, be at loggerheads with economic or cultural habits, or with the administration's way of working.

He has to learn to explain the new situation effectively. His opinions must be concise and to the point. The essence of his ideas will be lost if he gives a minister or a government service a report that is hundreds of pages long. The expert has to be able to condense and simplify his ideas because his audience is not composed of experts. This does not mean being too sketchy or oversimplifying - speaking personally, this has been my biggest problem. Last, the expert should never forget his role as a scientific expert. He is not the politician's counsellor, his only role is to express a scientific opinion.

Handling the mad cow crisis was made more complicated because public health, unlike agriculture, is not controlled at the European level. The media thought that the Ministers of Agriculture were dealing with the crisis, while actually the matter was completely in the hands of the national health authorities of each individual country. The man in the street did not know that.

The experts must never forget the importance of reconsidering work they have done in the past, revisiting their earlier decisions. A multidisciplinary committee, like the one I belong to, must constantly re-analyse the situation and offer the government work plans with guaranteed efficiency. The expert should also be careful that his statements do not contradict those of the decision-makers, the public authorities. On the other hand, if an expert committee feels that a decision, be it political or administrative, is harmful to public health, its duty is to convey this information. With the exception of this particular case, it is best to keep the two parties clearly apart.

Last, no discipline-specific expert should ever become a professional expert. He may be solicited from all sides, as is the case today, but if he accepts to wear several hats, he risks losing his place in the research world. He may become more skilled in the techniques of an expert, but he will lose the very substance that qualified him as a scientific expert to begin with.

Mr Duby, *Mathematician, Director General of SUPELEC, France*

Scientists working in the agriculture and agro-food sectors are undoubtedly those solicited the most by politicians and even by their fellow citizens. The agro-food industry, indeed, is relying increasingly on technologies that stem from fundamental research such as genetics, which means that the technological risk is becoming part of the food sector proper. The free movement of foodstuffs, even in the food chain, makes the problem worse - as we have seen in the case of meat-and-bone meal. On the other hand, developed societies are increasingly averse to risk. We have even gone as far as to proclaim a "right to health". Society wants zero risk while technological, economic, political and industrial evolution all contribute to increasing the risk element in foods. The political decision-makers respond to this demand by making rules and regulation and by calling in the scientists - not only in times of crisis.

All this contributes to altering the scientist's profession. He is commissioned to make expert evaluations which are quite different from his brief as a research scientist. First of all, it involves a different time frame. The scientist takes a decision when he is ready to, while the politician takes a decision when he is obliged to. As an expert, the scientist may have to work much faster than he does in his own profession.

It also requires a change of logic. The scientist's job is to construct an explanatory hypothesis, consolidate it and, in some instances, defend it against other hypotheses. The expert's job is not to defend a thesis, but to examine the facts in the light of all possible theses.

Expert evaluations also require a different mode of communication. To respond to social demand, the scientist has to grow accustomed to the demands of the media. Scientific publications use a review committee, with a strict code of ethics and methodology. The public media are much more difficult to control, and the propagation effects are totally unpredictable. It was surprising, for instance, that dioxin, which never caused anything worse that skin rashes, was stigmatised as a violent poison, while the media never really condemned methyl isocyanate, that killed thousands of people in Bhopal. The scientist has to learn to communicate via these media, whose basic logic is unfamiliar to him. This requires a different code of ethics. I am shocked when I see certain research scientists hold a press conference before having their article published in a magazine like "Science" or "Nature". Media, like the Internet, also create serious problems of subterranean literature. An article that appears on the Internet has not been read by any review committee. Furthermore, rumours that circulate on the Web can be dangerously exaggerated. There have been AIDS patients, for instance, who have decided to change their medical treatment because of unconfirmed information on various molecules they received through the networks. The consequences can be very serious. What would have happened if the Villejuif tract on food additives, that circulated in the 1970s, had been available on the Internet? In media of this kind, the scientist no longer has the protection that comes from the peer reviews that precede publication in scientific journals.

All these changes also affect the responsibility of the scientist. For a long time in France, laws on responsibility were closely connected to compensation, but now jurisprudence and legislation tend to look more and more upstream, and we are beginning to see responsibility levied on the scientists. Exonerating the developers from risk responsibility is also a question under scrutiny, especially in France.

Finally, since economic interests can be considerable, scientists may unwittingly be used by pressure groups. They can become accomplices, or even form their own pressure groups to influence the political decision makers. In the case of xeno-oestrogen, for instance, didn't certain scientists, in certain countries, knowingly try to alarm the politicians in order to win public contracts?

How can research organisations react to these changes? First, rather than trying to avoid problems, I think they should try to integrate them into their field of work. This was the attitude that made INRA start studying the mad cow crisis as a way to examine relations between scientists, politicians, the media, animal producers and manufacturers. Second, research organisations should reconsider their way of evaluating their scientists. Scientific results and industrial repercussions, the exclusive parameters of the past, should be combined with other parameters such as media communications, and violations of the moral codes should be sanctioned. Third, scientists should be given the resources (this goes beyond the material resources) to discharge their new mission as experts. Their responsibility, when they work for public research, should be protected in the same manner that it is in the private sector where civil responsibility devolves to the employer. Last, in their "watchdog" role, research organisations should act, rather than merely react. Politicians quite naturally rally to public opinion, but we all know that the social perception of a risk and the reality are worlds apart. And very often, the politicians' main interest lies in the risks the public are most aware of, not the most important ones. Public opinion, for instance, is very strong on the greenhouse effect, whereas there is less interest in the equally serious threats to plant biodiversity.

Mrs Hermitte, *research scientist at CNRS/GAPP, jurist specialised in the rights of living organisms*

There may be small bits and pieces of jurisprudence on the scientist's responsibility, and we may be able to find provisions in civil, administrative or criminal law, but we have too little hindsight to be able to make a sound judgement.

This problem is new because for a long time, discontented people and victims usually sought compensation first, which meant that the search for the responsible party focused on financially solvent entities. The manufacturers, because of product liability, were on the front lines. But now victims want to go as far as the criminal courts to establish who is at fault, regardless of whether it is a physical person or a legal entity. This means that the judges have to search for individual behaviour behind what would seem to be group behaviour, and condemn the action of one individual who favoured some decision or another, as well as the silence or reticence which, in certain situations, can be sanctioned under criminal law. Beyond the manufacturer's responsibility, we are looking for the responsibility of the European Union, the State, local communities, and individuals, be they politicians, experts, or civil servants. Jurisprudence should slowly but surely determine the personal responsibility of the individual.

The mentality of the victims was well revealed through the false information suit filed against the Academy of Medicine for a report on asbestos. This was a premier, because no complaint had ever been filed for a reason of that sort before. Legal experts and experts from insurance companies are organised, and have a code of ethics that lays down the rules of the art, unlike the scientific experts who participate in policy decisions - both in times of crisis and in normal times to establish a standard or a threshold of danger. This leaves them at a loss when brought before the court where the ordinary law on responsibility is applied. I have read various reports by scientific experts who were called upon when a natural disaster occurred and have noted that they had no, or practically no, idea of the rules of law concerning their legal responsibility and, therefore, made some glaring mistakes. What part of the rules that apply to the legal expert could be transposed to the scientific expert? The obligations are clear, even if they are not always easy to respect. The first obligation is for the expert to be totally independent of the interests at hand. For a legal expert, this obligation is clearly defined in the civil procedures code and the criminal procedures code: the expert must have "no connection", of any sort - family, amicable, professional, financial - with the implicated party. He can be challenged on all his motives under a procedure of objection. We might query whether a public research contract does not inherently contradict the expert's obligation of independence. This said, there are some interesting experiences, in particular in other countries, like the creation of special financial organisations to ensure the independence of the experts, or, the decision taken by the British Commission (in the case of the authorisation of genetically modified organisms) to have the experts sign that they had no connection whatsoever with the agro-food industries.

The second obligation for the legal expert is to respect the rule of cross examinations. If he is the only designated expert, he must listen to all the opinions and record them fully. This is where legal culture differs from scientific culture. The jurist seeks contradictory arguments while the scientists seeks consensus through debate. In a scientific evaluation efforts are made to end up with a common position, which is not the case in a legal evaluation.

The other major category of obligations for the experts concerns the constitution, presentation and circulation of the results of his research.

The first element, "constitution" is defined as "an obligation to undertake all the studies needed to remove scientific uncertainty". This obligation was marginally brought up in the case of the blood transfusions and the contaminated blood. It could be brought up in the BSE case since studies, apparently, could have been started sooner, when the first warning was given. The other obligation in information constitution is to hear substantiated contradictory evaluations, since complete objectivity is not possible. Even if he is very honest, an expert - if working alone - will tend to draw conclusions based on his personal convictions. One of the reasons Dr. Garretta was found guilty is that he systematically rejected opinions that were contrary to his own. Furthermore, the references of contrary opinions must always be kept on file, especially when information is conveyed to the public. Another rule for scientific experts should be to stay strictly within their field of excellence. If need be, they should call in experts from neighbouring fields. In legal examinations, the "super-specialists" are always called in.

When speaking about the circulation of information, a distinction needs to be made at three levels: information for 1) the public authorities, 2) the scientific community and 3) the public at large. In the contaminated blood case, Dr. Rous and Dr. Netter, who worked for the Ministry of Health, were found guilty of not providing the political leaders with all the available information as quickly as possible. What they transmitted was incomplete, deceptive, and watered down.

As concerns the distribution of information to the scientific community, if we believe the many articles on BSE that appeared in Nature, the English Ministry of Agriculture and the experts apparently were reticent about transmitting all the information they had. In as serious a crisis as this, the courts, some day, may well recognise the right of the international scientific community to have access to all the available information.

Last, the most sensitive issue, of course, concerns the information to be transmitted to the public at large. Dr. Garetta was convicted for not transmitting full information to the public at large, and, more significantly, to haemophilic associations, "through all possible media - written press, television, ect." Let me point out that criminal law defines these obligations, which become even more binding when public health is at stake. But in some cases, they contradict other obligations, such as professional or medical secrecy, whose non-respect is considered a criminal offence. Jurisprudence on expertise other than scientific expertise, purports that the expert can be relieved of his obligation to confidentiality and professional secrecy when collective interests are at stake, in particular, in the event of danger to public health, on the condition that he informs his mandator accordingly. A mandated expert, however, always signs a confidentiality clause, and his expert report is always the property of his mandator, who is the only person entitled to release it.

I would be tempted to make two proposals. One is completely utopic and consists of giving the scientific expert a protected status similar to the status which labour law applies to a company's staff representative. In the case of the mad cow disease, certain people in other countries were dismissed because of the position they defended. I don't know of any case of this kind in France, but some people may have lost their senior rank in a laboratory. This said, it would be rather difficult to grant a protected status in the present liberal context. My second proposal would be to start preparing, as soon as possible, a European, and, even better, an international code of good conduct that spells out the obligations of the scientific expert and how he is protected.

Mrs Lenoir

You clearly demonstrated how the former compensation-based logic has been replaced by fault punishment. We know the far-reaching repercussions of Mrs Dufoix's (former French Minister of Social Affairs and Solidarity) statement: "I am responsible but not guilty". Nowadays, we look for the guilty, and we sanction behaviour.

You proposed that a code of good conduct be prepared, and even that a protected status for the scientific expert be defined. This does not seem utopic to me. You referred to the status of company staff representatives. In a 1988 decision, the French *Conseil constitutionnel* recognised this protected status as a constitutional right because it meets the provision of the Constitution that requires workers to participate in defining their working conditions. In 1975, since the *Conseil* also recognised the right to health protection as a constitutional right (by a decision taken in 1975), I do not think that it would be inconceivable to define a protected status for scientific experts, even if it did not have any constitutional weight. In any case, independence and protection must be combined because protection should not be a universal pretext that entitles everybody, including the experts, "to hide under an umbrella". This would go against public interest.

Mr Leaver, *biologist specialised in biotechnology, Oxford University, United Kingdom*

I am both a scientist and a teacher, and as such, am sometimes called upon as an expert in biotechnology. Throughout our discussions yesterday, we heard that agriculture had a triple challenge to face in the future. It is expected to provide adequate quantities of healthy, inexpensive foods and use methods that ensure sustainable development. The scientists have already produced tools that meet these challenges; biotechnology is one of them. Most people don't understand them. I am not only talking about the man in the street, but also the politicians, the media and certain scientists themselves. This means that we have to educate both the public and the politicians. As we know, ignorance always encourages resistance to change. We are beginning to discover the possible risk of genetically modified organisms, but have we thought of the potential danger attached to crossing the street at night? We should no more think about the risk in the first case than we do in the second. But what bothers the public is that the scientists are very seldom absolutely sure of the facts - and if they are, it is usually because they are poor scientists!

We need to work hard on improving communications on biotechnology This is not easy, because it is very difficult to prove what is positive, and almost impossible to prove what is negative. The public is right in wanting genetic engineering techniques to be perfectly safe and wanting the results to benefit all of society. The fact that industrial companies, - which are often headed by people without any scientific training who think first of all about the stockholders' profits, - want a return on investment in as short a time as possible can be a problem.

This said, I cannot imagine interest in biotechnology waning, now that we know what its potentials are. Our work as scientists and as specialists is to give independent explanations, instruction, and information on past decisions. Unfortunately, the way the mad cow crisis has been handled is not very reassuring for our fellow citizens. Biotechnology can make a good contribution. We should put our heads together, whatever our origins, to think about the best resources to use, in particular through a multidisciplinary partnership that brings together not only scientists, but also jurists, politicians, doctors, religious leaders, and others. This is an extraordinary challenge.

SCIENTIFIC APPRAISALS: THE ORGANISATION AND RESPONSIBILITY OF EXPERTS

Marie-Angèle Hermitte
CNRS/GAPP
Paris, France

Can a scientist be held legally responsible for an expert appraisement that has damaging effects? This may sound like a surprising question because there are very few examples of such responsibility being admitted or even introduced by the courts. Up to now, the blood transfusions lawsuits have targeted company managers and government employees because of their decision-making powers, not because of their scientific decision[1]. But the question is clearly coming to the fore, as can be felt through a certain number of tragic cases like the second investigation on the blood transfusion case, the spread of the Creutzfeldt-Jakob disease that comes from the hypophyses, the Mont Sainte Odile airplane crash, asbestos, and the mad cow[2]. This question, indépendant of any crisis, is already causing some degree of concern because of genetically modified organisms, although laws, admittedly, are written more often because of a crisis and its legal repercussions than as the outcome of peaceful, protracted réflexion.

In the beginning, victims were not well organised as groups and mainly sought compensation by implicating a financially solvent legal entity, usually a large manufacturer. Civil (and then administrative) responsibility as the consequence of the use of a product made their task considerably easier, and since the victims did not have to prove that the manufacturer was at fault, they were systematically compensated.

Reassured in that vein, now victims are trying to establishment responsibility, in the everyday sense of the term. In other words, they want to know what succession of decisions led to the catastrophe. They want the fault to be recognised or even to designate the guilty party, not only as a way of expressing their desire for vengeance, which they expect will bring them certain appeasement, but more fundamentally out of a citizen's concern for avoiding further disaster. They are also inspired

1. But this was an underlying question in the case of Dr. Allain because of his qualification as a hemophilia specialist, and more indirectly in the case of Dr. Roux and Dr. Netter, because they were "in the know", although they were not experts.
2. A point of procedure is to be considered in this trend. Here we have the development of what is called group actions. Cf Cl. Lienhard, Le droit pour les associations de défense des victimes d'accidents collectifs de se porter partie civile, Dalloz 1996, chron. P. 314, and, J.Morand-Deviller, Les réformes apportées au droit des associations et de la participation publique, Rev. fr. de droit adm. 1996, p. 218.

by a desire for reform[3]. The expert is challenged as part of this move to lay bare the decision-making process.

In a criminal case, the individual person rather than the legal entity is now the most deeply involved[4] because the victims want the judge to expose individual behaviour buried in collective decision, and to condemn an individual's action in favour of some decision, or his silence or reticence, or simply his slowness which, under certain conditions, can bring about a decision that leads to the damaging consequences. In the future, public law should become more important in this type of affair because behind the manufacturer's responsibility the search now turns to the European Union, the state and its employees, and the experts who were counselling the "princes". This new mentality was clearly revealed in the asbestos case through the complaint for false information filed against a scientific report that had been accepted by the Academy of Medicine and was judged to be dangerously lax[5].

Under what circumstances, thus, could scientific experts be involved? Unquestionably when handling a crisis. But they could also be challenged every time a decision is taken concerning a threshold of danger, in other words, the daily effects of rules and regulations in many domains, *e.g.* chemistry and, more recently, genetics. But we have to understand that scientific experts only have a vague idea of their potential responsibility and their obligations. Unlike experts in other fields (motor vehicle insurance, agriculture) governed by set rules or more broadly by rules applied to a judicial appraisal, they do not even have a code of ethics to use as a guide when called upon for an expert opinion during or in specific relation to a trial[6].

It would be interesting to think about rules that should govern the behaviour of a scientific expert. They could be inspired by the main guidelines followed in judicial appraisements, which are based on long years of experience[7]. To do a complete job would require drawing fully on the rules applied to expert appraisements in civil, criminal and administrative procedures, without forgetting the importance of the principles in Article 10 of the European Convention on Human Rights[8]. The specific conditions connected to the status of each type of expert have to be carefully studied. The expert may be a private individual or, more often, a public sector scientist who makes an appraisement within or outside the framework of his normal employment. A scientific appraisement (expertise) is an activity carried out more often than not intuitu personnae[9]. Although the general principles on expert appraisals

---

3. M.A. Hermitte, Le sang et le droit, essai sur la transfusion sanguine, Seuil Publishers, p. 435 ff.
4. Despite the innovation in the penal code. Cf. François Franchi, A quoi peut bien servir la responsabilité des personnes morales ?, Rev. de sc. crim., no. 2-1996.
5. It was declared unreceivable because only the Parquet (public prosecutor's office) is empowered to prosecute this offence which was designed to apply to political affairs.
6. Rules of ethics for judicial experts who belong to a company affiliated to the Federation; the text was published in Expertises judiciaires, Delmas, 10th edition.
7. The 1852 Dalloz "Jurisprudence générale" devotes 187 pages to the question of expert appraisal which, in the legal field, dates back to Roman law, in particular for everything related to land surveying. The development of legal rules for expert appraisals may well have been closely connected to the trade guilds.
8. Court of Appeal of Grenoble, ch. acc. 25 April 1995 JCP 1996 ed. G no. 44/45, 2225.
9. Some basic reference books: D. Duprey and R. Grandur, L'expert et l'avocat dans l'expertise judiciaire en matière civile, LITEC 1995; P. Feuillet and F. Thorin, Guide pratique de l'expertise judiciaire, LITEC 1991; J. Voulet, La pratique des expertises judiciaires et de l'arbitrage, Delmas 1994. All these manuals on civil and penal procedures include the question of expert appraisals. Extensive information is also available; inter alia, in public law journals, specialised reviews, the French journal on bodily injury, and the journal of auctioneers.

are well defined - independence, objectivity, proper qualification, professional secrecy, cross-surveying - some obligations, which clash with the working conditions or the structures of the profession, may not be easy to respect. Apparently, for reasons that have been well explained and are linked to the special nature of the art markets, it has never been possible to ensure that expert appraisals of artworks be made according to the dictates of the Code de procédure civile (rule of civil procedure). It is difficult to guarantee independence of judgement when we know that the experts are also the vendors of the artwork or receive a percentage of the sale.

Here we are pursuing a twofold objective, viz. briefly outline the contents of a code of conduct for making scientific appraisals and, a contrario, indicate what wrongdoing could engage the expert's civil or criminal responsibility. In the absence of jurisprudence on scientific appraisement per se, reference will have to be made to germane fields at least as concerns the legal structure, *e.g.* medicine, accounting, art, construction work. We will see that the rules are rather general and are not greatly influenced by the field of application, except, perhaps, as concerns the evaluation of incriminated wrongful acts.

THE INDEPENDENCE OF THE EXPERT

The first rule in a legal appraisal is the independence of the expert vis-à-vis the interest at stake. The expert must not have any connection (family, amicable, professional, or financial) with any of the people involved. If evidence of any such connections can be produced, he can be dismissed through a procedure of objection (*procedure en récusation*). The expert, then, is charged with a legal mission; if he commits a wrongful act with regard to one of the parties he will be held responsible on the basis of liabilty for damage due to misfeasance (*responsabilité délictuelle*) under common law; in this case it is up to the wronged party to produce evidence of the wrongdoing. Conversely, an expert who is chosen to appraise an object, without any procedure being applied, is a mandatary who signs a contract and has an obligation to be loyal and proffer advice to the mandator. If he commits a wrongful act, he will be prosecuted for breach of contract. This connection with the mandator does not release him from his obligation to be objective, nor from a certain responsibility vis-à-vis a third party. The expert appraisement that interests us here relates to the second situation, and is performed as part of a decision-making process.

The first condition, independence, gives rise to a very immediate problem, because in the world of today, one may doubt whether a scientific appraisement can be truly independent, according to the strict criteria set out in judiciary law. It must be precise, of course, and must be carried out at the optimal scientific level. This means calling upon the top specialists. The Code de procédure civile, moreover, states that the expert must carry out his mission personally and not delegate it to any co-workers who are not under his control and responsibility. Furthermore, he must enrol the services of a specialist for any part of the evaluation that does not fall within his special field of excellence. In this situation, the second expert is known as a "sachant"[10]. But, except if the case only involves fundamental

10. The obligation to call upon a "learned person" (un sachant) is a way to reconcile two obligations: personal work and specialisation. Interconnecting these two obligations is severely sanctioned. The Cour de Cassation (supreme judiciary court), for instance, confirmed the deletion of a certain Mrs E, an expert in toxicology, from the list of legal experts because she made an expert study that involved a genetic prints technique that she had Codgène carry out and interpret.

knowledge, the specialist is part of a small group of people who usually have some financial interest in the economic issues at stake. We might even say that the reorganisation of public research is so slow that scientists are becoming obliged to develop economic links with industry. To a great extent, putting public research under (timebound) contracts, thus, is contradictory to the independence demanded of an expert[11].

Judicial experts are rarely faulted for lacking independence. The most revealing experiences in this regard, - apart from physicians who work for insurance companies - are the ones relating to works of art[12]. Because of hyperspecialisation, it is very difficult for art experts to make truly indépendant appraisements; the same holds true for scientific appraisements. There are very few people, in some cases only one person, who can authenticate a picture as the genuine artwork of a given painter. Depending on how he perceives his personal interests, he will tend to be the custodians of the purity of the work - and will find it difficult to accept the existence of previously unrecognised works - or, on the contrary, will be tempted to authenticate works of art, since he may well be paid a percentage of the sales price.

It is important to think of the problem of independence, which brings out a conflict between three obligations: technical competence and specialisation which involve scientists connected to private interests, personal realisation of the appraisement[13], and independence. This requires special attention because the research scientist working in a technological branch tends to protect its development, even if he does not receive direct financial support. It may be worthwhile making a comparative evaluation of the various systems used to pay for appraisements, both in France and abroad.

THE PRINCIPLE OF CROSS-SURVEYS (*le contradictoire*)

The second rule that can be derived from judicial appraisements is respect for the cross-survey. Jurists often consider this a principle of natural law[14]. In any case, it is central at all trials, and the Cour de Cassation ensures its total respect. It guarantees that no party in a trial will be judged before being heard, both as concerns his position relative to the facts and his position relative to the legal rules being applied. More broadly, it gives a third party the right to be heard at a trial between two other parties if his interests are indirectly called into question (right of opposition or intervention) [15].

By sub-contracting this work she violated the obligation to make the study personally. If this was the main part of the appraisal, she should have said she was not qualified and, if it was an ancillary part, she should have abided by legal procedures in entrusting it to a third party, lst civ. ch. 10 April 1996, Bull. no. 174.

11. It is intriguing that Dr Benveniste was reproached for having the Boiron Laboratories finance his research on highly diluted products while all the public research institutes strongly recommend that type of funding. Cf Kerry Allbeury, "Une controverse entre scientifiques" in Le droit saisi par la biologie, supervised by Catherine Labrusse-Riou, LGDJ, 1996, p. 355.

12. L'expertise dans la vente d'objets d'art,***, aspects juridiques et pratiques, under the supervision of M.A.Renold and R. Byrne-Sutton, published by Schulthess Polygraphishers, 1992, p. 116.

13 This should condemn attempts made by certain research institutes to develop a capacity for making expert studies under the collective responsibility of the institute rather than under the responsibility of a given scientist.

14. H. Motulsky, Le droit naturel dans la pratique jurisprudentielle,***, le respect des droits de la défense en procédure civile, Mélanges Roubier, Dalloz-Sirey, 1961.

15. M.A. Frison-Roche, Généralités sur le principe du contradictoire, Paris II, 1988; Mahmoud Sayed, Le principe du contradictoire, Rennes I, 1990.

All jurists agree that this right not all provides protection for people, and is even a human right, but that it is also a technique for bringing out truth and justice: the truth of facts that should be discussed and thus should benefit from the opinions of parties that have conflicting interests concerning how the facts should be presented, and justice, or rather equity, thanks to the solution that is retained, following a judicious choice and interpretation of the legal rule to be applied. The cross-survey is what justifies the decision here[16]. At first sight, the link between this procedure and truth may seem somewhat remote from scientific thought, which is known to be based on experimentation and demonstration. But this "distance" can be explained by our special vision of scientific truth, *i.e.*, in a well defined context, a single, confirmed scientific truth exists. All scientists recognise the relative nature of this moment of truth and the ever recurring share of uncertainty, but they have not yet weighed all the consequences. The question thus is whether, for scientific issues in these zones of uncertainty, the cross-survey procedure constitutes a valid way to handle the acquisition and disclosure of scientific knowledge.

This juridical conviction, which considers the soundness of a decision to come from the contradictory procedure, is applied to expert appraisements when they are needed to investigate a case. Since results are expected to be derived from formally organised contradictor procedures, it is understandable that an expert appraisement abides by an essentially formalistic procedure[17]. Similarly, the expert who hears a "learned person" must inform the parties to the case of what he has heard so that they can discuss it from all sides before the his report is submitted[18]. The same rule applies to the expert's "pre-report". It is becoming more and more common for the judge to have the pre-report sent to all parties to the case, who then have 15 days tò discuss it and convey their remarks, to which the expert has to provide a well-founded response in the final report[19].

This detailed information has been given in the hope that it will be transposed for use on sensitive issues covered by scientific appraisements. A lawsuit is organised around the parties to the suit; similarly the decision could be organised around the parties involved in defining the danger threshold. The proponents of each danger threshold being proposed would be called forward to justify their position, to discuss their opponents' arguments, and to respond to arguments set before them. This is the daily fare of judicial experts, and it is these arguments which serve as the basis for making judgements, with the outcome being the judge's obligation to respond to the conclusions of each of the parties and to justify his response. If this method were to be used, the debate would be very different from what it is now.

It is this interlinkage, *i.e.* cross-survey procedure, response to conclusions, statement of motivation, that is the most sorely lacking in scientific appraisals which present and purposely paint a picture of the consensus of the moment, omitting, minimising, or "flattening" the hypothesis that are marginal or contrary to general opinion. This is the area where legal experience would be the most useful in helping to reach the next step in preparing a charter on expert appraisements. E. Naim-Gesbert

16. On relations between truth, legitimacy and justification, cf. Xavier Lagarde, Reflexion critique sur le droit de la preuve, Bibliothèque de droit privé, LGDJ, 1994, and also Le Masson, La recherche de la vérité dans le procès civil, Nantes, 1991.

17. L. Cadiet, Droit judiciaire privé, p. 493 ff.

18. 1st cir. 19 Dec. 1996, Bull. civ. I. no. 475.

19. M. Oliver, Du prérapport et son usage en expertise judiciaire, Gaz. Pal. of 26 Jan. 1995.

has gone as far as to talk about the "right to a full-fledged appraisal" and has shown that certain international conventions already oblige experts to present all angles of their scientific hypotheses[20].

In the legal culture, opposition between two theses is not seen as a transitional nor as a pathological phenomenon, but rather as a normal state of interhuman relations. The decision taken by Justice brings a case to a conventional end, without claiming that truth, consensus, or reconciliation has been achieved, but merely that after all paths of recourse have been used, the case has reached a stage of relative "peace". It marks the completion of an organised cycle of argumentation, which may tend towards truth without ever being able to claim that it has been reached[21]. Conversely, in the scientific culture, the debate reveals an uncertainty that is meant to be temporary. It is a moment in a learning process. The tendency is to go beyond this stage, even if the next step is another zone of uncertainty. This explains the search for consensus which, when politics are involved, all too often leads to compromise.

But the principle of cross-survey goes farther. It involves a counter expert's appraisement and a special, specific way of presenting opinions if several experts, with varied or even contradictory opinions, are involved. In this case, each expert has to present his opinion and sign for it[22]. This method, which provides the judge with opposing opinions, signed by their authors, is visibly foreign to the scientific appraisement which is often organised around a compromise/persuasion dialectic between the case rapporteurs and the chairman of the commission. The system is completed by a judicious selection of experts. There is a certain confidence in the experts' general opinions about sensitive cases, their propension to resign in case of disagreement, and their sense of confidentiality as well as in group discipline.

THE EXPERT APPRAISEMENT MUST BE OBJECTIVE AND CAREFULLY DEFINED, AND MUST REFLECT THE MOST RECENT SCIENTIFIC KNOWLEDGE.

Article 237 of the Code de procédure civile requires the expert to carry out his mission conscientiously, objectively, and impartially. These terms are somewhat redundant and, to a great extent overlap requirements linked to the independence of the expert.

The idea of strictly defining the expert's mission is vital. The mission is essentially technical, and the expert is only to give his opinion on the specific question that has been submitted to him. Hence, this does not concern us here directly. (???)

But another, fundamental, requirement could be added concerning the dissociation of the expert's - technical - mission from the judge's - juridical - mission. To caricature the situation, we could say that the judicial expert is to examine the cracks in the metal but must make no remarks about the

20. E. Naim-Gesbert, Les fondements scientifiques du droit de l'environnement, Lyon III, 1997, p. 661. He cites the Espoo Convention of 25 February 1991 on this question.
21. And in some cases knowingly prohibiting its achievement, e.g. when a disavowal of paternity is unacceptable, even if the legal father submits proof that he is not the biological father.
22. Cf. R. Genin-Meric, cited above, Jurisclasseur, vol. 662 no. 186.

possible responsibility of the manufacturer of the elements that caused the damage. While risk sociologists claim that the scientific expert inextricably links his objective scientific appreciation to consideration of various economic, hierarchical, and profession-related constraints and that the texts on appraisement confound scientific results with opinions on the risk or the interests at hand, the law obliges the judicial expert not to answer questions and never to give any legal opinion (Art. 238 of the NCPC). It is crucial to delineate different spheres (appraisement and judicial decision) precisely. It leaves the judge free to take a decision that does not concur with the findings of the expert appraisement, and further, it levies an obligation on the judge that is just as binding as on the expert. Many judgements are repealed when the appraisement seems to include an opinion about juridical matters (this does not occur when the judge personally appropriates what the expert expressed when exceeding his brief, Cass. 3rd civ. 29 May 1985, JCP 1985, IV, 278).

The separation of these two spheres may be difficult, but it can be learned, and the situation described by the sociologists is not a fact of nature that cannot be overcome. It is actually a culture that stems from the existence of networks that closely intermix the political and the scientific sphere. Experience in judicial appraisement shows that expert appraisement can be a "profession" with its own rules, and not only an unorganised ramification of knowledge which, as roles get twisted, turns into power.

Jurisprudence on the experts' responsibility stipulates the obligation to use the latest techniques available. This means the most recent jurisprudence in the case of accountants, notaries, tax advisers[23], or the most sophisticated investigation techniques when dating a piece of art.

It should be noted, however, that when time-honoured professional customs are used, the courts agree that art catalogues may be prepared for public auctions without prior sampling or chemical analyses of the artworks[24]. As a result of this simplified procedure, catalogues rather often contain mistakes that are not discovered until later, when the artwork is thoroughly appraised. Casuistry plays an important rôle in these solutions; certain judgements recognise the especially difficult character of a case and pronounce "absence of fault" while others refer to indices[25] that should have attracted the attention of the auctioneer and his expert. Nonetheless it seems that in some cases, the diligence required of an expert may be decreased as a result of customary practices. This type of jurisprudence could greatly reduce the responsibility of experts who use simplified procedures, like those that have at times been introduced to cover the distribution of genetically modified organisms (GMOs). A solution of this type, applied to artwork, constitutes a very limited exception to jurisprudence which is marked by the contrary[26]: "it is not probable (that exclusion or limitation of responsibility can be derived from customary practices). Such practices, which have no more value, as far as liability is concerned, than administrative authorisations, constitute a minimum level of precautions in the interest of the public whose protection is their goal. If they are too weak and leave room for negligence, the fact that they

23. Cf. a certified public accountants' company was not found guilty of wrongdoing (Cass. comm. 12 July 1993, Bull. IV, no. 298) because its reports were based on the most recent jurisprudence, which was only challenged by a decision taken later, by the Conseil d'Etat.

24. Example: Court d'Appel de Paris, 15th chambre, section B, 31 May 1996, Jurisdata, 021901.

25. The notion of indices that should attract the expert's attention needs to be examined in a study on scientific appraisements, because the danger indices would probably show up as one of the main points in legislation on responsibility.

26. Ph. le Tourneau and L. Cadiet, Droit de la responsabilité, Dalloz, Action, 1996, no. 485.

have been respected does not grant exoneration from the consequences of this negligence". The Cour de Cassation uses clear language in regard to a polluting activity that is covered by regulations: "administrative authorisations are always granted, as long as they respect third party rights, which means that the Cour d'Appel, by taking this more indulgent position, has violated the law mentioned above"[27]. Actually, considering present day practices and attitudes on responsibility, the tendency would be to institute proceedings not only against the originator of the damage, even if he has respected the standards, but also against the author of the standard.

RULES ON THE CIRCULATION OF INFORMATION

Since the scientific appraisement not only interests the party that ordered it, but also the public, it is especially important to decide on the principles to be used in organising information dissemination. A distinction will probably be made between three possible users: 1) the ordering party, the public authorities and manufacturers, 2) the scientific community, and 3) the public. Unlike the doctor's person-to-person obligation to give information to his patient, the information in référence above is collective in nature and is not covered by a well-established body of rules. The 1905 law on fraud repression, amended in 1978, is based on a general obligation to disclose risks inherent in a product; we might query whether the scientific appraisement of such a product would not be influenced by this general obligation. In certain special fields concerning a major construction project or a technology, the law has been able to make "information for the public" obligatory, *e.g.* the dissemination of GMOs. And then there are the well-known obligations to provide information on the contents of impact studies, classified installations, etc.

But these scattered rules lack cohesion and, more importantly, the precision needed to be clear and efficiency, or even effective. Furthermore, the relation between these rules and rules that apply to professional secrecy, business secrets and medical confidentiality are not well known by the people responsible for carrying them out.

Information for the public authorities

The blood transfusion case has paved the way on this particular point. Drs Roux and Netter, employees of the Ministry of Health, were found guilty on the grounds of abstention, in order words, for not having given the appropriate information to the appropriate political bodies. They were blamed for transmitting incomplete, deceptively appeasing information, and for doing it too slowly. Dr Roux was especially found guilty for becoming discouraged when the political authorities repeatedly refused to listen to him and giving up during the last weeks of the crisis. The message is clear: if the case is serious enough, whoever has the information must continue transmitting it, even if no one listens to him. He is not expected to be efficient, but he is expected to keep going[28]. But we have no idea of the obligation which devolves on the everyday scientist who possesses information on a risk. Normally he would inform the head of his laboratory, or his institute. What should he do if the information they are transmitting gets blocked somewhere along the way?

---

27. Cass. 2nd civ. 28 April 1993, Bull. no. 156.
28. For details on the judges' argument, cf. M. A. Hermitte, Le sang et le droit, op. cit. p. 406 ff.

Information for the scientific community

For the scientific community, certain crises or technical projects serve as an opportunity to share information freely, - at least if the information not connected to industrial interests. This was the case in the early 1980s when doctors and biologists started working on AIDS. Note that as concerns basic data on hereditary diseases, international co-operation is standard practice, regardless of economic interests. On the other hand, the very unique structure of the mad cow crises led to a concentration of data in Great Britain, and these data, apparently, were not very well channelled between veterinarians in England and the British Ministry of Agriculture, nor between them and their colleagues on the continent. "Nature" repeatedly stressed the reticence and silence of the Ministry of Agriculture. Similarly, the slaughterhouses, knackers, bone-meal and gelatine manufacturers and others working in the same sector do not seem to have been forthright enough in handling the crisis. An open question is whether the obstacles could have been avoided through negotiations carried out under the aegis of the European Union or the International Office of Epizootics. Another question that is still unanswered is whether some legal reference could have been found to empower a judge, confronted with the double threat of an epizootic and an epidemic, to issue a summons for the documentation needed for the scientific evaluation. We are still puzzled, for instance, about why the EU veterinary committee was unable to obtain data about certain methods used in making gelatines.

Information for the public

The blood transfusions case has also provided certain relevant direction by highlighting the obligation of the doctors in charge of the (French) national blood transfusion centre (CNTS), who knew about the risk of contamination from antihemophilic products, to warn their patients through the latter's associations and the media. Nothing is said about what government employees should do if they disagree with the political authorities. When they have obtained a certain number of forewarnings about the threat, should they inform the public directly, or are they obliged to continue informing only the public authorities? There is a gap in the legal rulings on this matter where silence seems to favour the government employee's obligation of discretion in contrast to the manufacturer's obligation to provide information.

Except for some procedures especially established for this particular case, the public's right to information cannot be deduced from the blood transfusion example. Yet, the right to information, loudly proclaimed in environmental law (with limited effectiveness) also exists in another domain, the domain of historical truth, which in many ways is quite close. When the Première Chambre civile of the Cour de Cassation had to decide on the historian's responsibility in presenting controversial facts, it immediately established the principle of "the public's right to information" the author of a publication on history engages his responsibility vis-à-vis the persons concerned when the presentation of the thesis being upheld, by distorsion of facts, falsification or gross negligence, shows flagrant contempt of the truth[29].

The whole question of information dissemination brings out the problem of the relationship between the right to information vs the obligation to provide information on the one hand, and other

29. The case of the heirs of H. de Portes, a friend of Marshal Pétain, against J. J. Servan-Schreiber, 15 June 1994, Bull. civ. no. 218.

obligations related to 1) the government employee's obligation of reserve, 2) medical, industrial or professional secrecy, and 3) the protection of the individual's private life, on the other[30].

Here again a certain number of satisfactory solutions have been worked out in judicial appraisements. But they may not be easy to transpose. The dissemination of information, in this case, is limited to the ambiguous scope of the lawsuit which only concerns the parties to the case, although it leads to a public decision supported by information that is only partly disclosed, *i.e.* the part that is needed for the juridical solution.

As far as the expert is concerned, Article 244 of the NCPC clearly expresses the general idea that the technical expert must communicate all relevant information, but nothing more, and only if this information has been lawfully compiled. Furthermore, this information may not be used elsewhere than in the legal proceedings. Last, if he violates the confidentiality of his mission, he may be sanctioned by virtue of Article 226-14 of the criminal code. As for his informants, they are bound by a very general obligation "to assist in bringing out the truth", and by precise obligations concerning documents that the judge can order them to produce.

One conciliation principle that is often used is to entrust the secret information (if there is any) to an outside person who is authorised to extract from it whatever is relevant to the case. A judge, for instance, can order a doctor who, for reasons of medical confidentiality refuses to disclose certain information about his patient, to turn over the medical file to a specially appointed colleague. In that event, the second doctor will handle the medical secret in the litigation.

The practice of "shared secret" is often used. The Cour de Cassation, for instance, decided that "an expert duly mandated to establish the composition of a product was not divulging an industrial secret by speaking about it at a non-public meeting devoted to the expert appraisement"[31]. This conforms with certain other solutions used elsewhere. The courts, for instance, decided that disclosing the composition of a product to meet the needs of medical experimentation on the product, and for nothing else, in no way impeded the novelty requirement applied in patent rights.

Similarly, legal decisions, which are accessible to the public, can be drafted in a manner that protects the privacy of certain technical or economic information by replacing it with (Í). This means that the information needed for the judge to prepare his argument for the published decision is more precise than the information set out in the decision[32].

The technique of "shared secrets" may not be adequate, and, in that case, the information under dispute may have to be disclosed. Many decisions confirm this, *e.g.* in the vast field of responsibility incumbent on chartered accountants and statutory auditors, companies concerned have to provide all the justificatory information needed to establish or refute responsibility, and are not allowed to plead

30. On the linkage between the historian's work and respect for one's private life, cf.the case of the widow of the Emperor of China, JCP 1997 no. 4, II, 22765, note by J. Ravanas. Certain facts of private life, if disclosed by the deceased person at the time they occur, may be repeated by an outside party in his work as a historian, as long as he respects the truth.
31. Cass. civ. 2nd, 15 March 1979, JCP 1979, IV, 177.
32. This practice may have certain adverse effects Cf. C.J.C.E, 11 January 1996, The Zunis vs Commission case 1996 I.1, note M. A. Hermitte, Journal de Droit international 197, no. 2.

"business secret". It was also interesting that the Cour de Cassation used a pharmacist's illegal practices as the legal basis for its decision to order the DRASS (Direction régionale des Affaires sanitaires et sociales) at Nantes to communicate its report; the pharmacist wanted to sell her pharmacy, and the buyer, claiming that the turnover figures had been artificially inflated as a result of these illegal practices, asked that the selling price be decreased accordingly. The Première Chambre civile of the Cour de Cassation justified its decision on the basis of Article 10 of the Code civil which makes it mandatory to assist the courts in bringing out the truth and stipulates that this obligation applies equally to people in the public and the private sector. It also stipulates that disputed documents required to solve a litigation cannot be restricted for use only in a disciplinary procedure, where the investigation remains confidential.

Reference must also be made to a decision that directly concerns public health. At a television programme on AIDS, journalists asked medical professors what they thought about X, who gave patients a product whose composition he kept secret. Prof. Y violently attacked the effectiveness of the product, whereupon X sued him for defamation. The Cour de Cassation recognised the good faith of the Prof. Y, who was "known world-wide for his scientific and medical research and wanted to draw the attention of the viewers and AIDS patients to the danger of a drug whose composition, against all rules, was kept secret, and whose effectiveness had not been established. The judgement added that Y wanted to warn the public about the possible danger for patients to interrupt therapy prescribed by their normal physician, because of a chimerical hope of being cured, and that the violence of his words also came from the fact that X's treatment was not free of charge. Thanks to these findings and facts, the Cour d'Appeal, reversing the burden of proof, was able to infer that Prof. Y had proven his good faith"[33].

The judgement of the Besançon administrative court, which the commentator considered audacious, could also be linked to the right to information on the quality of life,or on public health, since the two fields are close connected. It annulled a prefect's order not to make public a private consultant's report on the effects of noise from the airport, prepared by a private consultancy bureau. The document was called for when the airport noise exposure plan was being revised, but was withheld on the grounds that it was an internal working document on local urbanism and on how people react to noise. In other words, the Direction Départementale de l'Equipement felt that it owned the study and therefore could decide whether or not to make it available. The judge did not agree, and said that the study was a document "whose contents would be useful in informing the public." Withholding the document, thus, "made the revision of the noise exposure plan irregular"[34]. This decision is interesting because it shows that the principle of the public's right to information actually exists but is rarely applied.

---

33. C. Cass. 2nd ch. civ. 17 March 1993, no. 109, conversely, the Chambre criminelle did not except "good faith" for a member of a municipal council who defamed a hospital doctor. This can be understood by considering the reasons behind the defamatory remarks; the aim was to expose mismanagement that "jeopardised local finances". The Court judged that "the intention to inform the public cannot, by itself, constitute justification of good faith". But what would the decision have been if the information had been disclosed for reasons of public health, since the malfunctioning of a hospital service can endanger the lives of the patients? Cass. crim. 12 Oct. 1993, no. 289.

34. RJE no. 2-1995, p. 455 note F. Mallol who points out, as a reminder, that the secret should protect the citizen not the administration.

CIVIL RESPONSIBILITY AND EXPERTS' WRONGDOINGS

The principle has been recognised at least as far back as a 1914 judgement[35], and there is no doubt about the expert's civil responsibility. An amicable appraisement involves contractual liability, but if the appraisement is ordered by the judge, parties to the suit must have no connection with the expert who then has delictual liability. A wrongful act must be judged according to the terms of common law, with référence to the standards of a qualified, conscientious expert, and a constant reminder of the expert's obligation to have made his best effort and employed optimal conditions (obligation de moyens). But hesitating to invoke the expert's liability, the judges often speak about grosse negligence[36] and seldom recognise the causal link between the negligence and the deleterious effects, *e.g.* losing the case, undue detention[37].

Mention could be made of certain errors admitted by the courts in fields that could be of interest in a scientific appraisement because the facts and the legal situations are similar[38].

Thus, the courts declare carelessness of the expert if he ascertains the authenticity of a work on the basis of statements by members of the creator's family[39]. The judge's motivation is understandable; he knows that it is in the interest of the heirs to increase the number of authentic works. The error of the expert, who, in principle is indépendant, is to rely on what is said by a person who is not. Certain scientists, by adopting the position of a pressure group, without indicating their sources, may also go adrift in the appraisement they make for political authorities. Giving the position of industry on a problem is justified, but it must not be the only opinion admitted.

The most common admitted error relates to the idea that an expert should not categorically affirm that a work has been produced by a given artist if he is not certain[40]. In case of uncertainty, there are several possibilities. First he can carry out all useful investigations, compile full information and

---

35. Chambre des Requêtes, 26 October 1914, Dalloz 1916. But going further back in time, Jurisprudence générale Dalloz (1852, verbo expert, no. 122) cites a 1812 decision of the Cour d'Appel in Rennes, and a decision of the Chambre des Requêtes that recognises the legality of the sentence condemning an expert land surveyor to pay damages and to reimburse the cost of a second appraisement since the expert had committed a wrongful act, 21 Nov. 1822.

36. On admission of a minor negligence which does not interfere with the proper execution of his mission, Cass. civ. 24 Nov. 1966, Bull. civ. II, no. 920.

37. Cf. the jurisprudence cited by Régine Genin-Méric, Mesures d'instruction exécutées par un technicien, Jurisclasseur procédure civile, vol. 660, no. 146. Famous mistakes in expert reports have been brandished regularly in cases involving poison. Progress in toxicology has revealed serious judicial errors. Cf. P. J. Doll, De la responsabilité des experts judiciaires, Dalloz, 1966, chr. p. 47 who, to find a list of such errors, quotes Prof. Muller La responsabilité des experts, archives of the legal medicine institute of Lille, 1954, p. 28. When artwork is involved, the judge cancels the sale and grants damages to the buyer for the moral injury he has suffered.

38. In the case of a judicial expert, responsibility is unique because it is "settled" by invoking error of procedure: the expert's opinion does not entitle the wronged party to compensation. If the appraisement contains no errors, the party to which it is not favourable cannot blame the expert, be it but out of respect for his "independence and his freedom of judgement", as the Cour d'Appel de Versailles decided in the case of a doctor who disclosed the fact that a father, who wanted custody of his child after a divorce, was using drugs (CA Versailles 1st Chambre section 1, 29 November 1988, Jurisdata 048192). If an erroneous conclusion in an expert appraisement has repercussions on the legal decision rendered, it is usually up to the Judiciary as an institution to make the necessary amends.

39. Cour d'Appel de Paris, 8th chambre section A, 18 January 1988, Jurisdata 020870.

40. From the legal angle, the Cour de Cassation has decided that, unless there is some reservation, selling a work of art that bears a signature is tantamount to establishing its authenticity and thereby excludes any risk in the contract JCP 1995 ed. G no. 51/52 IV 2770 and Dalloz 1995 no. 43 IR 266.

opinions[41], and consult the person who, in some cases, is the only specialist on the painter or sculptor[42] in question. If, after all these procedures have been used, there is still not enough knowledge to be certain about the identity of the object, the expert should not confirm the work's authenticity, and the conclusion will be to attribute the work on the basis of well documented indices. If this information is too uncertain, or if there are counter indications, it is better to withdraw the object from sale[43].

Similarly, all the judicial experts stress their obligation, when examining a claim, for instance, to make the judge clearly aware of everything that contributes to determining the related circumstances, and bring out all controversial points so that he does not "embark too surefooted on a debatable path"[44]. Many reports from legal experts conclude with a statement saying that it is impossible to find the cause of the accident or the fire. There is no problem, not even a psychological problem in declaring this lack of adequate information. On the other hand, one of the main difficulties in making a scientific appraisement for a politician is admitting ignorance[45].

A technical error is not exonerated of responsibility[46]. It is defined as "any error that a halfway conscientious, diligent, attentive, informed technical expert would not have made"[47]. The notion of responsibility for erroneous information is tending to develop, and connections can be found between this notion and scientific appraisement. In the case of AIDS, albeit a very specific sector, the Tribunal de Grande Instance (court of first instance) of Valence recognised the responsibility of a laboratory that had sent a doctor inaccurate results of a seropositivity test, thereby causing serious shock to the patient and his concubine. The patient, who has remained emotionally frail, was compensated by a payment of 70,000 francs; his concubine was awarded 20,000 francs for indirect damages[48].

There are limits to the notion of moderate degree of average diligence, which judges often use. The average standard depends on the scientific sophistication of the activity in question. The architect of houses in a residential district will not be expected to have the same technical skills as the contractor who has accepted responsibility for the safety of a suspended bridge that was built using vanguard

41. Cour d'Appel de Paris, 1st chambre, 22 January 1993, Jurisdata 021136 and, Cass. 2nd civ. 20 July 1993, Gaz . Pal. 1993, 2, jp. 301.
42. Cour d'Appel de Paris, 1st chambre Section B, 22 March 1996, Jurisdata 021515.
43. For the bronze sculpture errorneously attributed to A Rodin, the Tribunal de Grande Instance de Paris, 1st chambre section 1, 14 September 1995 Jurisdata 046198, and 14 March 1996, Jurisdata 042912.
44. Mr Klein: A propos de la fiabilité des informations nécessaires aux rapports d'expertise médicale, Rev. fr. du dommage corporel, 1991 p. 380; the author points out that in case of doubt, the expert should solicit additional information, and if, even then, some doubt remains, it should be clearly expressed. At the end of this article there is an interesting bibliography on doubt and uncertainty in the decision-making process.
45. Cf. the writings of M.Cl. Hennion and F. Charue, in La chimie dans la société, supervised by G. Bram et al., L'Harmattan-CNRS 1995, p.157 and p. 304, on the difficulty chemical engineers have in talking about what they don't know and what has not been confirmed.
46. Mention could be made of the decision taken by the commercial court in Romans which levied a 50,000 franc fine for damages caused by a computer server that transmitted erroneous information; the information included untrue commercial information, which alerted a certain company's bankers and "made the customers feel suspicious". The payment for damages was low because of the unintentional character of the error which came from a confusion made between two companies, JCP 1996, ed. G no. 44/45, 2227.
47. R .Vienne, JCP 1959, ed. A, II, 11374.
48. Tribunal de Grande Instance de Valence, 8 December 1992, Gaz. Pal. 1994, 2 July, jp.p. 411.

construction techniques. In the scientific appraisement involved here, "average diligence" will mean the highest level of scientific sophistication.

It was interesting that the Tribunal de grande instance de Paris, after pointing out that absolute certainty is often unattainable in the field of art, blamed an expert for committing a wrongful act by limiting his work to merely repeating statements of authenticity contained in the catalogue of a previous sale, without examining it critically nor checking it[49]. Scientific experts, whose work on a critical issue brings out unusual elements of appraisal, (as was the case at the beginning of the AIDS crisis) are often reproached for their lack of critical insight.

In a rather unusual decision, an expert examining poor workmanship in a building, was blamed for reporting a purely imaginary risk of a landslide; he had not made any calculations nor done any thorough research[50]. This is an arm that could be used against "marginal" experts accused of catastrophism, but it will be difficult to use it equitably because the people who decry a risk use arguments that are much less solid than those who deny its existence. This distortion, unquestionably, is the main difficulty in research on prudent procedures. Suspicion of new risks is based on hypotheses derived from fragmentary, emerging, little validated information that is often kept secret. The sound proof seems to appear on the "absence of risk" side, which makes the principle of cross-survey very difficult to apply here.

In the issue at hand, the question is whether the wrongdoing will be assessed using standards that apply to scientific method as viewed by scientists, or if the judge will add exigencies derived from the juridical model with its principle of cross-surveying and the obligation to express underlying motivation (obligation de motivation). There is also the question of the degree of simplification that can be accepted because of the need to make the appraisement widely understandable. As concerns the details of the argument and the experimental proof, reference could well be made to precise bibliographic data, but there may be some conflict about how to present and simply the scientific hypotheses in question, the provisional results and, more importantly, the scope of the results, in other words, the context in which they have been produced. Simplification should not be a pretext for omitting inauspicious signals.

Last, it is important to try to summarise jurisprudence on two points that seem to increase the general level of liability or can even be the decisive underlying cause. One is the judge's ascertainment of imminent danger, and the second is the use of new technologies. Both ideas are directly linked to the principle of precaution.

On imminent danger, consider the following two examples. In the first case, the Cour de Cassation censured a decision that did not retain the liability of the Club Med for an accident that happened to a tourist who dived in shallow water. Club Med should have used all possible means to prevent this from happening, in other words, "to bring a dangerous activity to an earliest halt and thereby prevent the risk of an accident whose imminence should not have escaped them"[51]. The second case involved a city mayor who was sentenced for not have acted within a reasonable length of time

---

49. Tribunal de Grande Instance de Paris, 1st chambre section 1 of 25 April 1990, Jurisdata 042952.
50. Chambre d'Appel de Rouen, chambre civ. 1, 20 February 1991, Jurisdata 040463.
51. Cass. 1st civ. 28 April 1993, no. 152.

after the first warning signs of pollution appeared subsequent to problems with the water purification system[52]. Neither of these cases involved a scientific appraisement, but they clearly evidenced the notion of warning, alert, and imminence which are so important in defining obligations connected to crisis management.

As concerns new techniques, especially in the medical field, the use of new techniques used to be considered as a factor that could contribute to attenuating the charges of wrongdoing, because of the uncertainty connected to them, but now the trend is the opposite. It is well known that the jurisprudence established by the Conseil d'Etat in the medical field has
been reversed[53]

CRIMINAL RESPONSIBILITY

Doctrine on criminal responsibility, (except as concerns corruption, which is seldom applied), covers indiscretion and lying[54]. It dwells on precise offences, such as violating professional secrecy, but not on offences related to the deleterious effects that consequences of the expert's behaviour could have on health or the environment, which are the ones that concern us here. The question, thus, is to find out what legal basis a victim could use to prosecute an expert who knowingly omitted alarming information because it did not seem sufficiently sound, or because he feared it might cause panic that was disproportion to the risk; these two reasons are often proffered by experts accused of having minimised risks.

In the most serious cases, it may involve an outright lie, covered by the 1905 fraud law since it refers to deception that may be the result of a lie, simple reticence or embezzlement. Might not omission of alarming symptoms in the provisions of services (and not only goods) also fall under this law? Furthermore, there is a connection between these appraisements and merchandise, for downstream of them, products are - or are not - put on the market. The moral element of the offence, so well evidenced in the blood transfusion case, may reside in accepting risk on someone else's behald, which clearly reflects the structure of what the victim is reproaching the expert.

The doctrine gives a very restrictive interpretation of this offence, and points out that the notion of the substantive quality of service-giving is still unclear[55]. Yet decisions of Justice exist and create specific obligations for each profession, *e.g.* "the obligation for a travel agent to personally check the veritable existence of the services" promised by the owner of holiday rentals[56].

The obvious question is whether the causal link between the manipulated, or incomplete appraisement and the damage has been well established. The scientific appraisement leads to opinions that the public authorities do not have to accept, of course. But the more vanguard the technical field,

---

52. RJE no. 3-1995.
53. Cass.1st civ. 6 March 1996, Bull. no. 128.
54. J. Pradel, La responsabilité pénale de l'expert judiciaire, Rev. sc. crim. 1986 p. 247.
55. J.H. Robert, Consommation, Jurisclasseur competition-consumption, vol. 1010, no. 20 ff.
56. Cass. crim. 17 March 1993, Bull.crim. no. 123; although the main part of the jurisprudence focuses on deceptive advertisement.

the more the expert's, or more generally the expert committee's opinion will be reflected in the legal decision. With this in mind, the reproach against the Academy of Medicine in the asbestos case was interesting. The complaint was not that the appraisement, which was deemed optimistic, had harmed anyone's physical health, but rather that its optimism convinced the public authorities to play for time rather than make them decide to act rapidly.

There was an interesting decision taken by the Swiss federal court which condemned an art appraiser, who had been hired by the vendor of the painting, of swindling. He was not reproached for lying as such, but for having kept quiet about facts he knew, facts that he should have told himself might influence the buyer's decision, if he (the buyer) knew about them. It has been said that the expert should always remember that the information he gives out, or the information he keeps to himself is not neutral, but may have a decisive effect on his interlocutor. This is one of the main problems with art appraisements; when an expert is contracted by an individual, he has a general obligation of loyalty to the co-signer, but going beyond the private relations that stem from the contract, he also has a responsibility to third parties and to the public at large. It is important to keep this in mind, because an expert is often faulted for reticence, or withholding information[57].

CONCLUSION

The question of expert appraisement is worth a thorough study that would cover the various types of appraisements according to the status of the expert, the various types of missions entrusted to him, the beneficiaries, and, more than all else, the interconnection between secrecy, discretion, and the right to information. This has never been done, although the law does contain parsimoniously scattered solutions.

57. L'expertise en matière d'oeuvre d'art, op cit. p. 55.

PLENARY SESSION

(FRIDAY MORNING, 28 NOVEMBER, 1996)

EUROPEAN AGRICULTURAL RESEARCH, WORLD FOOD SITUATION AND DEVELOPMENT

AGRICULTURAL RESEARCH AFTER THE WORLD FOOD SUMMIT

STEIN W. BIE
Research, Extension and Training Division
Sustainable Development Department
Food and Agriculture Organization of the United Nations
Viale delle Terme di Caracalla
I-00100 Rome, Italy

The World Food Summit is over, and on behalf of the Director General, Mr Jacques Diouf and FAO, I am pleased to be able to give you a report on the outcome of the Summit, with particular emphasis on matters that relate to research issues. You will appreciate that FAO does view research in a more global than European setting. FAO's immediate concern is with food insecure in poor countries rather than in Europe. Food insecurity in Europe is - with the possible exception of areas in Russia - currently a question of purchasing power at household level. However, the research paradigm that underlies European agricultural research is surprisingly similar to that which has guided tropical agricultural research over the last few decades.

The Summit essentially suggested that efforts in agricultural research that relate to developing countries should now be seen in the immediate context of food security for an estimated 840 million people who are currently malnourished or undernourished. The Summit recognized that we must face a growing global population challenge with a Kirst things first strategy. Unless we do so, the number of food insecure will continue to rise, as it has done during the 22 years since the last major world food meeting in 1974. We know that halving the number of food insecure by the year 2015 will be a true challenge, but both Mr Diouf and many other speakers at the Summit made clear that their ambitions go beyond this. And as a fellow agricultural research scientist I find it personally also unacceptable to have a prospect of facing 420 million food insecure in 2015.

It is very clear to me that the food insecure in the developing countries cannot and should not be fed by surplus production in Europe or elsewhere in the industrialized world, be it by gratis food aid or regular trade. They cannot, in fact, be fed in this way, because they have no purchasing power to acquire imported food, and they have no wish to become permanent relief clients. Whilst it may be true that countries in South-East Asia, which used to export food and import TV sets, now export TV sets and import food, this pattern will not be repeated elsewhere for considerable time, and not in the most food-insecure continent, Africa. Whatever opportunities for food exports Europe may find through the gradual introduction of World Trade Organization rules, I do not foresee a European agricultural research agenda significantly based on the need to feed the food insecure of the developing world. The European research agenda will be based on Europe's needs for efficient and ecologically benign production strategies in the context of human public health issues. But whilst that starting point is different from that of food insecure families in the developing countries, many of the research approaches will not be. And the role of research will be central to agricultural production systems both in the North and in the South.

Agricultural research issues played an important role in the preparation of the Background Technical Documents to the World Food Summit. The Summit had a solid technical backing in 15 documents that we are sure will stand for quite some time as the current awareness papers on food security issues. They will be read when diplomatic words are forgotten. They are available in hard copy and on the FAO Home Page on the WorldWideWeb, in English, French, Spanish, Arabic and Chinese, and make up 3 large printed volumes. There are many papers that relate to research issues, but two stand out in particular: No. 9 on the Role of Research and No. 6 on Lessons from the Green Revolution. Both these papers signal a significant paradigm development in agricultural research as it applies to developing countries, but it also reflects on paradigm shifts in agricultural science in general, and thus - as I have previously indicated - also to European agriculture. Let me briefly dwell on such foundations in science.

Science rests on paradigms. Every age has its own set of paradigms. Modern agricultural science grew out of four distinct traditions, all with solid European bases:

- one from the chemical laboratory of Liebig and his colleagues in Germany, where the chemical constituents of plant food were discovered,
- one from the Czech monastery gardens of Mendel that gave the key to genetic heredity,
- and one from Darwin discovered not the least through his voyage with The Beagle and the Galapagos finches, that made us understand forces of genetic selection.

The fourth had a different background, because it sprung from the fields of an experimenter, not a scientist. John Lawes, whose name will always be linked to the long-term field trials at Rothamstead in England, that put the ideas of the scientists within the reach of the farmer.

These four European traditions, all conceived during the second half of the last century, have formed the foundation for our modern discipline of agricultural science this century. The notion has clearly been that we can modify the environment to create better or optimal growth conditions for crop plants and domestic animals. This paradigm of modifying the environment underlies irrigation science, because we irrigate when it is too dry; behind soil fertility management, where we add chemical fertilizers, lime and manure to ameliorate unfavourable soils and use tractors to till heavy soils and break hardpans; in weed and pest management where we spray to keep enemies of our crops at bay, and in veterinary science where we medicate and vaccinate our farm animals to escape disease. And our successes in modifying the environment have invited plant and animal breeding strategies that have created crops, livestock, poultry and fish that respond with higher yields when conditions improve.

To these four traditions we have added the fifth: the new knowledge that sprung from the realization of the Double Helix and the intrinsic molecules of life, a discovery where Europe also played a significant part.

There is little doubt that this approach to farming has yielded enormous results, and enabled societies in the industrialized world to build value-added industries on the foundations of highly productive rural primary agricultural industries. The Green Revolution, with its package of seed of improved rice, wheat and maize, fertilizers, irrigation and pesticides, in a setting of infrastructure, credit and extension, not only averted a major threat of hunger in Asia in the 1960s and 1970s, but also offered countries of South-East and South Asia prospects of creating industrialization as had been bestowed on the industrial nations of the North. Let there be no doubt about the virtues of the Green Revolution: human suffering, and real loss of life, would have been much greater in Asia had the Green Revolution not happened, and the prospects for economic growth in the now emerging powers of South-East and South Asia would have been much less. Social inequities, marginalization of women, possible genetic erosion in major food crops, and damage to the environment are all part of this criticism. Whilst recent reviews have on balance exonerated The Green Revolution, there is little doubt that some of the critics have pinpointed both negative and unintentional effects on individuals and groups of urban and rural dwellers.

In many respects the Green Revolution brought to tropical crops what agricultural research had already contributed to European and North American farms. It was no different from the high-input - high-output technologies that European farmers could access through their infrastructures.

But the Green Revolution largely bypassed Africa: the strong external political interests in Asia was not mirrored in Africa. To that extent there has been a approach in international politics on Africa that has had implications for food security in that continent.

But from a scientific point of view a more interesting explanation of some of the criticism of the Green Revolution may lie in a shifting set of scientific paradigms, with origins both in environmental science and in the social and economic sciences. Emerging from several decades of debate is now a realization of shifting paradigms. Most simply the shift can be explained thus:

Previously agricultural science was based on modifying the environment through external inputs for optimal farming conditions, with genetic improvements in crops and livestock designed to utilize the improved growth conditions where available, and to give new plants and animals that would perform well even if the optimal conditions were not achieved.

The new paradigm accepts that in many instances farming conditions cannot be dramatically changed through large external inputs, because the environmental costs will be too high, or that farmers just do not have at present or in the foreseeable future physical or economic access to the required inputs. We require a scientific foundation that can give real prospects of increased food production also for those who are food insecure through lack of access to significant external inputs to their production systems, whether be it for environmental, logistical, economic or social reasons. It follows therefore that access to inputs is a constraint that is much less relevant to European farmers than to tropical smallholders.

One may of course put the blame on politicians and policy makers that have been unable to create the enabling environments socially and economically for high-input-high-output technologies to thrive in developing countries. Indeed there will be people who argue that it is irresponsible to shift scientific paradigms for political convenience. Instead we should unequivocally strive to create that enabling environment that will allow green revolution technology to produce in developing countries - anything else is to condemn the poor to eternal poverty. One may argue that it is politicans who empower the poor, liberate the women and protect against economic exploitation, not the agricultural scientist. It therefore follows that social responsibility in science should not be taken to the extreme of changing scientific paradigms.

Considerations of equity and social equalities, and the very fact of the 840 million food insecure, do nevertheless suggest that there may be scope for alternative approaches. The experience of hundreds and thousands of non-governmental organisations and farmers' associations in the heartland of the green revolution in South and South-East Asia suggests that there is room for alternatives. There is room for scientists to focus on methods and technologies that realize that socio-econmic deprivation is for real. And there is room for looking at other approaches to improved yields - or at least somewhat improved yields - in the light of increased understanding of the environmental intricacies where irrigation water, fertilizer use and pesticides and other agro-chemicals are variables that do play part in the greater equation of sustainability.

The World Food Summit made some very specific references to shifting paradigms.

Commitment Three of the World Food Summit Plan of Action has a whole paragraph devoted to research. We scientists would do well in listening to what 186 countries decided:

"Research in agriculture, fisheries and forestry will be essential to achieving the sustainable food production increases upon which the short and long term food security of a growing world population will depend. The combination of such research, and an enabling environment, can improve food security both at national and household levels. Equity issues and equality between women and men should be given appropriate consideration when setting research agendas for the future. Research efforts should clearly focus on poverty eradication and on the creation of more environmentally sustainable agricultural, fisheries, forestry and food production systems. This research should be directed towards low, as well as high, potential areas according to their specific research needs. Renewed efforts should be made to involve farmers, fishers, foresters and their organisations in setting research priorities and directions, and to make experimental findings accessible to them". And The Plan of Action spoke specifically of technologies that originated as counter-technologies to the green revolution and industrialized country technologies, *e.g.*:

- Appropriate attention to areas that are less endowed with natural ressources,
- Special emphasis on under-utilised crops,
- Enhancing the institutional framework to allow for the full participation of all interested parties, including indigenous people and their communities.. in the identification of research needs,
- Promoting suitable systems, *inter alia* participatory systems, for the dissemination and extension of research results,
- Ensuring that gender perspectives are integrated in research, planning and implementation,

- Promoting research and development leading to the use of appropriate technologies, relevant post-harvest and transformation techniques, and adapted plant and animal breeding that meet local needs.

There is little doubt that this requires an additional research agenda to the one we have traditionally pursued in the established research communities, whether it is tropical couintries, or in countries in Europe and other industrialized countries on behalf of tropical agricultural science. It does not mean that we should abandon the old agenda, rather that we enrich our tool kit. The expanded paradigm does not reject the validity of high input - high output agriculture under suitable environmental conditions or where inputs are available to the farming community. It does not even fundamentally question the validity of the Green Revolution research agenda, it just appreciates that the older paradigm does not apply to all farmers in the South, and in particular will not be helpful to many poor farmers for many years.

Before discussing the ingredients of research agendas based on the expanded paradigm, let us consider an important shift in the interpretation of the concept of sustainable development. The World Commission on Environment and Development (The Brundtland Commission) suggested in 1987 that sustainable development was a route that both rich and poor nations should follow. By UNCED in Rio de Janeiro in 1992 the sustainable development concept had matured, not only to focus on the protection of the natural environment (including still undiscovered butterflies of the Amazon jungle), but also to consider economic, social and cultural aspects of development through wise use of our resources. To-day we are struggling to put this enlarged concept of sustainable development into operation for agriculture, forestry, fisheries and food production. The early focus on natural capital and on narrow management of natural resources, has been superseded by a sustainable development concept where the Total Capital we strive to sustain within and between generations consists of four separate components:

- natural capital (land, water, air, genetic material, ecosystems..),
- human capital (knowledge, science, culture, health, nutrition),
- institutional capital (schools, universities, research facilities, infrastructure),
- social capital (democracy, good governance, civil rights, equity, social harmony).

In banking terms this Total Capital constitutes the wealth of nations. When we nurture development, we attempt to let the sum of components grow, also by trading between components. As developed industrialized countries used natural capital to create valuable human, institutional and social capital (they burnt oil to educate their children), so developing countries must develop sensible approaches to balance their use and their creation of component wealth.

If you accept this expanded understanding of the concept of the Total Capital as the welfare function to be optimized with sustainable development, it becomes easier to interpret the effects of invoking the Expanded Agricultural Research Paradigm for increased Total Capital. It also becomes easier to understand and interpret the critique of the original Green Revolution - after all, it was long before the concept of Sustainable Development, long before our expanded definition of the wealth of nations. Afterthoughts are certainly better than no thoughts at all, and we accept that the critique of the Green Revolution and of similar intensive production methods in Europe and other industrialized countries, has helped us to define the new and additional research agendas in terms of sustainable agricultural development.

Here are some examples where the Expanded Agricultural Paradigm differs from the traditional paradigm.

The Expanded Agricultural Research Paradigm relates to the component of natural capital by:

a) suggesting that nutrient management be based also on the increase use of nitrogen fixation from the air and phosphate and micronutrient mobilization through crops, agroforestry methods, microorganisms, worms and termites, the use of composts, manure and sewage, the recycling of plant nutrients from deeper layers beyond the normal crop rooting zone by deep-rooted tree species, and by viewing investment in less mobile plant nutrients (*e.g.* phosphate) as a capital investment rather than a recurrent expenditure. (You will recall that the traditional paradigm suggests significant reliance on chemical fertilizers.)

b) understanding the ecology of weeds and pests, not only in farmers' fields but the functional agro-biodiversity in the total farming ecosystem. Integrated pest management and integrated weed management offer intelligent alternatives to heavy use of fungicides, herbicides, insecticides and other and pesticides often advocated under the traditional Paradigm, and also reduces the risk of accidental poisoning of animals and humans.

c) appreciating that large genetic diversity in crops and livestock constitute an asset and not a hindrance to efficient production. To explore how farmers can become actively involved in selecting those populations of plants and animals that provide the best production potential for them, both in terms of actual yields and security of yield. Poor farmers often operate in natural conditions with major risks, and must behave in a risk-averse manner different from richer and risk-loving farmers in better-endowed areas. At the same time small and poor farmers are important curators of valuable agricultural germplasm that all farmers, rich and poor, in the North and in the South, will rely on for the foreseeable future. The Original Paradigm obtains large yields from genetically homogeneous populations of crops and livestock and the control of risks by external inputs rather than genetic heterogeneity.

The Expanded Agricultural Research Paradigm relates to the component of human capital by:

a) seeking participatory methods of developing relevant research agendas, closely involving also less privileged groups (the poor, women, youth) in setting the agenda, to exploit their knowledge and ambitions, to give them ownership of ideas, and access to external knowledge through new modes of communication. The Original Paradigm frequently bases research on centrally developed agendas by specialists, in efficient top-down designs.

b) looking at farming systems with mixes of plants and animals that provide for balanced nutrition throughout the year to improve human health, reduce reliance on exploitive forces outside their communities, and distil pride in their own culture and choice of pathways to the future. The Original Paradigm relies on surplus from monocultures to create purchasing powers that can be used to acquire food for a balanced diet.

The Expanded Agricultural Research Paradigm relates to the component of institutional capital by:

a) facilitating the formation of farmers' groups that can provide focal points and services where central government in practice does not reach, provide efficient interfaces to outside agencies, and mechanisms for protecting against unethical exploitation of their resources. The Original Paradigm often assumes the existence of centrally governed mechanisms.

b) putting emphasis on decentralized knowledge centres, local experimentation, small-scale developments of infrastructures such as feeder roads, simple irrigation systems easily managed and maintained and alternative and appropriate energy generating systems that can overcome chronic or seasonal energy shortages. The Original Paradigm emphasizes economies of scale and efficiencies of centralized institutions with critical mass.

The Expanded Agricultural Research Paradigm relates to the component of social capital by:

a) empowering less privileged groups (often women and poor people) by making them decisions-makers and architects of their own futures, increasing their status as producers and curators of resources, supporting technology that can yield greater equity between groups, and showing other sections of the society more clearly their role in nation-building. The Original Paradigm often assumes rapid trickle-down effects to benefit the less privileged and creating better equity.

b) building foundations for increased political influence and less exploitation by stronger groups, and enhancing the role of equal rights and democratic ideals, by developing research agendas that apply to the less privileged rather than more favoured sections of the community. The Original Paradigm often points to increased national wealth as a basis for the growth of democracy and equality.

The Expanded Paradigm is not intended to replace the original foundations for agricultural research. Both approaches play important roles in nation-building. But an agricultural research agenda based on the Expanded Paradigm will be significantly different.

Whilst we vigorously pursue the Original Paradigm, and fine-tune it to reflect our new perceptions of sustainable development, we owe it to the 840 million, and many other underprivileged, to simultaneously explore alternative pathways. The interesting aspect of the Expanded Paradigm is also that it offers alternative approaches also for agricultural production in developed countries. Ecologically produced food is a major growth industry here. There are elements in the application of the Expanded Paradigm that tend to rely on fewer external inputs than otherwise commonly used in industrialized agriculture. A picture is emerging that bring together important elements of technologies that by default or design rely on the use of lower but smarter inputs. Industrialized agriculture strives to become more sustainable and reduce its overcapacity in an environment already polluted. My prediction is that this type of agriculture will increasingly come to rely on technologies that apply equally to tropical smallholder agriculture that strives to become more productive in order to feed more people. In agriculture, as with other disciplines, there is no free lunch, and simple theromodynamics still rule. There must be nutrients and there must be water. Simplistic solutions have no place. But smart solutions will arise from within the framework of the expanded Research Paradigm. The new green revolution will be different.

The Rome Declaration also made it very clear that research in agriculture, fisheries and food is essential for future food security. The governments declared unanimously that we recognize the need to adopt policies conducive to investment in human resource development, research and infrastructure for achieving food security. In The Plan of Action paragraph 5 the governments said explicitly: Long term investment in research and in cataloguing and conserving genetic resources, particulartly at the national level, is essential. If I may quote from another somewhat lengthy passage, it is to specifically point to what governments see as objectives (paragraph 35):

"To take decisive action in cooperation between the public and private sectors to strengthen and broaden research and scientific cooperation in agriculture, fisheries and forestry..."

"To this end, governments in collaboration with the international and scientific communities, in both the public and private sectors, as appropriate, will strengthen national research systems,.. focus on interdisciplinary research .."

"Strengthen international research systems, in particular the Consultative Group on International Agricultural Research (CGIAR), and promote coordination and collaboration among international, developed country, and developing country institutions "

These quotations constitute a subset of many paragraphs where governments emphasised research as an important tool for food security. Representatives of 186 countries -- including 41 presidents, 15 vice presidents and 41 prime ministers -- travelled to Rome for the World Food Summit, and they subscribed to the Rome Declaration and the Plan of Action that I have just quoted from.

The World Food Summit was a major event for agriculture. It was attended by close to 10, 000 delegates, and representatives of non-governmental organisations (NGOs), UN agencies and other international bodies, journalists and support staff. NGOs, youth, parliamentarians, family farmers associations and the private sector held parallel meetings in Rome during the Summit and reported to Summit on their conclusions. They did have strong debates. But nobody quarrelled with the need for research. And in this context this means research both in the North and in the South. From the FAO end we are of course pleased to note that initiatives to link research institutions in Europe and developing countries get such support. The recent European initiative, including EU countries and Norway and Switzerland, is a most welcome manifestation of Europe's will to collaborate.

May I end with a few quotations from what the Director General of FAO. Dr. Jacques Diouf said about Food For All:

"We have the possibility to do it. We have the knowledge. We have the resources. And with the Declaration of Rome and the Plan of Action, we've shown that we have the will. I challenge the world to a race against time to achieve the Summit goal of reducing the number of hungry people in the world by half by the year 2015. This is not a maximum goal; it is a minimum goal. This is only the beginning of the fight to ensure that babies will not cry of hunger and that mothers will not be looking at children who have no hope. And I am convinced that what you have shown here during these days that are history is certainly the greatest indication that together we will win against poverty, we will win against scepticism, we will win against cynicism, we will win against egoism and that the best of human values will prevail in the relations among nations, among states and within communities."

This is my report on research from the World Food Summit. I believe it applies to Europe's efforts - it challenges Europe as it challenges all governments of the South and the North.

EUROPEAN AGRICULTURAL RESEARCH: WORLD FOOD SITUATION AND DEVELOPMENT

MICHEL PETIT
World Bank Group
1818 High Street NW
Washington D.C. 20433, USA

The globalisation of world economics and the revolution in international communications affect all aspects of economic, social, cultural and political life, and therefore also deeply affect agricultural research. Globalisation exists in a variety of forms. In agriculture, there are four aspects that merit special attention. The most obvious is liberalisation of trade in agricultural products, which was made abundantly clear in the recent agreements concluded in the GATT Uruguay Round and now administered by the International Trade Organisation (ITO). From the economic standpoint, a phenomenon that perhaps is even more important than this agricultural liberalisation is the free movement of capital. The result is a massive increase in international financial flows, which greatly holds back the capacity of the Nation-States to implement indépendant macro-economic policies. This was clearly shown in France, for instance, during the last presidential election, through the difficulty in discussing the harmonisation of ideas and on the possibility of implementing an alternative macro-economic policy. Agricultural research, of course, is also affected by accelerated information exchange resulting from technological advances in computerisation and communications, which make it vital for scientists to be interconnected. Everyone knows of the possibilities that progress in the Internet has brought.

The fourth aspect concerns global environmental problems, and perhaps more generally the need, recognised the world over, to "allow" the planet to survive by improving natural resource management. All these tendencies are deep-rooted and profoundly affect what agricultural research can do and on what society expects from research. The present upheavals constitute redoubtable challenges for European agricultural research, but these challenges are also opportunities, even if they are difficult to define. It is clear that European agricultural research will only be able to seize these opportunities if it adapts to the rapid, far-reaching evolution, taking place on the international scene.

There are two aspects of this evolution that merit special attention: the changes in the international markets in agricultural and food products, on the one hand, and the requirements for the development of the poorest countries, on the other. The aim of this paper is to explore the implications of these two aspects for European agricultural research. After considering the long-term perspectives of the world food situation and the need for agricultural development in the poorest countries, the third part of the paper will explore the consequences that this evolution will have on European agricultural research.

WORLD FOOD: MEDIUM- AND LONG-TERM PERSPECTIVES

The World Food Summit organised by the FAO in Rome in November 1996 emphasised the vulnerability of hundreds of millions of ill-nourished people. The goal of reducing their numbers by 50% between now and the year 2020 may not seem very challenging. This ambition is modest because it reflects the conviction that it will be difficult to provide adequate quantities of food, at the right time, and in conditions which enable the recipient populations - whose numbers are growing - to either produce or pay for it. This conviction is the result of the great uncertainty, as reflected in the major differences in the experts long-term views of the world food situation, and depends on how experts view the prospects for increased production. On the demand side, experts develop their - mainly similar - projections by examining the prospects of population increase and average per capita income for each of the main regions of the world. The consumption of agricultural and food products is expected to more or less double in the next 25 to 30 years. It is more difficult to count on projection models on production. They interpret and adapt trends of the past as a function of various parameters whose effect is also assessed on the basis of elements of the past. The result is that there is great uncertainty on aggregated productivity growth to be expected from continued technological advances. Annual productivity increases are estimated at 1% or 2%; either percentage will have a radical, although different, effect on the long-term equilibrium between supply and demand. Experts are far from agreeing on a figure. On this point, pessimists like Lester Brown make projections that are very different from the more optimistic ones of Dennis Avery, and also of the major institutions such as the World Bank and the FAO. These long-term projections do not provide enough information to clarify and settle the question. The World Bank reviews price fluctuations for the main cereals on the world market, in real terms, for the 1960-1965 period and for the first half of 1996. The FAO presents the monthly price fluctuations between January 1990 and September 1996. The price trends are interpreted quite classically as a function of supply and demand.

The first graph (Fig.1) shows, very clearly, that the real price of cereals has followed a downward trend. We also know that this trend preceded the period covered by data presented here, thus indicating the existence of a secular tendency. The trend was accompanied by sizeable inter-annual variations, with short, sharp price hikes, particularly in the early 1970s.

The question is whether this downward trend will continue, despite occasional brief increases. A look at the second graph (Fig.2) suggests that the recent increase, which reversed itself in the middle of 1996, was probably only a passing surge. We know that the price paid for agricultural produce, more precisely for grain crops, started going up towards the middle of 1993, after having declined throughout the 1980s. The food price index published by the World Bank rose by 56% between June 1993 and May 1996. The explanations for this price increase can be found by breaking down global production into categories for countries that have dissimilar characteristics. The result suggests that recent increases were nothing more than the result of timebound circumstances. Consider that five leading export countries - USA, the European Union, Canada, Australia and Argentina where seeded/harvested acreage shrank considerably. Then add the case of the former USSR countries, as it undergoes the upheavals of deep economic transition. Together they account for the main part of the production decrease. In the other countries production has continued to rise. The average annual growth rate for the 1980-1995 period was 2.4%. During this same period, consumption continued to go up regularly, except for a slight slump at the end of the 1980s and the beginning of the 1990s. This led to the idea that the other countries of the world should, in the main, be able to satisfy a notable part of their food

Figure 1: World grain prices (deflated by the Manufactures Unit Value)

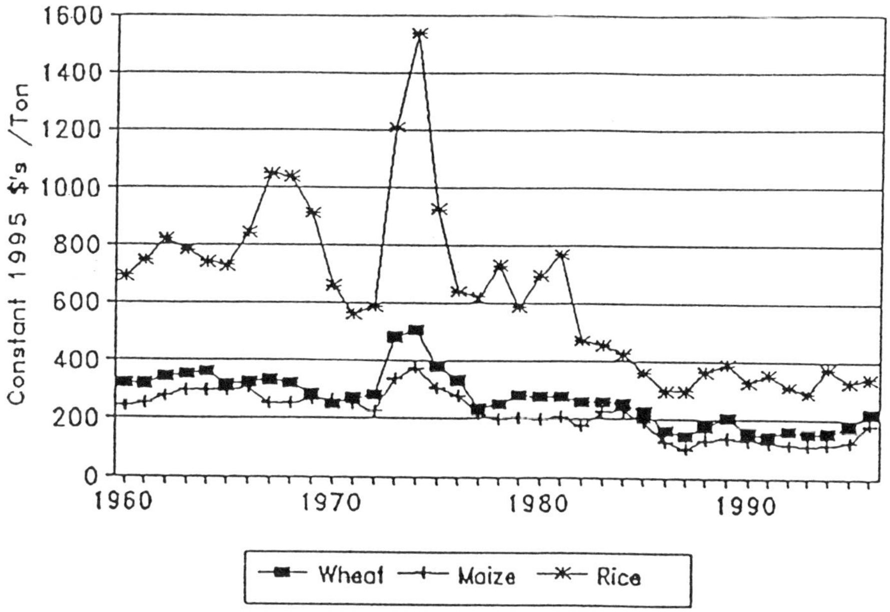

Figure 2: Grain prices (1990 - 1996, monthly)

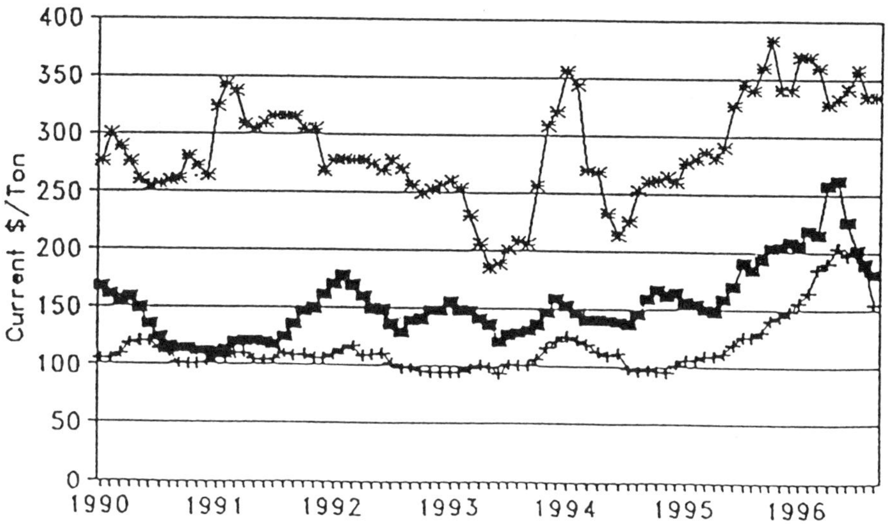

requirements themselves. Fluctuating world prices explain the variations in volumes produced by the five exporting countries. This conforms to the simplest economic theory as concerns the private sector, and also is relevant to decisions taken by public sector authorities. Take the European Union as an example. When international prices are low, the budgetary cost for export subsidies becomes high, triggering off heavy political pressure to reduce subsidies. Further, we know that as a result of political pressure from other exporters, in particular the United States, the European Union has promised to reduce its export subsidies. Yet we recently saw that when prices rose, farmers fought to limit the acreage left fallow, and hence, political pressure was exercised on the public authorities to revitalise production.This economic behaviour, *i.e.* a classical reaction by the exporting countries to prices is easy to understand, and does not justify long-term pessimism on these countries' production capacities. As for the ex-USSR countries, it is very difficult to predict how long the transition period will last. In the long run, economic reforms will probably enable these countries to use their very real agricultural potential to ensure a better food supply line for their home populations, and perhaps rejoin the ranks of agricultural and food product exporters they belonged to at the beginning of the century. Under communist rule, these countries were massive net importers of cereal grains; there is reason to believe that this situation was the result of price distortions which no longer exist.Consumer subsidies were huge and led to large-scale waste. All this suggests reasons for relative optimism in the medium and long term.

This said, the prospective demand need to be more deeply analysed. China is probably the best illustration. Population growth has been slowed down. But the population is so large that the most probably scenario - rapid growth in food consumption and more dietary diversification in favour of animal products - will lead to an accelerated, very high volume increase in the consumption of cereal grains. Many countries in East, Southeast, and maybe also South Asia, sooner or later, will follow the Chinese example. In sum, because of the dynamics of economic growth and the (slower yet noteworthy) population dynamics, the Asian market for foods will be very buoyant. The case of sub-Saharan Africa causes much greater concern. During the last few decades population growth has greatly outpaced agricultural growth thereby causing the average per capita food availability level to decline. There are real potentials for increasing production and productivity, even if, unlike the green revolution in South and Southeast Asia, there is no magic solution. But in many regions, the individual's safety, because of political unrest, is not secure enough to allow for increased, or even sustained production. Just think of Zaire, Rwanda, Sudan and Liberia at this very moment. In other countries, production has risen, but the conditions surrounding economic development, in particular the underdeveloped infrastructure and the institutional weaknesses are such that prospects for the future are not very encouraging. For its food supply, sub-Saharan Africa will probably continue to rely on exports from other zones, in particular the OECD countries, and on food aid.

Efforts in agricultural research must be pursued in order to ensure higher yields and, more generally, greater productivity. The same applies to everything that contributes to technical extension work and productivity improvement. There is a downward trend in public funding for the National Agricultural Research Systems (NARS) and in contributions from the international community. The Consultative Group for International Agricultural Research (CGIAR), in particular, went through a financial crisis a few years ago. The crisis is over, but the sustainability of the Group's financial capacity is far from guaranteed, yet the tasks that lay ahead are becoming more and more complex. Of course, there is hope that in many regions the private sector will increase its investments in research and development, but here again, prospects remain uncertain. In any case, the private sector will never be able to fully replace the public sector in agricultural research, a field in which much of the knowledge and services are what the economists call "public property". This quite naturally leads me to the second part of my paper.

EVOLUTION OF DEVELOPMENT-ORIENTED AGRICULTURAL RESEARCH

The conclusion that efforts in international agricultural research must be heightened becomes even clearer when attention is given to the development requirements of the poorest countries. The immediacy of needs is indisputable, and these needs will continue to grow. Increasing productivity is all the more necessary since other sources of agricultural production growth, *i.e.* increasing arable lands and, more specifically, increasing irrigated lands, cannot continue their role of the past.Potentially available arable lands are limited. There is still a large amount of land that is unexploited, in particular in the Amazon and Congo basins, but the economic and environmental costs, and the deforestation required to make lands cultivable are considerable. Such development is probably not advisable for the future of humanity. As for irrigated lands, expansion will be limited because of growing competition for available water resources resulting from urban and industrial growth. Productivity increases, moreover, should come from improved natural resource management.All too often the price for a straightforward per hectare yield increase is the deterioration of the natural resource base, which thereby jeopardises the sustainability of such an increase. Agricultural research ends up with very difficult technical specifications, and its brief becomes that much more pressing.

Unfortunately, agricultural research institutions in developing countries are still very weak, at present they are the weakest link in the global agricultural research chain, a system that is currently being established.Often, as part of the public sector, they suffer from the excessively constraining regulations of the public accounting system, bureaucracy, and from insufficient support from the state, both as concerns funding and the application of institutional reforms that are often "politically sensitive". The result is that in many countries, the number of research scientists has increased faster than the corresponding budgetary allocations, which means that their productivity is often inhibited by lack of sufficient resources. Furthermore, they have great difficulty in efficiently spending the small amount of funding that is made available.For this reason, it is absolutely essential that public research institutions be strengthened. The World Bank, the organisation for which I work, is striving to support this institution-building endeavour.

For European agricultural research, the most important international issue is probably the extent of present-day upheavals. The creation of the CGIAR, 25 years ago, consecrated the international research centres as an institution, and with it, the institutional innovation fostered by the U.S. Ford and Rockefeller foundations which had created the first of these centres, *viz*, CIMMYT and IRRI. Since that time, the CGIAR centres have become leading actors on the international scene. But since their creation, there has been deep-rooted turbulence which has led to the emergence of what we could call a global system of agricultural research. The main changes in the last 30 years have been:

- the considerable increase in the number of researchers in developing countries. Although, as we have seen, they are underemployed and often not very effective, they constitute an important resource that should be used;

- the veritable scientific revolution in biology, and the technological revolution taking place in the field of information and communications. These changes provide an opportunity for modernisation and considerably more collaboration between the traditional agricultural research teams and the more basic research teams working on biology, as well as for stronger professional relations through transnational networks;

- greater expansion in research. Thirty years ago emphasis was placed on increasing volumes produced, the size of "the cup of rice" so to speak. Nowadays, as we well know, natural resource management - which reflects the tradition of responsible agriculture and care for the heritage being passed on to future generations - is absolutely essential. This has resulted in broader horizons for research. Since most natural resource management problems occur at the local level, it is very important to use knowledge of the local situation; more often than not, this knowledge can be found in the minds of the local populations themselves. Participatory methods that bring together the farmers, the users and the communities that draw on the natural resources, be they forests or irrigation lines, need to be employed.

All these upheavals bring about an exponential increase in the number of partnerships, and intensified collaboration within them. This proliferation and intensification of partnerships suggests that a genuinely new global system of agricultural research is being constructed.

Everyone should find his due place. Recent changes are promising, because we can already see new partnerships being established, and, elsewhere world-wide consultations are beginning to be organised. The CGIAR has played an important role during the last few years, in particular through what is called its renovation process. It has also promoted and facilitated the creation of the Global Forum, now being established. Regional representatives of the NARS from the countries of the south will play a central role, and will receive support from the CGIAR organisations. The Forum is open to other partners, such as OECD, research institutions, private sector research organisations, associations, agricultural organisations, and NGOs.

CONSEQUENCES FOR EUROPEAN AGRICULTURAL RESEARCH

The globalisation process mentioned above obliges European research to face three major challenges:

1) **the challenge connected to keeping European agriculture and the agro-food sector competitive,** all the while meeting the increasing requirements of public health, the environment and the quality of life. As far as competition in the agricultural and agro-food sectors is concerned, technological and organisational capacities will become increasingly important. Agriculture and the agro-food sector are becoming more and more sophisticated, capable of efficiency, ensuring high quality products and providing consumers in various markets with products that have specific characteristics, and are adapted to a well identified and "conquered" niche. Along these same lines, we see that processed agro-food products are selling better than raw agricultural produce. When talking about markets, we have a great tendency - to which I have ceded during this talk - to focus exclusively on cereal grains and not recognise the importance of the processed and transformed products. The evolution of the agricultural and agro-food sector in Central Europe is very interesting to watch. The competitive edge in this sector is directly linked to the investment of multinational agrofood companies

that bring in capital and, even more importantly, technological and organisational know-how and market wisdom that enables them to target production to sales.

Clearly, European agricultural research, through deeper and renewed public/private sector partnerships, has a paramount role to play in ensuring competitiveness. To reach this goal without violating the ever stricter requirements applied to agricultural research, it will have to participate much more actively in the societal debate, in particular by making this debate more objective. The present debate on Genetically Modified Organisms (GMOs), badly documented and violent through the disputes it provokes, illustrates this need. This is a somewhat new rôle for research which most institutions have not yet fully recognised and are not yet well enough organised to play.

Last, I would like to point out that intelligent competitiveness is a requirement that will be based increasingly on the mastery of genetic resources, which means, first of all, access to these resources. The present free access regime is in an uproar. The potentials for genetic engineering are considerable, but there is great apprehension about developing them. There is no doubt that European agricultural research should become equipped to promote scientific development, exploit the possibilities it leads to, and communicate with public opinion and public authorities to be sure that these possibilities are exploited in a socially acceptable way.

2) The second challenge facing European agricultural research, as a result of present-day upheavals, comes from the need to **find its niche in the emerging global agricultural research system**. As I mentioned above, a Global Forum has recently been created. Institutions of the OECD countries are invited to join it. This is both an opportunity and a constraint, because going beyond just joining the Forum, the European institutions should organise themselves to speak the same language. There is a widespread, imperious need to forge partnerships, in particular with institutions from countries of the South. This would be invaluable since many genetic resources are still to be found there, and partnerships such as these will open up many scientific opportunities.

3) The third challenge to European research is **contributing** - if its leaders want to and can - **to progress in agriculture in the most resource-poor countries**. From the European point of view, this is desirable for humanitarian reasons, of course, and as a contribution to economic progress in the countries of the South. Such progress, in the long term, will also benefit the countries of the North. Furthermore, in the short term, a commitment of this type can serve European agricultural institutions. Confronting the realities of developing countries allows the scientist, individually, to benefit from a great range of experiences. At a more fundamental scientific level, many of the problems stemming from the need to increase productivity without jeopardising the natural resource base take us to paradigms that need to be developed, or, perhaps, still need to be invented. The multidisciplinary systemic approach of these paradigms comes from a macroscopic level which is probably of the same nature in both the North and the South. Scientists who participate in drawing up these approaches in countries of the South will be better equipped to do the same work in the countries of the North and thus contribute to solving the difficult problems mentioned earlier.

CONCLUSION

In conclusion I would like to appeal to the ambition of European agricultural research by inviting it to face the three challenges identified above and summarised below, namely:

1) contribute to making agriculture and the agro-food sector more competitive while respecting the growing requirements of the environment, public health and standards of living;

2) find the right niche in the world-wide agricultural system;

3) participate in the development of the countries of the South.

An ambition such as this would provide a new incentive for agricultural research; it would be a real contribution to satisfying the needs of the European society and, at the same time, to progress for mankind as a whole, without which durable, viable long-term progress for Europeans cannot exist.

This is a great ambition. But I feel convinced that a lukewarm attitude would prevent agricultural research from ensuring the permanency of its social usefulness.

EUROPE AND AGRICULTURAL RESEARCH FOR DEVELOPMENT

ALAIN DARTHENUCQ
European Commission, DG XII-B-4,
200, rue de la Loi
B-1049 Brussels, Belgium

ABSTRACT

In a world of global parameters which traverse national frontiers, where feeding a large percentage of the world's population and alleviating poverty are still major priorities, where public opinion is increasingly aware of environmental issues. Agricultural Research for Development (ARD) is evolving at international level in terms of organisation and partnership.

Europe has an important role to play, both as a donor and scientific partner, but this dual role is not evident and is insufficiently recognized owing to lack of coordination.

The European Initiative for Agricultural Research for Development (EIARD) has been launched to remedy the situation.

The European Commission is an active partner in the Initiative, especially via its ARD cooperation instruments.

There is every reason to believe that a European ARD system will be created, provided that the political will exists.

INTRODUCTION

The world of agricultural research is changing rapidly and faces many new challenges.

The world's population is estimated to reach 8 billion in 2025, which means that food supplies will have to double in the next 30 years. As we have just been reminded at the World Food Summit, agricultural research is one of the means which should enable us to meet these requirements. Also, as more people become city dwellers eating habits are changing, and this too must be taken into account.

The agricultural sector is very important in the economies of developing countries, in terms of both income and employment. The fight against poverty is therefore directly linked to the continuing improvement of agriculture in these countries.

Such development must also take account of environmental issues, of which international public opinion is increasingly aware, while we know that the amount of arable land available per inhabitant will decrease and most fertile land is already being farmed.

Furthermore, as we have been reminded, due attention should also be given to changes occurring in the world which will inevitably have a considerable impact on developing countries. These are the processes of globalization and internationalization and the emergence of new technologies, in particular biotechnology. Such changes may facilitate the integration of developing countries into the world community, but they could also widen the gap which separates them from the developed countries.

Thus in a context where different elements come into play - the internationalization of all forms of exchange, global climatic change, the fight against poverty, the need to meet food requirements, product quality - agricultural research for development faces new demands from farmers and professional organisations, consumers, governments and donors.
In these circumstances, efforts have been made at international level to improve agricultural research for development. Where does European ARD stand in this global system which is now emerging?

EUROPE'S ROLE IN INTERNATIONAL AGRICULTURAL RESEARCH DEVELOPMENT TODAY

At international level, Europe plays a dual role in the field of agricultural research for development: as donor and scientific partner.

Donor at two levels

Firstly, Europe - the Member States of the European Union, the European Commission, Norway (associated with the EU as part of the European Economic Area) and Switzerland, with which negotiations are underway concerning research - provides over 40% of the funding for CGIAR (Consultative Group on International Agricultural Research) international agricultural research centres co-sponsored by FAO, the World Bank and UNDP (US$ 129 million out of US$ 302 million, *i.e.* exactly 42% for 1996), plus a contribution in kind (researchers). However, each donor contributes separately, which gives Europe a relatively minor role compared to the other major donors, namely the World Bank and the United States.

Europe is also helping to strengthen the research capability of National Agricultural Research Systems (NARS) in the developing countries, through bilateral cooperation with the Member States or EU development cooperation programmes, such as the EDF for the ACP countries or the financial and technical assistance fund for cooperation with the ALA and MED countries. However, for a number of reasons, the funds allocated for this purpose are fairly limited.

Europe is also a scientific partner

Europe encourages activity in the field of agricultural research for development through:
- S&T cooperation developed individually by Member States;
- S&T cooperation developed at Community level by the European Commission.

Europe has a long-established tradition in this field, with many achievements to its credit. The system may, very briefly, be characterized by:
- its size, in human and financial terms.
- its diversity.

Some countries have specialized research centres (the KIT-Royal Tropical Institute, in the Netherlands, the IICT in Portugal, the NRI in the UK, the CIRAD and ORSTOM in France), but universities also play an important role. The private sector has played a very small part up to now.

Different focuses of interest are linked to Government policy. Different Governments encourage S&T cooperation with particular countries and regions.

There is also growing interest in scientific cooperation with the developing countries on the part of establishments not primarily concerned with this field (INRA is a good example). The training of students and trainees in European universities and research centres is still an important activity, for which there is much demand from developing countries.

Research, carried out in and for Europe, has become more relevant to developing countries. Agricultural research institutes have much to offer in terms of multidisciplinary approaches, modelling and advanced technologies. Their knowledge and experience put them, individually and collectively, in a very good position to play a key role in joint research programmes at world or ecoregional level, and to help NARS meet their priority needs, especially training in research management and administration, the definition of priorities, guideline analysis, and the creation of improved communication and information networks. Increasingly, regional organisations and the new Regional and Subregional Fora are encouraging the creation of partnerships involving collaboration, at the initiative of NARS representatives. These new types of partnership and constructive dialogue, stronger and more equitable, present new challenges which call for European commitment.

Despite the importance of these two roles, informal coordination at European level only concerned, until recently, CGIAR donors, and was limited to the exchange of information. In view of the general lack of dialogue regarding other aspects of European participation in ARD, it was proposed to extend the scope of the dialogue and get all interested Member States involved, not only CGIAR donors. This new system follows on from the previous one and its non-official character, and the principles of flexibility, subsidiarity and equal status of all participants have been maintained. It is called the European Initiative for Agricultural Research for Development (EIARD) and was presented by the Union's Presidency at the ministerial meeting of the CGIAR, held in Lucerne on 9 and 10 February 1995.

THE EUROPEAN INITIATIVE

Its role

On the basis of Europe's important contribution, the purpose of EIARD is not only to enable Europe to find its proper place and receive due recognition at international level, but above all to give it a more active role, as a result of increased coordination and dialogue. It is intended to be a forum for discussion and proposing strategies and new initiatives. The aim is to improve coordination between its 18 members and between research and development policies. EIARD also wants to strengthen partnerships and promote collaboration between all those involved in ARD, especially European Agricultural Research Organisations (ARO), NARS in developing countries and their Regional and Subregional Fora, and CGIAR international centres (IARC), with a view to achieving sufficient coherence to define common objectives, ensure complementarity, synergy and cost-effectiveness of activities and to optimize their operational and political impact.

EIARD's role is therefore to promote:

Coordination at various levels, namely:

Exchange of information within Europe, *i.e.* within the European Commission and within States and between States and the European Commission.

The exchange of information between all actors is of vital importance. Modern information and communications technologies offer numerous possibilities for facilitating such exchange, in the widest sense, which will also help mobilize Europe's research capability. A pilot study is underway to examine the feasibility of an information system which would allow optimum exploitation of existing data bases and information systems.

Concertation between European partners for closer cooperation between Member States and the Commission, greater synergy between the various policies concerned (research, cooperation, external relations) and more productive relations with other countries, international organisations and developing countries. Such concertation is now effective between European CGIAR donors, both as regards guidelines and strategy and allocation of funds to IARC, and takes place informally at regular intervals. If necessary, common European positions can now be adopted at CGIAR meetings.

Coordination of activities. This means facilitating the formulation of coordinated positions and expressing the point of view of all partners in the Initiative. An example is the study carried out on regional and subregional ARD cooperation mechanisms, how to make them operational and ensure sustainable financing. This study, presented in Washington during the last International Centres week at the end of October, already demonstrates the commitment of EIARD partners to adopting a coordinated position in international discussions concerning the establishment of a Global Forum and Regional and Subregional Fora for agricultural research.

Formulation of a common European ARD policy. This is the final stage of the coordination process and it will depend mainly on the political will of the partners associated in the Initiative and the development of the coordination process. It can be regarded as the Initiative's political objective.

Stronger partnerships

Partnerships must be strengthened at two levels.

European level

Initiatives have already been taken to structure concertation between European research organisations. Networks such as NATURA (29 members and 12 countries) for universities and ECART (6 members and 5 countries) for research institutes are steps in this direction. On the initiative of the European Commission, European thematic networks have been set up, such as ETFERN for tropical forests or the European group for integrated pest management (IPM-EWG). However, these are individual, sectoral initiatives and for European institutions to take their place in the partnership being built within the Global Forum, a regional forum for Europe needs to be created associating universities, research centres, NGOs and the private sector, as exists for Asia, Latin America, Africa, North America and Japan. This is something the partners in the Initiative will have to consider during the coming year.

The Global Forum and Regional and Subregional Fora

In the new Global Forum, all those concerned, namely NARS, ARO and IARC must have their proper place, which implies new forms of partnership. A major concertation effort is required at world level, which will call on the resources of all parties concerned, depending on their comparative advantages. Europe is already contributing to discussions within the Regional and Subregional Fora of developing countries and supporting their participation in the Global Forum. Three African subregional organisations, CORAF (West and Central Africa), ASARECA (East and South Africa) and SACCAR (South Africa), which represent the NARS in their regions, are receiving European funding to create a Subregional Forum and participate in the Global Forum.

Its organisation

To implement the initiative and ensure its continuity, a European Coordination Group (ECG) was set up in Vienna in 1995. This group follows on from the existing group of CGIAR donors, comprising 13 EU Member States, the European Commission, Norway and Switzerland, plus the remaining Member States which are not CGIAR donors, namely Greece and Portugal. The ECG thus represents the 18 partners of the European Initiative. It is an informal platform for coordination, where the research and development policies and the activities of EIARD partners are discussed. It is *not* a new organ of the European Commission or a mechanism governed by Community procedures applicable to Community programmes. In this instance, the Commission is a member just like the others and is accountable to the Council and the European Parliament, just as the other members are accountable to their Governments.

It was agreed that the ECG would have two representatives per member, appointed by their respective Governments. The Commission's representatives are appointed by the Commission.

I will not go into details regarding the ECG's administrative organisation, but given its informal nature, based on a voluntary and interactive approach, the future of the European Initiative will depend entirely on the political will of its members and its effectiveness on what each member contributes and receives.

I can confirm that the European Commission, for its part, is committed to being fully involved, as it has shown since the idea was suggested by France at the Council meeting in September 1994.

THE EUROPEAN COMMISSION AND ARD

As we have said, the European Commission, like the EU Member States, has a dual role to play.

In the implementation of cooperation and development policy, it manages technical aid and financial assistance funds, and in this way, can help strengthen the research capability of developing countries in terms of infrastructure and human resources and contribute to the financing of research activities as requested by each country. Examples of this type of intervention are the strengthening of the KARI and the CARDI by the EDF, and support given to research programmes such as the "Fallow" project in West Africa at the request of CORAF and the "Soil and water management" project in South Africa.

The Commission is also, like 13 EU Member States, a CGIAR donor; the allocation for 1997 is ECU 12.8 million.

However, to date the main Community instrument in the field of S&T cooperation on Agricultural Research for Development, is the S&T cooperation programme with the developing countries (INCO-DC) which follows on from successive STD programmes, the first of which began in December 1982.

S&T cooperation arrangements set up by developing countries or international development organisations can be divided into two broad categories: those which aim to transfer an assembled technological package to the developing countries and those mainly concerned with strengthening these countries' own capability.

The European programme for S&T cooperation with developing countries fits into neither of these categories. Limited financial resources do not allow it to attempt to strengthen capability. (The first STD1 programme (1982-86) received ECU 30 million for "agricultural research"). Neither did it aim to finance research simply in order to find technical solutions to given problems. The objective was, and still is, to facilitate cooperation, *i.e.* joint activity and partnership between European scientists and scientists from developing countries, on research topics of particular relevance to the South. The research is not done for them, but with them. It is therefore a means of mobilizing scientists from North and South around a problem relevant to the South.

This objective has been achieved. The STD3 programme (1991-94) statistics show that 790 different teams have been mobilized in 157 contracts in the field of agricultural research, representing 493 different establishments: 226 in Europe, 106 in Africa, 76 in Asia, 70 in Latin America and 15 in the Mediterranean region. In response to the last call for proposals under the current INCO-DC programme we received around 700 proposals on the subject of "Agriculture - natural resource management", involving the participation of over 2,100 different teams.

Another feature of the programme is that it has been, and is increasingly, an opportunity for scientists from developing countries to extend and diversify traditional partnerships. It has also forced European teams to learn to work together, or in some cases made them aware of one another's existence (the rule is that at least two European teams come from different countries). Similarly, the increased value of a project which associates several establishments from developing countries in the same region helps strengthen the regional cooperation which these countries are promoting through the creation of regional and subregional mechanisms.

The programme is open to all developing countries and favours a thematic approach rather than a geographical one. In this sense, it goes beyond Member States' bilateral initiatives which are always geographically limited and often fragmented and dispersed, and increases their value by "federating" them. Its scope can allow transcontinental consortia to be created on a particular theme.

In addition, it favours direct contact between the research teams and the Commission. The proposals are sent in directly by the proposers and contracts established directly with the participants.

Over the last three years, the programme has evolved, both as regards content and procedural arrangements.

As far as content is concerned, a programme which started off as essentially technological, consisting largely of mono-disciplinary projects, now takes account of the knowledge, practices, objectives and constraints of producers and other actors in the field, with more emphasis on human and economic sciences.

The environment has also become a major issue, to the extent that INCO gives equal importance to the sustainable management of natural resources and the improvement of agriculture.

The programme also tries to take account of the variability of physical and socioeconomic conditions and regional diversity, identifying different regional priorities by subject through dialogue with those responsible at regional level. The strengthening of regional coordination mechanisms in the developing countries should facilitate such dialogue and make it more productive.

Finally, partly because INCO-DC is a programme implementing Community science policy, procedures are followed (international calls for proposals, assessment by independent scientific experts, strong competition) which guarantee scientific work of a high standard.

This type of programme not only brings researchers in the South out of isolation, but also facilitates their integration into the international scientific community. The fact that it is part of the European Framework Programme also means that European researchers concerned with research for development are not isolated or cut off from the rest of the scientific community simply because they are working in this particular field.

Financing has increased from ECU 30 million for STD1 (1982-86) to ECU 125 million for INCO-DC (1994-98). Such an increase is uncommon enough to be worth mentioning.

However, the programme is not perfect and there is still room for improvement, especially as regards the balance of partnership (the Europeans are still more dominant in project planning); dialogue could be improved for defining regional priorities; potential partners in the south need to be better informed; the selection rate is relatively low (around 15%), which reflects the programme's success in terms of participation and project quality but can be discouraging.

CONCLUSION

The European initiative launched in 1995 is gradually becoming operational.

Today, the Commission and the Member States are discussing the place of science and technology in development cooperation policy. This will be the subject of a major conference in the Netherlands, in March 1997, when the Dutch take over the Presidency of the European Union.

Work has started on the creation of a Euro-Mediterranean area in the field of research and technological development.

As regards Community scientific policy, work on the 5th Framework Programme (1998-2002) has begun.

Hence, if the political will exists within the Union there is ample opportunity to demonstrate the importance attached to agricultural research for development and the emergence of a European ARD policy and a European ARD system.

CLOSING SPEECH

FRANÇOIS D'AUBERT
Secretary of state for research
Paris, France

An anniversary is an especially opportune moment to make a sincere evaluation of years gone by. It is a sort of intermission that encourages us to reflect on what has been achieved and make good resolutions for the future. I believe that what applies to you and me also applies to institutions and organisations.

This is undoubtedly why INRA is using its 50th anniversary as an opportunity to critically discuss, without any taboos, the new, difficult questions facing research on agriculture, agro-foods, and foods. The President of INRA and everyone who contributed to the high quality of this meeting are to be thanked.

These questions are difficult because they go straight to the heart of our fellow citizens' daily preoccupations, can lead to unimaginable progress and, at the same time, provoke fears, that are not all justified but which, whatever the case, must be treated in the hope that they may be dissipated.

Agro-foods, perhaps more than other field of activity, takes us to the question of attitude changes that I should like to discuss with you and that no one in this sector can ignore. I can see three basic trends. The first is the very idea of modernity, which has gradually become misunderstood by mixing a demand for progress with new scientific breakthroughs and a certain scepticism on the real improvements that could be expected in daily life. "Happiness can be found in the green fields" as the French expression goes, with thoughts drifting to life in the countryside and grandma's tasty cooking.

This search for what is natural, authentic and savoury does not impede progress, it brings out the relative position of its positive effects. It rings like a warning that we should stem the race for more speed and wealth. Progress should be consumed with temperance, progress which, all told, technological performance alone can no longer justify.

The second trend is connected to the first one. It focuses on everyone's attachment to the environment and to protecting nature. I say "everyone" because this is no longer a fashion or an ideological pretext. It is a requirement that the whole of society is imposing on everyone who, in one way or another, might threaten the great equilibriums, pollute the soils or deplete vital natural resources. There is increasing awareness of the fragility of our planet, and determination to protect its beauty: a field is a piece of landscape, and agriculture contributes to the harmony of this landscape. The concern for aesthetics is also new.

The third trend places health at the centre of our preoccupations. Good health has always been our most precious asset, but we are now more vigilant about it. We want to understand, we want to measure the risks of an unhealthy life style, in particular the risks connected to poor eating habits. Some of us turn to dietary products, others to natural regional foods. Consumption patterns are becoming increasingly differentiated as demand becomes increasingly diversified. Even our relationship with products has changed. In former times we ate what was grown in a vegetable garden nearby. The "traceability" of foods, as the professionals say, was very simple. Today, the two ends of the chain, the trajectory from field to plate has become extremely complex. Nearly all the products in our diet have gone through intermediaries and have undergone considerable processing. Our foods have become the composites of high technology. But the consumer would like to know and understand this trajectory. He wants to be informed - and know about a lot of disconcerting details. He wants to be reassured and be able to exercise his choice freely.

These trends, which could be called sociological trends, must be borne in mind by the agro-food industry and agricultural research. They will undoubtedly change certain perspectives, correct certain excesses and, at the same time, considerably expand the horizons of research. The field of study is not limited to the borders of the cultivated plot. It includes the atmosphere, the ecosystems, water, biodiversity and questions of food safety that henceforth require a global approach. Public opinion is increasingly concerned about the problem of food safety.

There are these major trends, and then there are two phenomena that lend them urgency. The first, of course, is the mad cow crisis, which shook the whole cattle sector. The second is the overpowering progress in genetics, in particular the highly debated arrival of transgenic plants on the market.

At this stage, what lessons can we draw from the mad cow crisis? Let me mention just three. The first is that, without pre-empting on scientific results concerning the nature and transmissibility of bovine spongiform encephalopathy to man, we have reason to wonder about the race for productivity being run by certain producers. When subjected to strong competition, increased productivity is the key to professionnal survival and job maintenance. But there can be no justification, whatever the case, for blind productivism that disregards the major equilibriums. The agro-food industries must think about this. The same is true for research organisations working on biotechnology and other vanguard processes. I want to underline that progress should be a source of service, not perversion.

The second lesson from this crisis is that research may be a long term action and usually part of a multi-year plan, but it must also be able to react in real time. This applies not only to matters of public health, but also to the citizen's social expectations. During the last few months, laboratories, especially those working with INRA, have shown that research knows how to cope with new challenges.

Last, in this type of crisis, it is indispensable to be open and above board with the public, and also to respect precautionary principles and adopt the responsible approach that guided the government's action and should be imposed on everyone. When public health is at stake, and indices - if there is no formal proof - are sufficiently alarming, protection should supersede all economic considerations.

The second accelerating phenomenon comes from research itself. I want to talk about the rapid development of genetics, in particular as applied to agriculture. We are all familiar with the debate - to which Greenpeace is party - based on the claim that genetically modified organisms are dangerous for the consumer. I think that we should abandon the strange idea that scientists are preparing a world *à la Frankenstein* somewhere deep in their laboratories. Just mentioning this reminds us that several years ago discoveries stemming from molecular biology had a decisive, positive impact on mastering the reproduction of farmed animals, veterinary medicine, plant health the seed industry. What is new is the great adventure of the genome now before us. It is difficult to predict the precise impact of the extraordinary amount of data that the scientific community will be acquiring in the next few years. But we can be sure that the plant and animal production methods will change considerably. Much has been said during the last few weeks about transgenic plants. The word "transgenic" might sound awesome, but here again, that would be too easy.

It is irresponsible to create a sense of fear, like the media have been doing during the last few weeks. Transgenic plants constitute a source of great hope. For the countries of the South, they may make it possible to adapt crops to drought or high degrees of salinity, and meet the countries' food needs better. As everyone knows, the challenge before us is titanic. In thirty years, there will be four billion more people to feed on earth. But feeding them will only be possible if plant genetic engineering is not only used to the benefit of tobacco, maize, tomatoes and rapeseed! Further, transgenic plants provide prospects for reducing the amount of input needed in agricultural production: fertilisers, insecticides and other plant health products. It also means new markets for agriculture as molecules with high added value are produced. Remember that the pharmacopoeia was born because a large variety of plants were used for therapeutic purposes. Just think of some of the recent discoveries like the use of vinblastine or taxol in controlling tumours. Transgenesis unquestionably opens up new possibilities for gaining access to this biodiversity and the use of these resources.

Promising research must be continued on the synthesis of plants for use in new drugs and vaccines, like the vaccine against rabies that is presently at the field trial stage. It is most important that transgenic plants demonstrate their real benefits for the environment and the consumer and not only for the people who develop them. Alongside these hopes there are still legitimate questions that can only be answered through research and the acquisition of new data. Discussions must be held on a sound scientific basis. In response to the often mentioned risk of developing resistant strains by cultivating transgenic plants that produce an insecticidal substance, a global ecological review will have to be made to compare the effects of this insecticide when synthesised by the plant and when applied to the crops using the traditional method.

More, intensive research should be carried out on genetic transfer between cultivated and wild species. Their effects should be evaluated: will protective barriers be needed, if so, what kind? We should never forget when we talk about changing a plant's genetic heritage, that this approach was used empirically for centuries for the advancement of agriculture and then more rationally when varieties were characterised and cross-breeding controlled. It is true that techniques used today differ from those used in the past, and that differences will be even more pronounced in the future. The whole issue is to evaluate them scientifically and get rid of imaginary, irrational notions aroused by the mere term "genetic manipulation".

As far as health safety is concerned, this aspect is taken into consideration as soon as the research project is evaluated. During experiments, tests are carried out, and if there is the slightest doubt about any potential toxicity or allergenic effect, the project is immediately re-examined. Obviously a new product has to have full safety guarantees before it will be authorised for sale. I want to emphasise how extremely strict the different commissions are in their work all throughout the various steps on regulations: the Genetic Engineering Commission to start a research project, the Biomolecular Engineering Commission to authorise that an experiment be carried out in the open air, the Standing Technical Committee as concerns breeding before a new variety can be commercialised, and, finally, the Public Health and Hygiene Committee to examine the products if they are different from the products synthesised by non modified plants.

At each step of development path an application for approval is submitted and carefully analysed. And here I am only talking about the national level. To complete the story I should add the European Commission's authorisation procedures prior to public sale. It would be hard to find the likes of it.

Although decisions are taken rationally without excess, and should be guided by the principle of vigilance, our fellow citizens must understand that whether we talk about new technologies or traditional foods, nothing is totally free of risk. This is exactly why we must keep careful track of a new product after it has been put on the market.

The consumer, quite rightly, wants to be informed. He wants to know about the product's origin and feel free to make his own choice. That takes us to the difficult problem of labelling. The first requirement is to be sure that the label meets the consumer's real demands, and establish what products should be labelled: "transgenic" foods if the composition is different from the so-called "natural" products, processed product if they contain even the most minute transgenic component, or if the products has been exposed to genetic engineering during processing? The label should not be an "alibi" that merely indicates that the product was "obtained through genetic engineering" or "contains chemical substances", as this would apply to nearly all the everyday products.

For food products obtained through genetic engineering, a nomenclature and perhaps codes like those used for food additives should be worked out, indicating the moment at which transgenesis occurs in the process and the function of the introduced genes, or else the type of genetic construction used. This is an important operation that should bring together research scientists, producers, manufacturers and consumers. Then and only then, would the label be useful. It should be accompanied with readily understandable instructive information on the methods, studies and validation tests carried out prior to commercialisation and on product monitoring thereafter. Last, labelling is only effective if some control mechanism is also used, which leads straight to the traceability of the DNA sequence.

The very question of labelling indicates that agricultural research is at an historical crossroad between technology and culture. You discussed the road that it should take, but thought must also be given to ways of strengthening it. I think that, on average, industry is not investing in agro-food research on a par with its economic weight; this applies to both the large groups and the small- and medium-size enterprises and industries, the SMEs.

That is why we firmly intend to help the SMEs become more active in co-operation programmes with research organisations. We recently launched an important programme on biotechnology together with other interested ministries such as the Ministry of Industry, and also ANVAR. It is designed to stimulate research-industry transfers by supporting development projects that bring state laboratories and enterprises together. The call for tender will last several months; we have already received 308 eligible proposals. The results are extremely promising: 37% of the projects concern the agro-industry, 37% the environment, and 12% marine biotechnology. A total of 305 state laboratories, 167 SMEs and 46 large groups have become involved.

I very much hope that the big supermarkets, whose turnover figures are the largest in all of the agro-food sector, will also invest in this indispensable research. The major groups, which are at the heart of the agro-food chain, should become more involved in this innovation and bring in their own added value, make their own contribution.

Further, we want to make a specific request to the state research organisations to think about ways to reinforce this research. One way would be to work together on the major themes we selected during the interministerial committee meeting on scientific and technical research held on 3 October 1996. As you know, at that meeting we emphasised the question of food safety, and in 1997 we will launch a research programme on food safety and hygiene. The brief will be to conduct research on microbiological safety and develop an analysis of related risks.

At the end of the day, whether the question is organising research or pursuing goals, the main idea is the same: what is needed is a global approach with, on the one side, more networking of research operations and, on the other, more attention to new parameters, because of which, this research is neither basic nor applied but rather research involved with society, in other words, research that is fully aware of the hopes and worries that will appear on the road to discovery and innovation. Moreover, this is the real meaning of the joint declaration that you adopted. It is more than a framework, for it expresses your will to pursue research for the citizen.

One thing is clear. In the world of today, there is no room for error or excess, and there is no experiment that can plead any degree of innocence.

Printing and binding: Druckerei Triltsch, Würzburg